GUERRILLA WARFARE

with a new introduction by the author

Walter Laqueur

GUERRILLA WARFARE

A Historical & Critical Study

Transaction Publishers
New Brunswick (U.S.A.) and London (U.K.)

Third printing 2002

This book is printed on acid-free paper that meets the American National Standard for Permanence of Paper for Printed Library Materials.

Library of Congress Catalog Number: 86-1913
ISBN: 0-88738-656-3
Printed in the United States of America

Library of Congress Cataloging-in-Publication Data

Bruner, Jerome Seymour.
A study of thinking.

(Social science classics series)
Reprint. Originally published: New York: Wiley, 1956.
Bibliography: p.
Includes index.
1. Thought and thinking. I. Goodnow, Jacqueline J. II. Austin, George A. III. Title. IV. Series.
BF455.B75 1986 153.4' 2 86-1913
ISBN 0-88738-656-3 (pbk.)

These problems (of guerrilla warfare) are of very long standing, yet manifestly far from understood — especially in those countries where everything that can be called "guerrilla warfare" has become a new military fashion or craze.

— B. H. Liddell Hart, *Preface to the second edition of* Strategy, 1967

Contents

Introduction to the Transaction Edition

Every book has its history and *Guerrilla Warfare* is no exception. Together with its sequel *Terrorism* and two companion volumes (*The Guerrilla Reader* and *The Terrorism Reader*) it was part of a wider study such as had not been attempted before—to give a critical interpretation of guerrilla and terrorism theory and practice throughout history. It did not want to provide a general theory of political violence nor did it give instructions on how to conduct guerrilla warfare or terrorist operations. It was read in various countries and established itself as a text. I hope it contributed to the clarification of certain issues not widely understood at the time among the public at large. The book certainly failed to bring about greater semantic clarity; there had been a widespread tendency in the media and our public discourse to equate great war with small war, guerrilla, terrorism, civil war, and banditry (social and asocial), and this has not changed since this book first appeared in 1976. On the contrary, unsuspecting readers consulting major public libraries for guidance on the subject of guerrilla warfare will find under this heading books on the theater, on business practices (especially sales strategies), on education, and on many other fields of human endeavor. The term guerrilla has become very popular; but in the same measure as the application has multiplied, the meaning has become even more diluted. In addition there has been the widespread use of the unfortunate and misleading term "urban guerrilla" as a euphemism for terrorism. Guerrilla has a positive connotation, by and large, whereas terrorism has not, hence the misapplication.

While the word guerrilla has been very popular, much less attention has been given to guerrilla warfare than to terrorism even though the former has been politically more successful: Most terrorist groups have failed; many guerrilla movements have succeeded. The reasons for the lack of attention are obvious: Guerrilla operations, in contrast to terrorist, take place far from big cities, in the countryside, in remote mountainous regions, or in jungles. In these remote areas there are no film cameras or recorders. This fact has been recognized early on by some guerrilla leaders who decided, unwisely in some cases, to transfer their activities to the big cities. As an Algerian militant put it, if his fighters killed thirty soldiers in a village, this would be reported in a few lines on the back page of the world press, whereas the noise of even a small bomb in a big city would reverberate throughout the world and make headlines.

Has there been more or less guerrilla warfare during the last two decades? The days of the classic, major guerrilla wars seem to be over, no new Mao or Ho Chi Minh, no Tito and no Castro have appeared in our age. The gradual liberation of territories, the establishment of counter institutions and the transformation of guerrilla bands into regular army units has been rare in our time. The war in Afghanistan is probably the only major exception; the list will also include Chechnya (1994-95), but one cannot think of many others. On the other hand, there is a long list of guerrilla wars which have not led to victory: This includes the Kurds in Turkey, the Karen in Myanmar, the rebels in Tajikistan and Southern Sudan, the Tamil in Sri Lanka, the EZLN (the neo-Zapatistas) in Mexico, Sendero Luminoso in Peru, Hizbollah in Lebanon, the various guerrilla groups in the Horn of Africa, various small Punjabi and Kashmiri groups, to mention a few. Some insurgents have used both guerrilla and terrorist strategies; this is true, for instance, with regard to Algeria. In other cases, regular armies have not been strong or well equipped and for this reason were compelled to apply, on occasion, guerrilla tactics as in the former Yugoslavia. The same is true with regard to civil wars as in Rwanda and Zaire. In fact, "pure" guerrilla warfare has become fairly rare. If Mao or Tito were to reappear they would probably not approve of the actions of their descendants, but it is not clear what advice they would offer to cope with conditions quite different from those in which they operated at the time; there is no big war in progress which provided welcome cover for the guerrillas sixty years ago.

A review of strategies and the fate of guerrilla movements during the last two decades shows certain common features. In their great majority they consisted of nationalists fighting for independence whether against foreign occupants or against another ethnic group in their own country. Communism or Maoism did not play a key role as in the 1940s; true, Sendero Luminoso and the Mexican EZLN are social movements, but at the same time they are ethnic rebellions of Indians against the white upper and middle classes. The only exception is the Colombian FARC which at one time gave up armed struggle, but following its electoral defeat in 1992 it resumed it. However, the Colombians became involved with some of the drug cartels to improve their financial condition and while this might be justified from their point of view by pragmatic reasons it has nothing to do with the Marxist-Maoist tradition. Virtually all other conflicts are ethnic in essence except those carried out by Islamic fundamentalists (such as in Algeria) or against fundamentalist regimes (as in Sudan).

Only in two major cases have guerrillas been successful, in Afghanistan and in Chechnya. Victory was achieved because the rebels faced an enemy which had the power to smash the guerrilla but not the political will. These two guerrilla wars happened to coincide with the collapse of the Soviet empire; the Soviet Union had, of course, the power to destroy its enemies but the war was as unpopular as the Vietnam War was in the United States, nor was the regular army prepared for a guerrilla war. Topographical conditions in Afghanistan favored a guerrilla war and in addition there was a tradition of such a war in this country. The war had lasted for years until it was ended by an armistice which provided for the evacuation of all Soviet troops in 1989. However, there was no common front of Afghan resistance fighters which could have effected a smooth political transition. In these circumstances, facing a population tired by war and antagonized by the ambitions of rival war lords, the fundamentalist Taliban, with massive help from Pakistan, occupied Kabul (in October 1996) and large parts of the country.

In most other instances guerrillas were less successful; they caused considerable harm to their enemies (as in Sri Lanka, Turkey, Algeria, Israel, and some other countries) but failed to gain decisive victories. In some instances guerrillas transformed themselves into political parties (as in Nicaragua and San Salvador) without, however, making significant political progress. The PLO and the African National Congress were more successful but they had not been primarily guerrilla organizations; to the extent that it had been a guerrilla organization the PLO had been totally defeated and compelled to leave Lebanon for Tunis. But it still made a political comeback not as the result of terrorist attacks but of mass action.

Most guerrilla movements had substantial help from abroad and it is doubtful whether they would have been able to continue their struggle but for this assistance in money, arms, and by other means. This refers to the Afghan rebels and the Palestinians who received large amounts of money from the Arab states. The Tamil Tigers were supported by Tamils in India; the Chechens received Turkish help; the Algerian fundamentalists were helped by fundamentalists in the Arab world and Iran. Hizbollah in Lebanon receives its budget from Tehran. Taliban received not only arms but also money from Pakistan. Not much is known about the financial sources and resources of the Latin American guerrillas; they did not have much foreign help and were among the proletariat of the global guerrillas. But even they probably did get help from abroad even though they got most of their money from internal sources.

Historical experience shows that guerrilla movements have prevailed over the incumbents only in specific constellations: If Chiang Kai-chek's government would not have been weakened by the Japanese invasion, if the Germans in the Second World War had been able to deploy more regular units in Yugoslavia, neither Mao nor Tito would have entered history as the great victors. There are exceptions such as Cuba under Batista where a regime in time of peace collapses as the result of its lack of popular support. But these are the exceptions, and in the current state of the world there have been few such exceptions. While the weapons of the guerrillas have only marginally improved, modern means of observation have become more sophisticated and effective. They still do not offer panaceas, whatever the advocates of modern technology may claim; much still depends on the terrain, the fighting spirit of the guerrilla, and other factors which cannot be quantified. But while small units can move as effectively as they could fifty or a hundred years ago, it is far more difficult for bigger units to assemble and operate. They have become more vulnerable than in the past, hence the great difficulties guerrilla movements have faced in seizing and holding territory and establishing a standing army. It could well be that the classical model of guerrilla warfare is no longer applicable in the modern world; even Mao admitted that what could be done in China might not be feasible in Belgium. Now it might not be possible in China.

All this does not mean that tactics akin to guerrilla warfare have no future, but it signifies that they will probably be modified in accordance with technical developments. We may witness a combination of political warfare, propaganda, guerrilla operations, and terrorism, and, in some cases, this could be a potent strategy not in the most developed and densely populated countries but quite possibly in the rest of the world. It is too early to write off guerrilla warfare, despite the trend of urbanization all over the world, and the range of possibilities is almost endless. We may witness small groups of sectarians, religious or political, directing their aggression not inwards, that is to say, committing collective suicide, but outwards, against the rest of the world. We have certainly not seen the last of separatist guerrillas, provided topographical conditions favor this kind of warfare. A new criminal guerrilla cannot be ruled out for instance in collaboration with drug producers especially in Southeast Asia and Latin America. Such guerrillas may eventually have at their disposal very sophisticated weapons, including weapons of mass destruction, which would open new and dangerous prospects.

These are trends and possibilities, not certainties. But it all means that in some parts of the world, at the very least, small wars will continue, even if big wars have become too expensive. And it is useful to remember that historically and etymologically guerrilla means precisely this—small war.

Preface

The present volume is the first part of a wider study which, I believe, has not been attempted before — a critical interpretation of guerrilla and terrorist theory and practice throughout history to the present age. This book deals with guerrilla warfare; it does not aim at presenting a universal theory, for such a theory would be either exceedingly vague or exceedingly wrong. The author of an excellent book on the Cuban revolution noted some years ago that in view of its unique character the events in Cuba were a subject for the historian rather than the sociologist. The same is true of most guerrilla wars: a tank is always a tank, but guerrilla wars differ greatly from one another. Throughout history unconventional warfare has been affected far more by indigenous political, social, cultural, and economic factors and, of course, by geography than has conventional warfare. But though one eschews sweeping generalizations there still are common patterns to be found and it is to the search for these patterns that this study is devoted. Hence the comparative approach, which has been used in this study in full knowledge of its limitations.

My original intention was to write a general essay on the subject without dealing with details of guerrilla theory and practice. But the more I became interested in the subject, the more I became aware of how much spadework remained to be done. Excellent monographs exist on some guerrilla wars, others have been neglected hitherto, and there is no critical review of guerrilla doctrine available.

The facts of guerrilla warfare have been covered by a vast overgrowth of mythology, and I have regarded it as one of my main tasks to distinguish the facts from the myths. The present study is therefore an attempt to demythologize guerrilla warfare, without belittling its importance. Fiction should not be disparaged; Balzac's *Les Chouans*, Tolstoy's *Hadji Murat* and Hemingway's *For Whom the Bell Tolls* perhaps convey a better picture of what it meant to be a partisan than many volumes of military history. But these novels deal with the fate of individuals; for an understanding of guerrilla war as a political and military phenomenon, fiction is of little value.

A study of this kind faces considerable difficulties. It was necessary to go back to the sources, a laborious task since some of the material is no longer available even in the world's leading libraries: copies of Carlo Bianco's *Trattato*, of Le Mière's pioneering work, of J. Most's *Revolutionaere Kriegswissenschaft*, and other important sources quoted in this book are not to be found in either the British Museum or the Library of Congress. The search for rare books is arduous but it has its compensations. In any case, this was not my main problem. The guerrilla phenomenon, unless one regards it as a mere technique of warfare, has innumerable aspects and facets; so that even proposing a definition of the subject is exceedingly difficult.

The term "guerrilla" was originally used to describe military operations carried out by irregulars against the rear of an enemy army or by local inhabitants against an occupying force. More recently it has been applied to all kinds of revolutionary wars and wars of national liberation, insurrections, peasant wars, and terrorist acts (such as hijacking airplanes or kidnappings). It has been applied to happenings in the theater and the arts, to certain activities in universities and even in kindergarten. In short, the term has become almost meaningless, partly as the result of indiscriminate use but also because there are, in fact, some real difficulties. Regular armies may use guerrilla tactics, but so may bandits; guerrilla movements have transformed themselves into regular armies but the opposite has also happened. Not all unconventional warfare is "guerrilla" war, nor should it be used as a synonym for revolutionary politics, civil war (such as in Lebanon and Angola), or terrorism.

The tactics of guerrilla warfare are not very complicated, nor are they shrouded in mystery — they have been more or less the same, with slight variations, since time immemorial. Typical guerrilla operations include harassment of the enemy, evasion of decisive battles, cutting lines of communications, carrying out surprise attacks. Guerrilla tactics are based on common sense and imagination; they vary from country to country, are affected by geographical conditions, by social and political processes, and also change as the result of technological innovation. These developments are examined in some detail in the following pages. It is in the analysis of the political background to guerrilla warfare that most difficulties are encountered.

Over the last two decades there has been a tidal wave of guerrilla literature. Some has been straight propaganda or hagiography bearing little, if any, resemblance to reality. Other books provide useful advice on how to conduct successful guerrilla warfare or, alternatively, how to combat it effectively. But they are of little help in explaining why guerrilla wars break out in certain circumstances and not in others, why some wars succeed and others fail. On the other hand, there have been a great many sociological and psychological studies of guerrilla motivation and behavior, often with a ponderous emphasis on methodology. This approach may perhaps one day produce some interesting new insights but that day has not yet arrived and it is not certain that it ever will. Thus, after a great many detours, the student of the guerrilla phenomenon finds himself back at his starting point: in order to explain guerrilla warfare one has to write its history.

According to widespread belief, guerrilla warfare is a new way of conducting unconventional war, discovered by a stroke of genius by Mao in the Yenan period, and later successfully applied to other parts of the world by left-wing revolutionary movements. Observers with a longer memory point to T. E. Lawrence as the great pioneer of modern guerrilla warfare; some go even further back recalling the Spanish resistance against Napoleon. In actual fact, guerrilla warfare is as old as the hills and predates regular warfare. Primitive warfare was, after all, largely based on surprise, the ambush and similar tactics. But too little is known about the subject and I have not burdened myself with an attempt to search for the roots of guerrilla warfare in prehistoric times.

The political context of guerrilla warfare has been and continues to be the subject of much confusion. Thus it has been asserted that before the 1930s guerrilla movements were usually parochial, not nationwide in character, that they had little more than nuisance value, and that they were ideologically conservative. Recent guerrilla movements, on the other hand, are said to be revolutionary, no longer spontaneous outbursts but part of a nationwide (or international) political movement, and it is this that gives them greater cohesion than in the past. There is some truth in these observations, but it is certainly not correct that until recently guerrilla movements were all of local importance only. Nor are "wars of national liberation" a twentieth-century innovation. Guerrilla movements of the early nineteenth century were indeed "right wing" in character, intensely patriotic, monarchist, religious-fundamentalist, whereas modern guerrilla movements do appear more often than not to be left wing, revolutionary in inspiration. But on closer inspection it transpires that the issues involved are not that clear-cut. It is not difficult to detect strong populist, antiaristocratic elements among the nineteenth-century guerrillas in Spain, Ireland, Italy, Latin America, and even in the Vendée. On the other hand there were and are many movements which simply do not fit into the obvious categories of "right" and "left." Quite frequently their ideology has encompassed extreme "left" and "right" components (as among the "Fighters for the Freedom of Israel" [the Stern Gang] and Dr. Habash's "Popular Front for the Liberation of Palestine"). The IRA and the Macedonian IMRO at various times in their history had connections with both Fascism and Communism (or Trotskyism). Latin American guerrilla movements manage to combine a bewildering multitude of conflicting ideological attitudes, and even in Communist-inspired guerrilla movements, nationalism has almost always been the single most important factor.

How much importance should be attributed then to the political orientation of guerrilla movements? Or, to put it differently: are there perhaps certain basic nationalist-populist-revolutionary impulses underlying their political programs and slogans, a free-floating activism, which may turn "right" or "left" according to political conditions and the changing fashions of the *Zeitgeist*? As a first step in the right direction a moratorium should be declared on the use of terms such as "Marxist" with regard to guerrillas in Chad and Zaire (even if their leaders have attended courses at the Sorbonne) and with regard to terrorists in places as far afield as Berlin, Beirut, or San Francisco.

The old term "guerrilla warfare" has been used in this study because there is no better one. Newer theoretical concepts such as "modern revolutionary warfare" or "people's war" can be of use only with regard to a few countries and applied elsewhere they are misleading; not all guerrilla movements are led by a monolithic political party, or a Communist party, or are either a people's war or a war of national liberation.

The term "urban guerrilla" poses something of a problem for the student of guerrilla warfare. It is a misnomer and its widespread use is to be regretted. Insurrections and revolts have occurred in towns and so, of course, have acts of terror, but urban guerrilla warfare only on the rarest of occasions. The essence of guerrilla warfare lies in the fact that the guerrilla can hide in the countryside and this, quite self-evidently, is impossible to do in a city. The distinction is of more than academic importance; there have been guerrilla units of ten thousand men and women but an urban terrorist unit seldom, if ever, comprises more than a few people, and urban terrorist "movements" rarely consist of more than a few hundred members.

What is now commonly called "urban guerrilla" warfare is, of course, terrorism in a new dress. We shall encounter it from time to time in these pages because it has been advocated and practiced as an alternative to or in conjunction with guerrilla warfare. But there was no room in this study for a detailed and systematic analysis of terrorism; I shall return to the subject in a different context. The political importance of urban terror has frequently been overrated, perhaps in view of its highly dramatic (or melodramatic) character and the fact that, unlike guerrilla operations, urban terror usually has many spectators. The attitude of the media towards the "urban guerrilla" reminds one of T. E. Lawrence's description of his Arab levies — they thought weapons destructive in proportion to their noise.

I owe a special debt of gratitude to the Library of the Royal United Services Institution in London. I have also visited or received the assistance of the following collections: the Military Library in Helsinki, the Royal Library in Copenhagen, the National Library in Madrid, the Lenin State Library in Moscow, the Library of the Ministry of National Defense in Paris, the Library of Congress in Washington, the Widener Library at Harvard, the Princeton College Library, the Asaf Simhoni Library at Tel-Aviv University, the British Museum, the Wiener Library, and specialized collections in Stanford, Vienna, Madison, Bern, Florence, Aargau, and Berlin.

I received advice from colleagues and friends who provided suggestions and criticism and read parts of the manuscript. I am grateful to Aviva Golden, Bernard Krikler, Kimbriel Mitchell and Freda Morrison who provided editorial and research assistance. Dr. David Abshire, chairman of the Center for Strategic and International Studies, and my other colleagues there, in particular, Mrs. Ethel Eanet, my assistant, have helped me in the course of this study;

so did those who work with me at the Institute of Contemporary History and Wiener Library in London. Generous support given by the Ford Foundation enabled me to complete a study that in view of the many difficulties involved may well have otherwise remained unfinished. Parallel to this volume I have edited an anthology on the historical development of guerrilla doctrine and the problem of terrorism will be covered in the third part of what was intended to be an essay and became a trilogy.

London–Washington, January 1976

GUERRILLA WARFARE

1

Partisans in History

Irregular forces and guerrilla tactics are mentioned, perhaps for the first time in recorded history, in the Anastas Papyrus of the fifteenth century B.C. Mursilis, the Hittite king, complains in a letter that "the irregulars did not dare to attack me in the daylight and preferred to fall on me by night." While peeved, Mursilis obviously lived to tell the tale.

Guerrilla tactics, of course, predate recorded history, as indeed they predate regular warfare. In Melanesia, the chosen practice was for the warriors to attack when the enemy was at its sleepiest and most unwary; the same approach was used by the Kiwai in New Guinea. The southeastern Indians of North America liked to be pursued by the foe so that they could lure him into the hollow of a crescent formation. The mock retreat and the ambush were also known to many other tribes; a classic description is in *Joshua* 8.

Generally speaking, primitive people had an aversion to open fighting.[1] But surprise and deception have their use in every military conflict and there are basic differences between primitive and guerrilla wars. Far more often than not, the former consisted of sporadic, unorganized sorties, hit-and-run raids, the object being either to plunder or to seek vengeance for some grievance such as trespass, personal injury, or wife stealing. Primitive warfare evolved in small tribal social groups who had no capacity for any sustained effort such as protracted war; the scope of movement was quite restricted, and ideological issues were certainly not involved.

The Bible mentions guerrilla leaders such as Jiftah and David. Of David, it is said that "everyone that was in distress, and everyone that was in debt, and everyone that was discontented gathered themselves unto him; and he became a captain over them." (*Samuel* 1:22) He used the Judaean desert (Ein Gedi) as a temporary base, engaged in forays in the Hebron area, imposed tribute on the rural population and, with a force of between four hundred men (the Caves of Adullam) and six hundred (the Battle of Ziklag), fought the Amalekites and other enemies. Gideon's night attack at Ein Harod (*Judges* 6–8) is a good example of exploiting the element of surprise. Of the twenty-two thousand men who were with Gideon, three hundred were chosen and, though there were many more Midianites — "like locusts, and no one could count the number of their camels" — they spread confusion in the enemy camp. The sound of trumpets and the sight of torches (hidden in jugs until the very last moment) created pandemonium and the Midianites turned against each other. To make the attack even more effective, it was launched at the time when the guard was about to change.

The Maccabaean revolt in 166 B.C. made use of guerrilla tactics in its early phase; Mattathias and his sons went into the mountains and, having no arms, Judah had to pick up the sword of the enemy soldier who fell first. They lived in secret places in the wilderness, "as wild animals do." Judas Maccabaeus launched his attacks mainly by night (*Maccabaeans* 2:8), while his brother, Jonathan, the second leader of the rebellion, launched raids from his base in the Judaean desert, harassing the enemy and evading frontal encounters; his strength lay in the mobility of his troops and a superior intelligence network.

In the Jewish war against the Romans, guerrilla units did not play a decisive role; the country was densely populated, and there were no inaccessible mountain ranges or wild forests which could have provided shelter for the insurgents. Josephus describes how the retreat of Cestius Gallus, the governor of Syria, from Jerusalem ended in a rout. Cestius's legions were heavily armed which greatly impeded their march, whereas the Jews who attacked them were traveling light. While the Roman forces moved through open terrain they were relatively safe, but eventually they had to cross a narrow downhill pass where their horses stumbled and the attackers all but annihilated their columns.[2] John of Gischala, a military leader in Galilee and a rival of Josephus, is described as little better than a robber — crafty, cunning, motivated not so much by patriotism as by the lust for power and spoils. The initiative in these wars was usually with the Romans, and the internal conflicts in the Jewish camp made an effective use of guerrilla tactics almost impossible.

The Bar Kochba rebellion (A.D. 132–135) was the last stage in the Jewish war against Rome; caverns and subterranean passages provided the insurgents with hiding places. They avoided open battle and from their footholds in the mountains undertook devastating raids upon the country. When Julius Severus, one of Hadrian's most capable generals, was appointed to suppress the revolt, he quickly realized that there would be no open engagement and that the rebels had to be hunted from their hideouts one by one.[3]

There are not many examples of guerrilla tactics in Greek military history. One of the few exceptions was Demosthenes's campaign with three hundred Hoplites in mountainous Aetolia. Thucydides, our main source for this disastrous invasion, relates that the Aetolian forces were scattered and lightly armed, having no defensive armor. They attacked the Athenians with their javelins from a safe distance without much risk to themselves, retreated when the Hoplites advanced, advanced when the Hoplites retreated, and, generally speaking, demoralized the enemy by constant withdrawal and pursuit: in the end the Hoplites got lost in pathless woods and stumbled into ravines from which they could not climb out. The Aetolians ringed them with flames and a hundred and twenty of the Hoplites ("the best soldiers of Athens") fell before their commander decided to retreat.[4]

Instances of guerrilla warfare are more frequent in Roman military history — in North Africa, Gaul, Germany, and, above all, in Spain.* Tacfarinas, the elusive Numidian chieftain, caused a great deal of aggravation to the Romans. According to Tacitus, he had served in the Roman army, deserted and collected a group of bandits with whom he undertook his expeditions. Was it a case of "social banditry," a "war of national liberation against the imperialist enemy," or perhaps a mixture of both? The Numidians could not resist the Romans in open battle but they were excellent practitioners of the art of small war; whenever the Romans attacked, they retreated, only to return and harass the Romans when these had ended their onslaught. In the words of Tacitus (*Annales*, 3:22), the Romans got very tired as a result of these frustrating experiences and it was only when the Numidians left the desert and tried their luck near the coast that they were beaten. Despite various setbacks, Tacfarinas continued his struggle; at one stage he even sent ambassadors to Rome, much to the mortification of Ti-

* The Romans used *quadrillage* and, generally speaking, had fairly developed techniques of counterguerrilla operations as described, for instance, in Sallust's account of the Jugurthan war. For an excellent summary, see S. L. Dyson, 'Native revolts in the Roman Empire,' *Historia* (Zeitschrift) (1971), 239–274.

berius. It was not until seven years later, when the Romans changed their tactics to using lightly armored soldiers who knew their way in the desert, that they succeeded in defeating Tacfarinas decisively. They came upon him at daybreak with his troops unprepared, horses unsaddled and not even any guards on duty — and that was the end of Tacfarinas and his band (*Tacitus*, 4:25).

Arminius the Cheruscan, like Tacfarinas, had served in the Roman army; he was a Roman citizen, a knight and an ally of the Romans. As Varus set out with three legions and auxiliaries, constituting a total of twenty-seven thousand men, for the forest of Teutoburg, he had every reason to suppose that Arminius was a friend. But Arminius was motivated by anything but friendly feelings when he persuaded Varus to move his headquarters from a Rhenish fortress to the Weser. He assumed, quite rightly, that this move would force the Roman leader to disperse his troops and his lines of communication would become vulnerable.[5] The first two days of Varus's march having passed uneventfully, on the third he reached the forest. The undergrowth was thick, there were no roads, the constant rain made the ground slippery, and leadership in the Roman camp was deplorable. Suddenly Arminius attacked, there was a mass slaughter and, while in the subsequent eight years the Romans avenged themselves in a series of battles, large and small, they never again succeeded in firmly establishing their rule east of the Rhine.

The Battle of Teutoburg Forest was an example of a successful large-scale ambush, whereas Julius Caesar, in his conquest of Gaul, encountered something more akin to full-scale guerrilla warfare. Vercingetorix, king of the Arverni and military leader of the tribes that rebelled against the Romans, had no military education and his soldiers were untrained. Having been defeated by Caesar's cavalry, he changed his tactics; Caesar quotes him addressing his fellow chieftains:

> We have to conduct the war quite differently, to cut off the Romans from their food and supply, to destroy isolated detachments. All the open villages and farms from which the Romans can get their provisions will be cut off and the Romans will starve.[6]

Vercingetorix realized that, while his own infantry could not face the Romans in open battle, it had the advantage of speed over Caesar's legions which were hampered by their huge baggage trains. His tactic was to tempt Caesar to pursue him through difficult terrain, to tire him out and thereby compel him to disperse his forces. He applied this strategy with considerable success for almost six years against Rome's greatest military genius. It was unfortunate that he had to cope not only with the Romans but also with the impatience and foolishness of his own countrymen, who time and again forced him to commit errors against his better judgment: insisting on his giving battle, for instance, or retreating after a battle into a prepared fortress. In the year 52 B.C. Vercingetorix was enclosed by the Romans in Alesia; Caesar beat off a relieving army and took the fortress by means of circumvallation. Eventually the Gallic leader had to surrender and was executed.

The elements of guerrilla warfare, such as the evasion of battle, the attempt to wear down the enemy, to attack small detachments in an ambush by day and larger units by night, were not of course a novelty to the Roman generals. They had applied them, more than once, in their own operations. Fabius Cunctator, who had been made dictator after the disaster at Trasimene Lake (217 B.C.) had employed the same expedients against Hannibal. He camped on high ground, nibbling away at the Carthaginians' rear guard, avoiding battle. While Hannibal ravaged the Campania, Fabius used his strategy of exhausting the invaders to good effect and this despite the mounting criticism of his fellow Romans thirsting for a decisive battle. After yet another and even greater Roman disaster (the Battle of Cannae [216 B.C.]), Fabius's conduct of war became official strategy until the Romans were strong enough to pass on to the offensive.

The examples mentioned so far refer to regular armies using 'small war' tactics because they were not strong enough to apply any others. Guerrilla warfare in the strict sense of the term was endemic in Spain among the Celtic-Iberian and Lusitanian tribes. Inevitably it has raised the question of why. The Roman historians saw the insurgents as mere street robbers and highwaymen (*latrolistes*); their leader was *latronum dux* or *listarchos*.[7] There was a great deal of unpolitical banditry (*bandolerismo*) in Spain in ancient times, as in later centuries, but this hardly explains the emergence of whole armies of thousands of men who defeated numerous Roman legions. Economic and demographic reasons, such as the distribution of land and the density of the population in certain parts of the peninsula, have been adduced to explain this phenomenon. For a long time Spaniards served as mercenaries in the ancient world; to take up arms served as a safety valve. In Hannibal's army which traversed the Alps, Spanish soldiers seem to have outnumbered the Carthaginians. The Iberian and Lusitanian tribes had every reason to hate the Romans for the heavy tributes imposed on them by several generations of *praetors* and the atrocities committed by Roman troops; above all, the treacherous massacres carried out by Servicius Sulpicius Galba (151–50 B.C.) and Lucius Licinius Lucullus. Thousands were killed; thousands of others were sold as slaves to Gaul.

The fighting in the peninsula, which involved many Roman legions, reached its climax in the war between Viriathus and Rome (147–139 B.C.). Born in the mountains of the Sierra de la Estella, between the Tajo and the Duero, probably of poor parents, Viriathus had been a shepherd and hunter in his youth, and later a small-time *bandolero*. Roman historians portray him as a man of very strong physique, quick, impervious to heat and cold, to hunger and thirst and without apparently needing sleep. He escaped Galba's massacre in the year 150 B.C. and, three years later, became the supreme military leader of the tribes — in Mommsen's words, the "chief of the guerrillas." Soon after Viriathus had been elected commander in chief, his soldiers were surrounded by the Romans but he achieved a getaway by use of a strategem that he was to apply many times in later years.[8] This is how Appian described his tactics:

> He [Viriathus] drew them all up in line of battle as though he intended to fight but gave them orders that when he should mount his horse they should scatter in every direction and make their way as best they could by different routes to the city of Tribola and there wait for him. He chose 1,000 men only, whom he commanded to stay with him. These arrangements having been made, they all fled as soon as Viriathus mounted his horse. Vitelius [the Roman *praetor*] was afraid that those who had scattered in so many different directions, but turning towards Viriathus who was standing there and apparently waiting for a chance to attack, would join battle with him. Viriathus, having very swift horses harassed the Romans by attacking and thus consumed the whole of that day and the next, dashing around on the same field. As soon as he conjectured that the others had made good their escape, he hastened away in the night by devious paths and arrived at Tribola with his nimble steeds, the Romans not being able to follow him at an equal pace by reason of the weight of their armour, their ignorance of the roads, and the inferiority of their horses.[9]

A few days later he attacked Vitelius from an ambush, killing four thousand of his ten thousand men; the rest were no longer capable of giving battle. Vitelius's successor, the *praetor* Plautius, fared no better; he was defeated twice by Viriathus, who again pretended that he was about to abscond when he was in fact preparing for an attack. Plautius was so weakened that, in the words of the Roman historian, he "retired to his winter quarters" — and this in the middle of summer. Viriathus also defeated another Roman *praetor*, Claudius Unimanus, but, after the year 145, his position became precarious inasmuch as the Romans had realized the seriousness of their situation on the peninsula and, after the victory over Carthage, dispatched larger forces to Spain. A consul was sent — Fabius Maximus Aemilianus, Scipio's brother — with an army of seventeen thousand preponderately new recruits. They were at first more successful than their predecessors; Viriathus continued to attack small detachments as the Roman general preferred to train his troops before launching a major campaign. In 144, Viriathus lost two cities in southern Spain, but his luck changed when Fabius was called back to Rome and when several tribes, hitherto in league with Rome, joined him in his struggle. A new Roman offensive was undertaken after the arrival of Fabius Maximus Servilianus in the year 141; but the forces at his disposal (twenty thousand men, ten elephants, and some African cavalry) were not sufficient to defeat the enemy.

Despite several setbacks and the loss of some more cities, Viriathus attacked the Romans almost without interruption. The following year he cut off and encircled the main body of the Roman forces and disaster seemed inevitable when, instead of pressing home his advantage, he engaged in negotiations that led to a peace treaty, with the Romans for the first time recognizing Viriathus, who became *amicus populi Romani*. The reasons for his seemingly inexplicable behavior have been the subject of a long and inconclusive debate on the part of latter-day historians. According to the Romans, he was guided by generosity; others argued that his army was simply tired — and he was under pressure to bring the war to a speedy end. Yet others saw it as a psychological enigma; Viriathus could have put the whole enemy army to the sword and thereby ended the war, for the Romans would never have retrieved such a loss.[10] Against this, it could be argued that Viriathus must have known that, in the long run, he could not maintain himself against Rome, that the Romans would not accept defeat, that new and stronger legions would be sent against him, and that formal recognition and semi-independence was the most for which he could hope. In any event, the treaty remained in force for less than a year; the war party in Rome had been opposed to it from the beginning and tried to provoke Viriathus into breaking it. When this did not succeed, Caepio, the new Roman leader in Spain, renewed the conflict. There was, however, no victory for him in the military sense, the war ending only after the assassination of Viriathus in his tent by three Roman emissaries who had been sent to renegotiate with him. The Romans were glad that their dangerous foe was dead; they were less than happy at the manner of his dying. Caepio was not awarded a triumphal return and Viriathus's murderers did not receive the promised reward. Tantalus, Viriathus's successor, capitulated soon afterward.

Viriathus's conduct of war was, in all essential respects, identical to the campaign waged in the peninsula nineteen hundred years later.[11] He made optimal use of the terrain and established *foci* in inaccessible places. He provoked the Romans to pursue him until he had maneuvered them into an ambush; he cut off their supply lines, harassed them ceaselessly with minor attacks. His principle was to attack quickly and to retreat with equal speed so as not to give the enemy an opportunity to discover the strength of the attacking party. While he preferred arms to be used from a distance, the weapons used by his troops in close battle were no less effective — the *paraxiphis* in particular, a lance with a strong barbed hook which caused terrible wounds.

The Roman historians praised his personal integrity and his military leadership; he was a just and unselfish man who had nothing but contempt for pomp and ostentation. Thus the *dux latronum* turns into the *vir duxque magnus* (Livy) and the national hero of Portugal, glorified in Camoes's patriotic epos.[12]

Fifty years after Viriathus's death there was a revival of guerrilla warfare in the peninsula. The insurgents were led by Sertorius, a Roman nobleman of Sabine extraction, who received his military training in Marius's army, was wounded and lost one eye. Sulla barred his election to the tribunal and later, when appointed governor of Hispania Citerior, had him proscribed. Sertorius escaped to Mauretania, where Lusitanian emissaries reached him and asked him to become their leader in the struggle for independence. At the time there were two thousand Romans and seven hundred Mauretanians with Sertorius; after his return to the peninsula his little army had grown to eight thousand and although Sertorius had to face the greatly superior armies of Pompeius and Metellus, he was never defeated. However, his deputies were not of the same caliber; on one occasion, in the midst of the war, he dissolved his army because he was dissatisfied with its performance and enlisted and trained new units to continue his campaign. He attacked the supply lines of the Roman legions and devastated the villages in their rear, compelling the legions to retreat to Gaul in the winter. He had an alliance with the pirates who attacked the Roman forces from the sea. Plutarch, who notes his quickness and dexterity, commented upon Sertorius's strategy:

> . . . Sertorius vanished, having broken up and dispersed his army. For this was the way in which he used to raise and disband his armies, so that sometimes he would be wandering up and down all alone, and at other times again he would come rushing into the field at the head of no less than one hundred and fifty thousand fighting men, like a winter torrent.[13]

Sertorius was a master both of regular and guerrilla warfare; with inferior forces he kept many legions constantly on the alert. Like Viriathus, he was assassinated: he was killed during a meal in the home of Perperna, one of his aides. This, to all intents and purposes, was the end of the revolt against Rome, for Perperna's qualifications as a military leader, in contrast to those of a conspirator, were nonexistent.

Sertorius no doubt more than once regretted his decision to become head of the Iberian tribes. He could rely on his soldiers only while the going was good; in adversity they were easily discouraged and they lacked staying power. There was a romantic streak in his character; at one stage he planned to sail to the mythical Isles of the Blest. He had far-fetched political ambitions and, for all one knows, planned, like Caesar, to conquer Rome and establish the rule of his party, the moderate democrats. He tried, without much success, to conclude an alliance with Mithridates and other enemies of the Roman Empire such as the pirates. Yet deep down he remained a Roman patriot — one of the most gifted and most attractive personalities of his time, a tragic figure who, owing to the intrigues of his enemies, had to die far from his native Rome for which he longed forever. Mommsen says of him that he was one of the greatest, if not *the* greatest man produced by Rome up to that time, one who, in happier circumstances, might have been the leader of a moral and political revival.[14]

The Roman legions were superior to the enemies facing them in the west and the east owing to their discipline, their military prowess, their sound tactics based on the maniples and above all their engineering feats (road networks, fortress building and siegecraft). On the other hand, these forces were dispersed over a wide area and their lines of supply and communications remained often unprotected. It is therefore surprising that guerrilla tactics were not used against the Romans with greater success. There was not that much to choose between the arms used by the Romans and those used by the "barbarians"; no major war industry was needed to keep either army supplied. But while Roman rule provoked resentment and, on occasion, active resistance, there was no unity and no common purpose among Rome's enemies. The Romans were fighting tribes, not nations, and in the heyday of the Roman Empire a policy of "*divide et impera*" helped to maintain the *Pax Romana* and military power had to be used but rarely.

The armies of the nomadic or seminomadic peoples, who brought about the downfall of Rome, applied new tactics, putting greater emphasis on cavalry and the use of missile weapons. Guerrilla warfare played little if any role in the early Middle Ages during the campaigns of the Vandals and the Ostrogoths, of the Huns and Byzanz, of Muhammad and the Mongols. In all these wars the fate of peoples and continents was decided in battles between large armies. Thus the Middle Ages are, on the whole, an unrewarding period for the student of guerrilla warfare. There was as much fighting as in former or later ages, but its character changed. In Europe, Christianity constituted a common religious and cultural framework with the Pope as ultimate arbiter. The Church did not seek to suppress war but to humanize it. According to the *Treuga Dei*, promulgated by successive church councils, it was forbidden to attack women and children, to rob or kill peasants, to use the arbalest against Christians, because its radius of action was a hundred and fifty meters and it was therefore considered too murderous. Fighting was legally permitted only for some one hundred and twenty to one hundred and fifty days throughout the year. These decrees were not, of course, always observed and were certainly not adhered to in the wars with non-Christian armies. But in principle, feudal warfare was diametrically opposed to the concept of guerrilla fighting; reading *Don Quixote* will help one to understand why. The knights jealously saw to it that the art of war remained their and their vassals' (the squires, *les écuyers*) monopoly; they would not tolerate any other fighting men next to them.[15] True, swords for hire appeared in Italy by the eighth century but this remained largely an isolated phenomenon, which became widespread in Germany and France only in the thirteenth and fourteenth centuries. In northern Italy, the *condottieri* extended their activities — the Visconti in Milan, the Scala in Verona, the Este in Ferrara, the Malatesta in Rimini — but the tactics used in their depredations and raids were very different from those used by guerrillas; they certainly did not have the support of the local population. In France and Germany small private armies were organized by the feudal lords and gradually became a major nuisance. In order to get rid of them, the popes tried, quite unsuccessfully, to dispatch them on a new crusade, but they preferred greener pastures nearer home.

Elements of guerrilla tactics can be found in the peasant uprisings of the Middle Ages and among the bandits roaming Europe's border areas. The war of the Swiss against the Austrians, however, and the Dutch war of independence against Spain were of a different kind. The Swiss were not, as is often believed, a peaceful tribe of peasants and cowherds; they had a long martial tradition. In their battles (Morgarten, Sempach, Nancy), they used regular army tactics; they were superior in number to the other side, but, in contrast to the Austrian knights, they thought it a waste of time and effort to take prisoners. In the Dutch struggle against the Spaniards, guerrilla warfare was well-nigh impossible for geographical reasons; Dutch resistance was concentrated in the cities, there were no mountains and forests to retreat to, while naval power played an important role.

In the Hundred Years' War between Britain and France, originally a dynastic conflict, which developed by degrees into a people's war, guerrilla tactics were occasionally used by the French, above all by Bertrand Duguesclin, Constable of France. He refused to attack the English, but, fighting in distinctly unchivalrous fashion, raided them by night, ambushed their convoys and in general engaged in a *guerre d'usure*, with all its familiar vexations. The English gradually lost what they had without so much as an opportunity of meeting the foe on the field of battle. Duguesclin is first mentioned in the chronicles at the Siege of Rennes (1356–1357) when, at the head of a hundred men, he penetrated the enemy camp and seized a wagon train of two hundred carts loaded with food and war supplies.[16] Duguesclin was among those who pioneered the technique of using the cavalry as mounted infantry; quickly transferred from one place to another, the soldiers dismounted once the battle had started. This was the French answer to the longbow used by the English. Subsequently the technique fell into disuse, only to be rediscovered in the eighteenth century. Duguesclin and his raids were mentioned more than once as a shining example of partisan warfare in the manifestos of the French Resistance in World War II.

The history of warfare in early modern times offers a great many examples of the technique of *petite guerre*. These were actions intended to damage the enemy while avoiding major battles. The element of surprise often played a central role, as in Prince Eugen's capture of Cremona (1702) which ended in a defeat, or the victory of the Austrians over Frederick the Great at Hochkirch (1758). This battle took place in wooded country with the Prussian king laboring under the delusion that he was facing merely another by now almost routine nightly raid by Austrian irregulars, whereas in fact Marshal Daun, the Austrian commander, had moved thirty thousand men to within two hundred yards of the Prussian forces, which were in consequence completely routed. Even before, in the Thirty Years' War, Wallenstein used raiding parties of Croats to pursue Mansfeld, and Gustav Adolf had special light brigades of about five hundred men for similar purposes. The main practitioners of the art of raiding parties were Johann von Werth in the last phase of the Thirty Years' War and the Austrian commanders Trenk and Nadasdy with their Croats and Pandurs who caused considerable trouble to Frederick the Great. The units consisted of anything from two hundred to two thousand horsemen and in their raids, which sometimes lasted for weeks, they covered hundreds of miles to the enemy's rear. But they almost always acted within the framework of a regular army.

Between the fourteenth and the seventeenth centuries, peasant revolts periodically swept throughout Europe (Flanders in 1323–1327, the Jacquerie in 1358, Wat Tyler's rising in 1381, the war of the Taborites in Bohemia and the German Peasant War in 1525). It was not so much the absence of a clear political program, inefficient leadership or inferior armament that led almost invariably to the quick defeat of the peasant armies, but chiefly the fact that peasants were incapable of engaging in a sustained effort; the Jacquerie in the Paris region, for instance, was crushed within a mere two weeks. The peasants would congregate for a big battle or a major campaign, presenting an easy and convenient target for their opponents. They lacked organization and discipline: after a few weeks or months they would disperse and their armies would simply melt away. There was nonetheless the occasional exception: while the French peasants were defeated with such ease, the struggle of the Tuchins in the Auvergne lasted for twenty-one years. It was an early and successful case of social banditry; they robbed travelers and attacked small English and French detachments.[17] The rebellion of the Remensas in Catalonia went on for over ten years (1461–1472) and there was another upsurge in the 1480s. The Taborites, more a national than a social movement, effectively used a new military technique (the *Wagenburg*) which, though known in the early Middle Ages, had been totally forgotten in the centuries between. The Taborites did not, however, evade confrontation with their enemies; they could more than hold their own on the battlefield and their military exploits do not therefore belong to the history of irregular warfare.

The Balkans were the main scene of banditry from the fifteenth to the late eighteenth century. Although they had been occupied by the Turks in the fifteenth century, local resistance had continued, sometimes in the form of a struggle for national (and religious) independence, but more often as banditry. The most successful of the opponents to Turkish rule was George Castriota-Skanderbeg. A Serbian by birth, he had served with distinction with the Turkish army for many years and was governor of a *sanjak*. In 1442 he moved to Albania and at the head of some three hundred local warriors raised the banner of national revolt. He proclaimed himself a Christian and, temporarily uniting the Albanian tribes, defeated countless Turkish generals such as Fizur, Mustafa, Hamza, Balaban, Yakub Arnaut, as well as the great sultan himself. Skanderbeg was a military leader of remarkable talent. Never defeated in battle, he mastered both regular warfare and guerrilla tactics. When the sultan attacked him with a strong army (a hundred and thirty thousand men according to the chronicles), he wisely evaded battle and, instead, intercepted the sultan's convoys, cut off his communications, wiped out small Turkish detachments, harassed the Turkish flanks and rear by day and by night, and so fatigued them by constantly changing ruses and stratagems that the sultan decided that the game was not worth the candle and retreated from this inhospitable country. After Skanderbeg's death (1468) his movement collapsed. He has become the national hero of Albania, the subject of countless histories, novels and even a film, which usually gloss over the exceedingly cruel character of these wars. Pardon was never granted, treaties and promises were broken, women and children were indiscriminately slaughtered; it was, in brief, a war of extermination which was far removed from the restrictions imposed by feudal chivalry on medieval Europe.

Successive generations of Balkan freedom fighters could resist the Turkish armies for comparatively short periods only, and then only if they had the support of some outside power. Thus Skanderbeg was assisted in his struggle by the Venetians and the Pope (Paul II) with men, weapons and money. The Serbians were supported by the Austrians; the great Serbian revolt known as the Insurrection of St. Sava lasted for thirteen years (1593–1606). When the Austrians made peace with the Turks the rebellion disintegrated.[18] The Greeks in their freedom struggle had the support, moral and material, of the whole of Europe.

Alone among the Balkan peoples, the Montenegrins continued their struggle from the fifteenth to the eighteenth century almost without external help; but even they could not have fought forever had they not had the supply line from the harbor at Cattaro, on the Adriatic, to their capital, Cetinje. In later years they had the active support of the Russians. They were further favored by the inaccessible nature of their Black Mountains (*Chernagora*). The country was exceedingly poor, roads were all but nonexistent; thus it held altogether no particular appeal for the Turks. A small Turkish army constituted no danger to the local inhabitants and supply difficulties prevented a large army from staying for any length of time in so uninviting a land.

The Montenegrins, like Skanderbeg and the Greeks, committed acts of great brutality, though it has been argued that this was the natural reaction to the barbaric oppression to which they were subjected. However, while the Turks were not exactly paragons of humanism, it is difficult to find in the annals of Ottoman history up to the nineteenth century many incidents like Danilo's massacre of the Turks in 1702 or the mass murder of Turks (and Jews) in Greece in 1821.[19] In some cases there was the deliberate intent of provoking the Turks to countermassacres; without the atrocities in Morea, Philhellenism would not have attracted so many supporters. The Turks, nevertheless, had a bad press in Europe; a British nineteenth-century traveler, with no axe to grind, noted, with some surprise, that there had been a mistaken tendency in Europe to invest the southern Slavs with almost supernaturally noble qualities. In actual fact, the Turks, "whether it be from a consciousness of their own decrepitude or some other reasons," behaved on the whole better.[20] He also noted that the Montenegrins greatly exaggerated their successes against the Turks, and that "they forget that they owed their present position not to their own prowess, but to foreign intervention."[21] Such exaggerations frequently occur in the reports about their struggle against the Turks; the Christians usually did not face Turkish elite troops and, more often than not, they had numerical superiority.

Brigandage was no less widespread in the Balkans in the seventeenth and eighteenth centuries. Many ballads celebrated the daring exploits of Haiduks and Klephts — both terms meaning robbers. It has been said that, given the severity of Turkish rule, a young Christian villager could either become a cleric, emigrate or turn to robbery. This is a somewhat simplified analysis of the opportunities open to the Balkan Christians, quite apart from the fact that the Turks and Albanians were at least as numerous among the Haiduks. Paul Ricaut, an attaché at the British embassy in Constantinople, who chose the overland route on his way home, reported that even the strongest convoys were attacked by the Haiduks in the forests, and that rocks thrown in narrow passes did as much damage as an artillery salvo.[22] According to the Haiduk sagas, many of them were shepherds, underpaid and bored, who knew the terrain extremely well and were lured by the romanticism and the material rewards of brigandage. There was also a fair proportion of escaped criminals among them. The gang was usually led by a *voivode;* the second in command was the *bairaktar,* the standard-bearer, in charge of accounting, which meant in this context the distribution of the booty. Their work was seasonal; since there were hardly any travelers to rob during the winter, these months would be spent preparing a camp and storing provisions at, if possible, a reasonable distance from their summer field of action.[23] Judging by accounts, they kept their weapons (rifles, pistols, *yatagans*) in excellent condition. They did not rob the peasants in their immediate neighborhood, since they had to rely on them for intelligence, and they also needed their help for a quick escape in an emergency. As it was mostly strangers they robbed they became the object of something like a Robin Hood or Rob Roy cult, not only on the part of the village population but, in later years, among town dwellers as well. The orthodox clergy blessed their weapons and gave them absolution for murder. As the popular poets saw it, the Haiduk was the only free man in an empire of slaves. Whether they were really rebels against Turkish oppression is doubtful, but it is certainly true that, while they did not discriminate between Christian and Muslim insofar as robbery was concerned, Christians had a better chance of escaping alive whereas Turks, including women and children, were killed without compunction. The Greek Klephts, particularly those in the Morea, showed less national spirit; they lived mainly at the expense of poor Christians and "rarely ventured to waylay a rich Greek primate, still more rarely to plunder a Turkish aga."[24] To kill compatriots, to burn a Greek village, caused them not a twinge of conscience. Koloktronis, who was the leading Klepht of his age, later became commander in chief of the Greek army and a member of the government. Even the Philhellenes were appalled by the savagery of the Klephts and the Greeks in general. One of them wrote: "Whatever national or individual wrong the Greeks may have endured, it is impossible to justify the ferocity of their vengeance, or to deny that a comparison instituted between them and the Ottoman generals . . . would give to the latter the palm of humanity."[25]

Large-scale banditry was not, of course, limited to Europe. To provide but one example: during the last two decades of the Ming dynasty, from 1628 to 1644, guerrilla warfare was endemic in northern China, particularly in Shensi, which was the birthplace of two of the main gang leaders of the period, Li Tsu-cheng and Chang Hsien-chung. With the cooperation of many smaller bands of army deserters and peasants, they virtually divided the country among themselves. Gradually these bands became armies, headed by warlords. The central government, far too weak at the time to defeat these internal enemies, tried to buy them off, or integrate them into the Chinese army. The generals sent to fight them usually preferred to let them escape. On the whole it was a fairly lackadaisical guerrilla war, despite the fact that it profoundly affected Chinese history and that a great many people perished in the recurring combats.[26]

Partisan warfare played a notable part in the American War of Independence. It was certainly not a crucial factor in the eventual outcome but it did have a delayed influence on military thinking; Clausewitz and the many theoreticians of the *petite guerre* frequently referred to it.[27] In some important respects it was different from previous wars. American soldiers did not have a good reputation; Wolfe, at Quebec, had called the American rangers the worst soldiers in the universe. They had little training and no discipline; another contemporary European observer wrote about a "contemptible body of vagrants, deserters and thieves" fighting on the American side. A recent historian has drawn attention to the fact that the American settlers were anything but noble savages — they belonged to a wealthy and sophisticated society with a high standard of living; they were willing to defend their homes but it was difficult to make them serve outside the militia area.[28] They were not very useful as regular soldiers. Circumstances made them choose a kind of warfare more congenial to their upbringing and more suited to their territory. They acquitted themselves well in the small enclosures and narrow lanes near Bunker Hill. This became the general pattern of the war; the Duke of Wellington was one of the first to point out that military operations were impractical by large bodies in North America (a thinly populated country, producing little food) except where there were navigable rivers or extensive means of land transport. Though their discipline was lamentable, the Americans were good shots and skilled horsemen and had learned about scouting from the Red Indians, whom they had fought for years around Fort Ticonderoga, at the Susquehanna River, and elsewhere.

Partisan warfare began in earnest in 1780, after the fall of Charleston; this was the time of Benedict Arnold's conspiracy, when Washington wrote about the "inevitable ruin of our cause." Threatening plunder and confiscation, Cornwallis tried to compel South Carolina to take the oath of allegiance. Among the patriots who retreated to the swamps and the mountains of the interior was one Francis Marion, aged forty-eight at the time, who became the most successful and most famous partisan commander of the war.[29] Marion was of Huguenot parentage, a small thin man, hardy and taciturn, whose favorite drink was a mixture of water and vinegar. He wore a scarlet outer jacket and leather cap. In the beginning he commanded a mere twenty men, among them a few boys and several blacks. Dressed in rags, miserably equipped, their appearance caused great hilarity among regular troops. They shot with bullets of pewter, buckshot and swan shot. Their swords had been fashioned from saw blades; they had neither tents nor baggage trains. "We had not seen a dollar for years," Marion was to write later. As the raids continued, the number of men under his command grew to a hundred and fifty, and eventually to several hundred, but there was a constant coming and going; he seldom had the same people serving under him for as long as two weeks at a stretch.[30] With the beginning of winter many recruits returned home. The raiders had continuously to be taught the essentials of partisan warfare. Some of them were desperate characters who were out for plunder and would change sides from time to time. "Many who fought with me, I am obliged to fight against," Marion sadly wrote on one occasion. Nevertheless, Marion constantly engaged platoons vastly superior to his own, broke up British recruiting parties, cut off Cornwallis's supplies and forever eluded Tarleton, who had been sent out to catch and destroy his forces. The mobility of his units was impressive; there were days on which they rode sixty miles and more. He instructed his men in the skills of night attack and in the techniques of small war. Whenever a bridge was to be crossed near the enemy camp, it was to be covered with blankets, so that the clatter of hooves on loose planks would not warn the foe. His scouts, frequently posted in the thick tops of trees, were taught a distinctive shrill whistle that could be heard at great distance. Sometimes his men went into action with only three rounds of ammunition apiece; they obtained rifles and cartridges from their prisoners or the enemy soldiers killed. He never informed his men of the anticipated length of a raid. When he established camp in the midst of the swamps, he had all boats in the vicinity burnt but his own. When the British forces burned the houses of the patriots, Marion retaliated by shooting enemy pickets, a practice that was considered unethical according to prevailing military convention. By his tactics of constantly retreating, he exasperated the British, who called him a robber and challenged the "Swamp Fox" to "come out and fight like a Christian." Marion did not deign to reply. He suffered defeat and setbacks: on one occasion, one of his recruits betrayed him; on another, the British discovered and devastated his main base at Snow's Island. Toward the end of the war, his raiders came under increasing pressure by the British, who had meanwhile mastered his tactics. But by now the tide had turned in favor of the colonists; Marion returned to civilian life, married a woman of means and became a state senator.

Among the other generals who engaged in partisan warfare, Andrew Pickens, Williams and, above all, Thomas Sumter ought to be mentioned. Sumter, like Marion, was a man in his forties when the war broke out. He had been a sergeant in the campaign against the Cherokees in 1762 and subsequently a captain in the mounted rangers. At the beginning of the war, he raised a militia force in Carolina. Unlike Marion, who took infinite pains over detail, he tended to neglect preparations for an assault, failed to reconnoiter and to coordinate his forces, and was forever becoming embroiled in quarrels with fellow officers.[31] A courageous and imaginative soldier and a man of great physical endurance, he would attack when the odds were overwhelmingly against him, sometimes against Marion's advice. He lacked strategic sense and, more often than not, his campaigns ended in defeat. The practice established in his units — "Sumter's Law," according to which his militia was recruited for a period of ten months and paid in plunder taken from the Tory loyalists — did not work too well; there was periodic discontent among his troops. He retired from military life before the end of the war and died in 1832 at the great age of ninety-eight. General Daniel Morgan was a greater strategist than either Marion or Sumter but he was not primarily a partisan leader. He employed irregulars of the militia in conjunction with the territorial army — with considerable success, as in the Battle of Cowpens.[32]

It has been argued that Washington did not win the war but that, owing to the terrain rather than to the enemy, Britain lost it.[33] Washington had realized early on that the war must be defensive in character, that the colonists — even with the help of the French — were not capable of facing the British in open warfare. Thus, the Americans developed their own style of conducting war, and partisan operations became frequent in the South, where the brutal behavior of the British forces had antagonized the local population. Furthermore, only relatively small detachments of cavalry could be deployed and there was only limited scope for artillery in the punishing terrain — wild and mountainous, with impenetrable woods and swamps, with no roads or negotiable rivers.

The leaders of the Southern irregulars were almost all veterans of the Cherokee campaign of 1761; in the war against the Indians they had learned the importance of taking cover, of moving silently, shooting accurately and other essentials of the *petite guerre*.[34] Their main weapons were the firelock and the saber; except for a short time in 1780, they had no artillery. Marion, considering it a burden, preferred to be free of it to retain his mobility. On the other hand, they occasionally used wooden dummies with good psychological effect. The partisans were mounted although the actual fighting was usually done on foot. In Marion's hit-and-run attacks, frequently carried out at night or at dawn, some of his men charged on horseback, but most would use their firelocks and muskets after dismounting.[35] The irregular forces used multiple-pellet loads (buckshot) for their smooth-bore weapons because there was always the chance of hitting more than one man with one discharge.[36] But the Americans were also among the pioneers of accurate, aimed shooting, a practice that was not yet widely accepted in the military manuals of the period. Marion applied tactics which Viriathus had so often used in his wars against the Romans: a small detachment would attack the superior enemy, then retreat in apparent disorder and lead the pursuers into an ambush. Furthermore, especially in the early days, the irregular forces would scatter in every direction after an engagement, making effective pursuit quite impossible, and would then meet again after a few hours or days to prepare a new attack.

Partisan activities in the South caused considerable losses among the British and weakened their resolve (which was not very strong anyway) to pursue the war. It has been argued that the Southern partisan forces were the salvation of the American cause since, if the British and Tories had had only the Continental regulars to contend with, "they would have won a complete victory in the summer of 1780."[37] But this argument ignores the fact that, by 1780, the British had already given up the North, their aim thenceforth being to preserve what they could of the Southern regions of their former colonies. Moreover, the decisive battles in the South, such as Cowpens and Guildford, were fought mainly by regular troops. Lastly, the tactics of the Southern irregulars were used more than once by their British and Tory opponents. While partisan tactics played an important part in the American War of Independence after 1780, they by no means won it.

THE VENDÉE

Fighting in the Vendée broke out in 1793 and, after some initial setbacks, the Vendean army was defeated with relative ease by the forces of the Republic. But this was not the end of the affair, for the second phase of the revolt, the *Chouannerie,* lasted for three more years and there were further, albeit short-lived risings in 1799, 1815 and 1832. The fighting affected large sections of western France, the *marais* and the *bocage,* the marshes and the forests on the left bank of the Loire; it also spread to Anjou and Haute Poitou. The rising has entered the annals of history as a classic manifestation of a counterrevolutionary movement, consisting of the most backward, ignorant and fanatical elements among a population that had not yet broken the shackles of their feudal masters, and the clergy, obscurantists unaware of the benefits of the revolution. The army, as the Republicans saw it, consisted of "deserters from all European armies, smugglers, gamekeepers and poachers."[38] These men, living in darkness, were manipulated by the royalists and the Church, which had joined forces in a giant conspiracy against the forces of reason and progress.[39] This, very briefly, is the traditional interpretation of the Vendean rising and it is, of course, correct in so far as the movement was directed against Paris and the new revolutionary authority. Religious inspiration was strong, stronger, in fact, than royalist influence. But there was no conspiracy; the risings were largely spontaneous, and had more to do with the unwillingness of young people to serve in the army and with the traditional conflict between town and country than with the speeches of Robespierre and the program of the Jacobins. The peasants bitterly resented the attempts of the bourgeoisie to dominate their communes. Aristocrats were prominently represented among the military leaders of the rising but there were even more commanders of very humble origin, more, actually, than among the generals of the Republic; the "nobles," moreover, were not dukes and viscounts but usually mere country squires. Finally, there was the resentment of local people against foreigners speaking another language, heirs to different traditions. The Vendée uprising was, in short, a bloody civil war, cruelly fought on both sides; it devoured about a hundred and fifty thousand victims, more than French losses in Russia.

From a military point of view, the campaigns are of considerable interest because the Republican army, itself the practitioner of revolutionary new tactics, had to face a new mode of combat which disconcerted it greatly. "Amid fire, skirmish lines, the exploitation of difficult terrain, rapid concentration of force, unhampered by any logistic straitjacket."[40] According to Joseph Clemenceau, who was captured by the Vendeans, their generals

> could never form the Vendeans into a permanent army or keep them under arms; it was never possible to make them remain to guard the cities they took; nor could anyone make them encamp or subject them to military discipline. Accustomed to an active life, they could not stand the idleness of the camp. They went to battle eagerly, but they were never soldiers.[41]

There had been a wave of unrest in western France, inchoate, without clear direction, even before the Revolution; a first small-scale armed rising took place in August 1792, near Chatillon, but the general attack started on 10 March 1793, when the tocsin was sounded in six hundred villages throughout the Vendée. At first the Vendean generals, such as Cathelineau, Bonchamp, Stoffet and d'Elbée, succeeded in making some headway against the Republican forces, of which there were not many in western France. They were beaten, however, at Lucan and Cholet in late autumn. By the beginning of 1794 they had lost their best officers and soldiers as well as most of their war material. Instead of avoiding direct confrontation and the siege of big cities, the Vendeans committed all the obvious mistakes; instead of retreating after their defeats into the interior of Brittany, where the Republicans could have followed them only with the greatest difficulty, they again went to battle against superior forces equipped with artillery which they themselves lacked. If, despite the capable leadership of Hoche, Berthier, Kellermann, Marceau and other famous generals, the armies of the Republic did not defeat them more quickly, the main reason was that the troops at their disposal were untrained and of inferior quality. Even more to the point, there was no unified command, political commissars sent out from Paris interfered constantly and gave orders which were, at best, unhelpful.

Paris had assumed at first that the Vendeans would be defeated in a matter of days, whereas the generals on the spot soon realized that they faced a mass insurrection and that pacification would be at best a long drawn-out undertaking. Kléber bemoaned that the Vendeans were always much better informed about the movements of his forces than he was about theirs, that they were constantly sending out patrols and attacking small Republican detachments. From the very beginning, Hoche stressed that pacification was a political rather than a military problem: "For the twentieth time I repeat," he wrote to the Directoire in Paris, "if one does not grant religious tolerance, one has to give up the idea of peace. This country needs civil administration — military administration does not suit it."[42] And, on another occasion: "If you are not tolerant, we shall go on killing Frenchmen who have become our enemies, but the war will not end." The Paris authorities were loath to show clemency to the enemies of the Republic, nor were they as yet fully aware of the extent of the revolt. Instructions were that all rebel leaders and soldiers were to be executed as well as anyone trying to evade conscription or who was found bearing arms. Since there were no game laws in the Vendée, everyone had a rifle, and thus could be shot without trial. Subsequently more lenient orders were issued. Only the leaders of the revolt were to be executed, a ruling more honored in the breach than the observance. When the Vendeans, for instance, killed their prisoners at Cholet, Westerman countered by killing prisoners and civilians and, from 1793, there was a vicious circle of terror and counterterror. In punitive raids women were raped and tortured, children killed, houses and crypts systematically burnt.[43] Relatives of Chouans were seized as hostages and executed. All over the west of France "traitors" and "enemy agents," however innocent, were sought out and arrested. Far from breaking the popular movement, such measures made it only more popular. The manifestos issued in Paris, claiming that the insurgents were creatures of the British, never gained credence.

The generals and the political commissars had reported to Paris as early as October 1793 that the "Vendée no longer exists." Such confident reports were correct to the extent that the rebels were no longer able to raise an army of fifty thousand, as they had done in the beginning. But for all that, they remained in effective control of the country and their guerrilla tactics made it far more difficult to attack them; Hoche needed more than a hundred thousand men, including the whole army of Mayence, to suppress the rebellion. Three years later, in July 1796, he could report with greater justification that the Vendée had been pacified. But even this was not final victory, for though all the major leaders had been taken prisoner and executed, and their troops decimated, unrest on a small scale still continued. Two thousand peasants attacked Nantes in 1799. The Vendean army was royalist in its sympathies, but the last thing the Comte d'Artois (the future Charles X) wanted was to accept the generalship that was offered to him. The leaders of the insurgents were a mixed lot; Chouan was apparently the nickname given to the four Cottereau brothers (Jean being the best known), who were smugglers at the little city of Laval, but they played no leading role in the war. The first generalissimo was Cathelineau, a wagoner and church sexton and son of a mason. He was a native of Anjou; intelligent and fearless, but no great master of strategy. He was killed in the very first months of the uprising and was succeeded by d'Elbée, a former captain of the cavalry, and La Rochejacquelein, a very young lieutenant, who is mainly remembered for having admonished his followers: "Let us find the enemy. If I retreat, kill me; when I advance, follow me. If I am killed, avenge me."

Bonchamp, also a former army captain, was killed early on in the campaign. One of the two principal leaders of the revolt was Stoffet, who had been a corporal in the army and subsequently a gamekeeper, an Alsatian by birth and son of a miller. A good leader of men and a capable officer, he had nothing but disdain for the nobles and always stressed that he fought for religion and the Church, "which makes all men equal."[44] He had the reputation of being a cruel man and many atrocities were ascribed to him. Mercier du Roche, who fought against him, wrote that he would have been a good general of the Republic. The other important military commander was Charette, a former naval lieutenant, who, like Stoffet, was hunted down in early 1796, captured and executed. Napoleon is said to have thought highly of him. To repeat, the notion that the Vendean revolt was a movement inspired and commanded by feudal chiefs is not borne out by the known facts. Caillaud was a locksmith, Forestier, the son of a shoemaker. There were not a few adventurers in their ranks, the outcasts of all classes. By and large, it was a popular movement with anticapitalist undertones; the operations of the Chouans struck terror among the bourgeois of the cities.

The popular character of the rebellion has to be stressed because it provides the key to an understanding of the roots of the *Chouannerie*. It was a movement of national liberation of sorts, even though its ideology was diametrically opposed to the ideals of the French Revolution. The enthusiasm of the Chouans surprised and mystified Republican observers. One of them commented: "One goes to battle like to a fête — women, old people, children aged twelve to thirteen, I have seen them killed in the front line."[45] Hedonville, who commanded the Republican forces in the later stages of the rising, reported to Bonaparte that the local population invariably gave the army wrong information about the whereabouts of the Chouans. There was no such solidarity among the Republicans. "We never lacked ammunition because your soldiers sold it to us," Coquereau wrote mockingly to the leaders of the Convention in Paris. Contemporary observers noted the prominent role played by women, both in the preparation of the rising and the actual fighting; they urged their sons and husbands to go to war, and many accounts have it that they were the most fanatic proponents of a *guerre à l'outrance*.

The tactics of the Chouans have been described by many eyewitnesses. Thus Joseph Clemenceau:

> They fought without order, in squads or crowds, often as individual snipers, hiding behind hedgerows, spreading out, then rallying, in a way that astonished their enemies, who were entirely unprepared for these manoeuvres; they were seen to run up to cannons and steal them from under the eyes of the gunners, who hardly expected such audacity. They marched to combat, which they called *aller au feu*, when they were called by their parish commandants, chiefs taken from their ranks and named by them, centurions, so to speak, who had more of their confidence than did the generals chance had given them; in battle as at the doors of their churches on Sunday, they were surrounded by their acquaintances, their kinfolk and their friends; they did not separate except when they had to fly in retreat. After the action, whether victors or vanquished, they went back home, took care of their usual tasks, in fields or shops, always ready to fight.[46]

Hoche, in a letter to Aubet-Dubayet, noted that the Chouans had friends and agents everywhere and always found food and ammunition whether it was given voluntarily or taken by force. Their principal object was to destroy the civilian authorities, and to that end they intercepted convoys, assassinated government supporters, disarmed Republican soldiers and even tried to foster unrest among the town dwellers. Their tactics were to disperse silently behind hedges, to shoot from all directions, and at the slightest hesitation on the part of the Republican forces to attack them while shouting their war cries. Encountering stiff resistance, they would retreat and renew their attack on another occasion. A Chouan leader addressed the following warning to the Republican authorities:

> The Chouans will demolish all the bridges, intercept all communications, destroy all mills which they do not need, cut the knuckles of all cows and the hocks of the horses employed in transporting supplies, kill every state employee performing his duty, kill everyone obeying requisitions; they will force the population under threat of the death penalty to follow them . . . they will not lower their arms until they have made France a great cemetery.[47]

The Chouans were organized on a local basis with assembly points in every village. The peasants kept their own arms and brought food along for three to four days when they went into action. Some of them wore uniforms — green coats and pantaloons with red waistcoats and a white cockade. They frequently attacked at night and they were past masters at ambushing small detachments in a country which favored such operations. They were less inventive when fighting involved larger units on each side. Kléber observed that their plan of battle was always the same:

> It consists in greatly extending their front so as to envelop us and throw disorder in our ranks. . . . When they advance on us they generally take care to dispose their army in three columns whatever the character of the ground . . . their right column is always the strongest and made up of the best men. The center column, strengthened by a few cannons, moves forward, while the other two open up into skirmish lines along the hedges, but the main effort almost always comes from the right.[48]

Republican commanders noted that it was the practice of the Chouans not to attack unless their force was greatly superior. When forced to retreat, the Chouans would rally at a distance of a few miles and immediately counterattack the unprepared enemy. On certain occasions they would appear on the battlefield with Republican hostages in front of them; at other times they would wear

Republican uniforms so as to surprise the enemy. They were men hardened by long winter nights, men who could jump over hedges and ditches, could see in the dark and hear the slightest noise.[49]

In the jaundiced view of their Republican enemies, the Chouans were little better than untamed animals. In actual fact their approach was quite sophisticated, when engaging in psychological warfare, for instance. Early in the war the Catholic army was given orders not to rob and to show leniency toward prisoners. These were released and given special passes on condition that they gave their word of honor not to fight again in the Vendée. Among the Chouans there were detachments of French, Swiss and German deserters; these were not very strong, but their very existence helped to demoralize the Republicans. In a leaflet signed "Les Brigands," the Chouans addressed the Republican soldiers as "*nos amis et frères.*" The city dwellers who had grown rich (it said), were those culpable for the bloodshed. The Republican soldiers were mere dupes, exploited (like the Chouans) by the common enemy.[50] On more than one occasion the Chouans attacked, caught, and executed gangs of bandits to dissociate themselves from criminal elements and to help the peasants who had suffered from them. In the end the revolt of the Chouans was put down because the Republic concentrated vastly superior forces against them and used effective counterguerrilla tactics both on the military and the civilian level. A network of entrenched camps was established which the Chouans could not bypass. The Republican soldiers systematically combed the area, seizing the peasants' cattle until they surrendered their arms. All suspects were arrested in the course of these operations, which were carried out in great secrecy with the help of the local police. In the instructions to his officers, Hoche always stressed the importance of establishing a good intelligence service and of deceiving the enemy about their intentions.

But Hoche also understood that this war could not be won by arms alone. He insisted on strict discipline on the part of his troops; severe punishment threatened those who assaulted civilians and did not respect their property. In an appeal to the civilian population, he announced that, unlike the rebels, the soldiers of the Republic would not answer cruelty with cruelty, terror with counterterror: "You will find in these soldiers zealous protectors just as the brigands will find in them implacable enemies. Peaceful and honest citizen, stop believing that your brothers want your ruin; stop believing that the fatherland wants your blood, it merely wants to make you happy through its beneficial laws." In his correspondence with Paris, Hoche emphasized the importance of treating the clergy well. He saw the priests as the main instrument for regaining the confidence of the population. They could be made to understand that the continuation of the *Chouannerie* meant war without end, with all the human suffering and the material losses that involved. A victory by the Republic, on the other hand, would restore peace and religious tolerance. Hoche's moderate course of action did not make him any more popular among the Chouans, while in Paris he was suspected of being too soft and thus ineffectual. A tribute was paid to him only in 1796 after the pacification had been completed, when Carnot, the president of the Directoire, expressed the thanks of the Republic to the "Army of the Ocean" and its commander.

The Chouans received some supplies from England but this assistance was at no stage decisive. Of the nobles who had left France, only few were inclined to return and fight. Those who did despised the peasants and their conduct of war. They wanted to fight according to time-honored tradition — and were quickly wiped out by the armies of the Republic. Since the Chouans did not hold any major port, supplies were neither substantial nor regular and in the last resort they could rely only on their own resources. The Chouans had no guerrilla doctrine; they were simply adapting their war to local conditions and, after their initial setbacks, fought the only way they possibly could.[51] The Russian partisans under Denis Davydov were regular soldiers, engaged in raids to the enemies' rear. The Spanish guerrilla fighters received substantial help from Wellington and the various local juntas, which also provided political leadership. The Chouans, in contrast, had no government, no political advisers. Of all the early guerrilla movements, it was the most spontaneous, the most isolated, and thus, in many ways, the "purest" specimen of the lot.

SPAIN

The Spanish war against the French (1808–1813) not only produced the term *guerrilla* but remained for many years the guerrilla war *par excellence*. Napoleon had badly misjudged the situation, assuming that with the defeat of the Spanish regular armies the war in the peninsula would come to a speedy end. But popular resistance continued and tied down substantial French forces. Taken in isolation these activities were mere pinpricks, but they had a cumulative effect. Furthermore, they provided inspiration to the anti-French forces all over Europe.

Spain at the turn of the century was a backward country ruled by the largest and most useless aristocracy in Europe.[52] The small-town bosses, the *poderosos,* were an important factor, often employing armed gangs or cooperating with them. Spain had a long tradition of banditry and guerrilla fighting of sorts. In a proclamation, the Empecinado, one of the leading guerrilla chieftains, invoked the memory of Sertorius and Viriathus.[53] The Spanish establishment and the liberals were on the whole pro-French; the former because they had realized since the lost war of 1783–1785 that resistance against their powerful neighbor in the north was hopeless, the latter because France was the country of the revolution. The attitude of the liberals gradually changed, largely because in Spain Napoleon supported the most reactionary forces — Ferdinand and Godoy. Sections of the Church were more popular in character than they were usually credited with being; the lower clergy was quite poor, their contact with the peasantry was close — most of them were, in fact, of peasant origin. The peasantry, or at any rate a militant minority among them, turned against the French with the same slogans that had been used previously in the Vendée, and later in Russia: in defense of Church and monarchy — though not necessarily the king. There was a genuine patriotic upsurge provoked by invasion and conquest — the Spanish regular forces were beaten, but, in marked contrast to the political and military tradition of the age, the people would not accept the fact. There was also a great deal of xenophobia — foreigners had no business to be in Spain. Thus it was essentially a war carried on by *le petit peuple* with an admixture of students, monks, local notables, a few officers, doctors and lawyers. But the backbone of the movement was rural, very much in contradistinction to Garibaldi's "Thousand," among whom there were no peasants at all. As in the Vendée, there was a strong populist bias, and the French mistakenly assumed that this would drive the men of property into their camp. It was the "mob," not the establishment, which in Saragossa demanded arms and patriotic resistance and the same happened in many other places. The performance of the Spanish army had been disastrous; true, there had been the victory at Bailén, and Saragossa resisted for ten weeks — to the admiration of all enemies of Napoleon. But these were followed by the calamitous Battle of Ocaña and the Spanish flight at Talavera. "I have never known the

Spaniards do anything, much less do anything well," wrote Wellington.

By 1809 the Spanish regular armies had virtually ceased to exist and it was in that year that the guerrillas first appeared on the scene. Nominally, they were subject to the supervision of the central junta which had retreated from Madrid to Cadiz via Seville, and the various provincial juntas — in Castile, Asturias and elsewhere. But these bodies were often paralyzed by internal dissension, and they helped the guerrillas only in a small way, monetary for the most part. Since they were chased by the French from town to town, they could not act as an alternative government. The first major guerrilla bands appeared in Aragon and Galicia; in the beginning the insurgents were ill trained, shot poorly, kept unreliable watch and were given to panic. Only by trial and error did they learn caution, and also learn not to fight except in the most favorable conditions. Soon much of Castile and León was in their hands, which made it impossible for the government to collect taxes. Originally the French had stationed garrisons only in major cities such as Burgos, Valladolid, Segovia, Guadalajara, but the *guerrilleros'* activities compelled Suchet, the French commander, to leave some units in every town and major village. Thus, the French army was forced to spread its forces very thin on the ground. Every messenger had to be given a substantial escort and the smaller garrisons were by no means safe from attack. The central junta had called for a people's war (*corso terrestre*) in a decree published in Seville in April 1809, promising awards for special feats of heroism and support for windows and orphans of fallen soldiers. It called on the population to give food, supplies and information to the *guerrilleros*, who, however, had not waited for official appeals. There had already been attacks against French soldiers by isolated individuals who had hoped that others would follow their example and that as a result the country would be saved from the invaders.[54] Gradually small bands of guerrilla fighters came into being, to increase in strength over the years. The most important were those headed by Espoz y Mina and the Empecinado.

Espoz y Mina was born in a Navarrese village in 1781, "the son," in his own words, "of honest farmers of that province."[55] He raised sheep and cattle, having taken charge of the family's farm after his father's death. "I lived in deepest peace," he wrote in his autobiography, until the convulsions of 1808 put an end to the rural idyll. He enlisted as a soldier in Doyle's batallion in February 1809 to fight the French and later joined a small guerrilla band which had been formed by his nephew Xavier, nicknamed "the Student." The younger Mina engaged in a number of daring actions, such as storming the town of Tafella, but was captured by the French in March 1810, whereupon the seven remaining insurgents elected Espoz their chief. Since, again as recorded by Mina, no one belonging to the rich classes appeared on the scene to hoist the banner of resistance, he rallied the patriotic elements, notably the youth. Among his first operations was an attack against a rival gang led by one Echeverria who had become the scourge of the villagers, robbing, plundering and in general making himself very unpopular. Though his own forces were inferior at the time — Mina had only four hundred men under his command — he attacked Echeverria and had him and three of his aides shot.

In April 1810 Mina began his raids which were to harass the French for the next four years. One of his major accomplishments was to attract large sections of the French Ninth Corps to Navarre — about twenty-six thousand men according to his own account, eighteen thousand according to other sources — a force badly needed by Masséna at the battle of Salamanca:

> Mina's lot during this period was no enviable one: he was beset on all sides by flying columns, and was often forced to bid his band disperse and lurk in small parties in the mountains, till the enemy should have passed on. Sometimes he was lurking, with seven companions only, in a cave or a gorge; at another he would be found with 3,000 men, attacking large convoys, or even surprising one of the blockhouses with which the French tried to cover his whole sphere of activity.[56]

Mina was chased simultaneously by several French generals — Reille, Caffarrelli, d'Agoult and Dorsenne — with forces five or six times as strong as his own. In recognition of his services, the Regency made him commander in chief of the *guerrilleros* in Navarre. The French, on the other hand, were exceedingly frustrated by the tiring, costly and ultimately pointless marches through almost impassable territory; they burned villages and shot all captured guerrillas. Mina retaliated in kind, threatening to execute four French officers for every Spanish officer, and twenty French soldiers for each Spanish one. When the French shot four members of the Burgos provincial junta, the Curate Merino, another guerrilla commander, had eighty French soldiers shot. In October 1811, after the French governor in Pamplona executed several civilians, including some priests, for having helped Mina, Mina avowed "war to the death" to every French soldier and officer, including the French emperor.[57] But in 1812 both sides, by mutual agreement, stopped this indiscriminate slaughter.

The most important military action conducted by Mina took place in March–April 1812. About thirty thousand French troops were massed in pursuit of the guerrilla leader who at that time had some three to four thousand men under his command. Caffarrelli invaded the Pyrenean valley of Roncal where Mina's hospitals and supply depots were located. Mina escaped into Aragon, much to the chagrin of the French who were, however, consoled by the thought that even though he had again escaped, he would need many months to recover from the blow that had been inflicted on him. But a mere two weeks later Mina was back in action and scored one of his greatest successes, attacking and destroying a great convoy in the Pass of Salinas. Five thousand Polish soldiers were killed or wounded, four hundred and fifty Spanish prisoners were freed, and immense booty fell into the hands of the *guerrilleros*. Once again major French forces were sent out to capture this elusive man, whose boast was that he was never surprised, but who, on 23 April, found himself nevertheless encircled.[58] Malcarado, one of his chief lieutenants, had betrayed his whereabouts to the French general Pannetier. For once the element of surprise was on the side of the French; five Hussars appeared on Mina's doorstep. His men were dispersed through the village of Robres; some had already been taken prisoner. Mina counterattacked with a few orderlies, rescued some of his officers and men, and, having escaped in a hazardous march to the Rioja, executed the traitor on the way. Like most *guerrilleros*, Mina was given to magnifying the extent and the importance of his exploits; it is unlikely, for instance, that the French, as he claimed, suffered forty thousand casualties in the battles against him, and that he took fourteen thousand prisoners (total French casualties in the Peninsular War from 1808 to 1814 were ninety thousand). But there is no doubt, as he maintains in his memoirs, that at a critical time for Wellington's army he distracted large sections of the French army of the north which, if employed elsewhere, might have been of decisive significance.

Mina received two heavy siege guns in late 1812; they had been landed on the Biscay coast and were put to good use in the siege of Pamplona in February 1813. Mina proceeded to expel the French from Navarre. In late March of that year he all but destroyed two French batallions in open battle. A last effort to pursue and destroy his forces was made in May 1813 by General Clausel with the help of four French divisions. Mina countered with the stratagem used so often before when facing superior forces: he dispersed the units until his command, and the small detachments easily filtered through the enemy net. But Clausel also attacked Mina's chief supply base, gambling on the assumption that the guerrilla leader would not surrender his vital strongpoint without a fight. The gamble paid off in part; one thousand guerrillas were killed or captured in the battle that ensued and Mina had to escape posthaste to eastern Aragon. Once here, however, he rallied some of his scattered troops and enlisted fresh volunteers. During the following months, Mina's forces gradually became part of the new Spanish regular army; they were among the first units that entered France in pursuit of the French troops. In this last phase of fighting on Spanish soil, Mina no longer evaded battle as in his early guerrilla days. General Buquet wrote Berthier: "Mina's troops are now so hardened that, when forces are equal, they accept battle quite gladly. He has now at least 7,000 men ready to give fight, including a cavalary of 1,000, which is not at all to be despised."[59]

Next to Mina, the Empecinado (Juan Martin Diaz) was the most notorious guerrilla fighter of the Peninsular War. The forces serving under this Castilian peasant were smaller than Mina's, but his daring attacks harassed the French, gave fresh courage to his countrymen, and made him a legend in his own lifetime.[60] Diaz was born in September 1775 in a village near Valladolid; his parents were peasants and he himself, in the words of his biographer, "acquired great bodily strength working in the fields." At the age of sixteen he ran away from home and joined the army, but his parents persuaded him to return. Later on he rejoined the army and, after serving for a while as a private, returned to his farm work and married. Then in March 1808 he cajoled two neighbors, one a lad of sixteen, into going along with him in a challenging enterprise; they intercepted first one, then another French courier, and killed them both. Several more acquaintances volunteered their services and a small guerrilla band was formed which became active southeast of Madrid. The Empecinado claims that his unit killed six hundred Frenchmen in the summer of 1808, well before Mina appeared on the scene; the figure is no doubt greatly exaggerated.

In these early days the Empecinado gave no quarter to prisoners because there were no facilities for transporting and detaining them. Like other guerrilla leaders, he offered his men both daily pay and a share of the plunder, but he also claims that, when on one occasion early on in his guerrilla career he seized a great sum of money, the larger part of it was given to the junta. His contemporaries described him as a man little above middle stature, firmly knit and of muscular frame, with a dark complexion and black and animated eyes. He was calm but could react quickly when circumstances demanded it. Steadfast in adversity, he was, according to his friends, a modest man who sought no rewards for his activities.[61]

The Empecinado was arrested by Spanish government forces in November 1808 following a denunciation. He broke his handcuffs and escaped from jail just as the French entered the village in which he was being held. Pretending to be a stableman, he waited on the French dragoons, picked one of their best horses and made off on it. His second guerrilla career began with small operations in the Madrid and Salamanca region carried out in collaboration with his three brothers, one of them only fifteen at the time. By the spring of 1809 his little band numbered forty-eight men and he received some money from General Moore to keep it going. The Empecinado (the name is derived from the black soil of his native region) attacked gendarmes and small French garrisons. Flying columns were sent out to destroy him, and the villagers were threatened with dire punishment if they helped him; even his aged mother was arrested and faced with execution by the French. In order not to endanger the villagers, the Empecinado launched sham attacks against them to obtain food and other supplies. On several occasions he abducted the *alcaldes*, the village elders. Like Mina and other guerrilla leaders, he engaged in almost constant hit-and-run assaults; he suffered numerous defeats but was never decisively beaten. With a hundred and twenty men he entered the fortress of Pedraza, then withdrew to the Avila mountains where seventy of his men deserted. Made captain of cavalry by the junta, he harassed the troops of Soult and Ney at the time of their temporary retreat from Galicia in 1809. Shortly afterward, he briefly enlisted four thousand peasants armed with firelocks and fowling pieces. He attacked Salamanca, but some of his soldiers again deserted and he had to restrict his activities for several months. On one occasion he fought a duel with a French commander, whom he killed. The junta dispatched him to the Guadalajara area, where he evaded a trap set for him by twenty-five hundred French troops. By this time he was treating prisoners according to the rules of war; like Mina, he had detachments consisting of Polish, Italian and German deserters. There were even a few German officers among them. According to the Empecinado's own account, he induced about six thousand men to desert, again no doubt a greatly exaggerated figure. But it is true that the French commanders regarded these defections as a serious problem.[62]

At one stage of the Peninsular War robbers became scarcely less of a menace than the French occupiers, and the Empecinado, "scrupulously attentive to the interest of the laboring class," destroyed the gang of one Don Bernardo Mayor.[63] By 1810–11 he commanded a thousand infantry and four hundred horsemen; he went through the pretense of forcing young villagers to join his units so as not to subject their families to French acts of retaliation. Like Mina, he had to face treason and mutiny in his own camp and he suffered more than one setback. The provincial junta gave him bad advice and two hundred of his men were taken prisoner in a French surprise attack while in church on Good Friday, 1811. Hugo, the French general who had been sent out to capture him, offered a deal: why would he not enter King Joseph's service? The Empecinado's answer was, "Would you kindly stop writing me. . . ."

In May 1811 he received three pieces of artillery, but there was renewed trouble among his officers and he was beaten by the French at Cuenca. Having commanded forces almost equal to a divison, his effective strength following these reverses was again down to four hundred men. The new junta, however, helped him rebuild his forces and in lower Aragon he took "thousands of prisoners" over a period of several weeks. After these successes, there was once more a sudden defeat; attacked by renegade Spaniards, he was surrounded and could save himself only by jumping down a precipice. Severely wounded, he was out of combat for some months. By March 1812 he had recovered, and he entered Madrid with Wellington; Guadalajara surrendered to him.

The setbacks encountered by the Empecinado were the result of inattention to orders, to disobedience and, perhaps above all, to lack of punctuality. But, as the chronicler of the Empecinado's campaign notes, the strict observance of military discipline was impossible to enforce in a corps formed amidst the bayonets of the enemy. The *guerrilleros* were officered by people whose minds and habits were not those of professional military men.[64]. Furthermore, there was always the danger that those who had been punished would desert. It was only toward the end of the war that the rudiments of military discipline became firmly implanted in the Empecinado's little army. He claims that he never let his men be idle; when he seized funds that had been extorted from the townspeople by the French, they were returned to the owners. Even with the French, relations were almost friendly toward the end; on one occasion General Suchet handed over to him twenty Spanish officers who had been taken prisoner.

Espoz y Mina and the Empecinado were but two of a whole host of guerrilla leaders. There was Merino, a goatherd in his youth who became a curate, but who spent more time hunting than attending to his clerical duties. His base of operations was in the Burgos and Soria region where he had his own musket factory. Among his early recruits there were many young law students; one of them, Santilan, later became Minister of Finance. At the end of the war he had three to four thousand men under him. Merino, too, was made a general by the junta but he usually wore his black surtout, which he preferred to his general's uniform. Those who knew him intimately described him as a man of small build but iron frame, of vindictive and cruel disposition, but truly disinterested; he never claimed any part of the booty.[65]

Among the other guerrilla leaders, Julian Sanchez deserves mention; a former professional soldier, he collected some hundred horsemen, and attacked French convoys in the neighborhood of Avila. Later his forces became part of Wellington's Anglo-Portuguese army. The French thought more highly of Sanchez (whom they referred to as "*chef de parti*") than of Mina and the Empecinado who in their eyes were no more than bandits. Camilo Gomez, a wealthy Castilian farmer, and Lucas Rafael, a young clergyman, gathered small bands around them; in each case members of their families had been killed or molested by the French. In old Castile a guerrilla leader nicknamed El Medico ("short, ferocious, dark eyes, lots of hair") raided the countryside up to and including the suburbs of Madrid. In Zamora Province, el Capuchino (Jean de Mendietta) was the main guerrilla leader. At one time he captured the French general Francheschi, but very soon released him. When he was in turn in French captivity and fell ill, he appealed to Francheschi's wife for help and obtained it. El Marquesito was made captain general in Asturias and Francesco Longa was another prominent leader of the insurgents.

Reference should also be made to the *Miqueletes* and the *Somaten* (*Honrados*), local Catalan militias, which had existed since time immemorial. The Somaten, divided into *partidas* of a hundred men, were more loosely organized; the Miqueletes consisted of smaller units of mountain peasants. The Miqueletes took a prominent part in the defense of Gerona, while the Somaten were mostly employed on guard duty.[66]

The subsequent fate of the guerrilla leaders was not, on the whole, a happy one. The war years had whetted their appetite for power or adventure or both. As far as they were concerned, fighting did not end in 1814. The younger Mina was executed as a rebel by Spanish forces in Mexico in 1817; his more famous uncle survived him for many years, a not very effective general in the Carlist wars. Most of his later life he spent in exile — in England, Switzerland and, longest of all, in France; the very country against which he had declared "war to the death" provided a shelter. Merino, the curate, and the Empecinado were on opposite sides in the Carlist wars, the former allying himself with the French party, the latter joining the constitutionalists. Neither distinguished himself in the fighting; it was one thing to attack foreign invaders, but another, much more difficult task to face one's own countrymen, who knew the countryside equally well. The Empecinado was made prisoner and hanged. The Marquesito was executed by the Spanish authorities in 1814. The Capuchino plotted against King Ferdinand and was hanged one year later, in December 1815.[67]

While most guerrilla leaders were eager to be recognized by the central junta, and to receive military rank, their contact was mainly with the provincial juntas; and even more often with the local communes. When financial assistance failed to arrive, the guerrillas sent out their own tax collectors, much to the chagrin of the French. Mina had a yearly income of some three hundred pounds from a customs house he kept at Irun, near the French border; the French authorities, needless to say, greatly resented such impertinence, but gradually a *modus vivendi* nonetheless developed between the customs officers. Mina's excellent spy network throughout the Pamplona region was based on the information received from government officials who kept him informed of French troop movements. The *guerrilleros* expected such help from the local population. When it was not forthcoming, they were not slow in displaying their anger. On one occasion Mina had three *alcades*, who had not warned him of the presence of enemy units, hanged.[68] Mention has been made of the fact that the major guerrilla bands had their own hospitals, workshops, ammunition factories and mobile workshops for the manufacture of clothing and saddlery. Mina maintains that he never imposed contributions on the town population (excepting only "collaborators" with the enemy), and that he took nothing from the peasants but bread, meat and wine, and barley for his horses. The rural populations nevertheless suffered from the guerrillas; the small bands often plundered indiscriminately, and the larger ones needed provisions which the peasants could ill afford to spare.[69] But this, as the guerrillas argued, was inevitable in a popular war of national liberation. On the whole, they could count on the sympathy of the local inhabitants, even though they were as much feared as loved.

Their relations with the Spanish regular army officers were not too amicable. To the professionals, Mina and the Empecinado were mere peasants, lacking military training, experience and discipline — little better than brigands. The guerrillas for their part, hardened in countless skirmishes, made short shrift of such criticism. "I was several times wounded, had four horses killed under me, still have a ball in my thigh," wrote Mina, but to the professionals "I remained a mere *guerrillero*. . . . To be sure, we learned the art of war not in the military academy, but on the field of battle."[70] The guerrillas had little respect for the personal courage of the regular army officers whose record against the French had not been impressive.

The attitude of the British commanders toward the guerrilla was ambiguous. Wellington hardly ever mentioned them in his dispatches, Napier wrote that these "undisciplined bands" had been as dangerous to their own country as to the enemy, while Sir William Vane asserted that the rural population was more afraid of them than of the French because they plundered everyone who fell into their hands. In reality the guerrillas helped Wellington's army in more ways than one. The French had a four-to-one superiority in regular soldiers on the Peninsula and it was the guerrillas' diversions that prevented a mass concentration of these forces against Wellington, with what well might have been fatal results.[71] They intercepted many messages between the French commanders and delivered them to Wellington's headquarters. Dispatches of political interest were sent to the free Spanish press in Cadiz for publication, providing grist for the mills of psychological warfare. At the time of the Battle of Salamanca the French armies were situated a mere fifty miles from each other, but, owing to the activities of the guerrillas, they failed to realize it and did not unite against the enemy. By attacking the forces of Soult and Suchet in Andalusia and Valencia in the spring campaign of 1812, the guerrillas prevented them from sending reinforcements to Marmont for his battle against Wellington. The following year, Wellington, by coordinating his actions with those of the guerrillas, kept two French armies apart in central Spain, defeated one and forced the other to retreat to France. Britain supplied Spain with millions of pounds and tens of thousands of rifles, guns, powder and uniforms; much of this went to regular Spanish units, but a substantial part was either allocated to or fell into the hands of the guerrillas.

In purely military terms the importance of guerrilla warfare has been overestimated both by the guerrilla leaders and by some other writers. Mina may have insisted that he was never surprised and the Empecinado that he never lost a battle. ("When the French pressure gets too hard, I retreat. . . . The French pursue me and get tired, they leave people behind and on these I jump. If a man remains one step behind the rearguard he is not seen again.")[72] But in the event, the guerrillas were beaten by the French forces time and time again. Toreno claims that by 1810 the French needed some 108,000 men to keep their lines of communication clear.[73] The figure is, of course, exaggerated; furthermore, these were in the main not French units but inferior troops from other parts of Europe. It is also true that most attempts to coordinate the struggle of the guerrilla bands, let alone to integrate them, failed completely. But what matters in the last resort is that the guerrillas had a considerable nuisance value; even the most determined attempt to stamp them out in early 1812 came to nothing. Oman describes the situation at this critical turning point in the Peninsular War:

> Large forces have been put into motion; toilsome marches have been executed over many mountain roads in the worst season of the year; all the bands of the insurgents had been more than once defeated and dispersed. But the countryside was not conquered: the isolated garrisons were still cut off from each other by the enemy, wherever the heavy marching columns had passed on. The communications were no more safe and free than they had been in December. The loss of men by sickness and in the innumerable petty combats and disasters had been immense. The game had yet to be finished, and the spare time in which it could be conducted was drawing to an end.

The guerrillas achieved no brilliant or major victories, but this was never their intention. Their purpose was to weaken the French army by incessant unpredictable harassments. At the start a small detachment was enough to protect a French courier, later a company or even a batallion were not sufficient. A Prussian officer serving with the French recorded in his diary an observation that reflects what successive generations of regular soldiers were to discover from their own bitter experience of fighting guerrillas: "Wherever we arrived, they disappeared, whenever we left, they arrived — they were everywhere and nowhere, they had no tangible center which could be attacked."[74] The French always had to be prepared for an attack; the mood is perhaps best depicted in a picture by Colonel (later General) Lejeune who, for a time, was one of the guerrilla's prisoners. His canvas, *Attaque d'un convoi par une guerrilla,* was exhibited in the 1819 salon.

The guerrillas could have been destroyed by the French had the area to be controlled been smaller, the mountains more accessible. In the plains of southern Spain there were no guerrillas in this campaign, nor in earlier nor subsequent wars. Without Wellington's army and Napoleon's defeat in Russia, Spain would not have been liberated. The real achievement of the Spanish guerrillas was political and psychological, the fact that popular resistance continued and that it constituted a serious problem for the occupiers. The French image of invincibility suffered as they proved incapable of imposing their will on the Spanish nation. Thus the little skirmishes and ambushes of Mina and the Empecinado ultimately had repercussions far beyond Spain. They provided comfort and inspiration to the enemies of Napoleon throughout the length and breadth of Europe.[75]

THE TYROL

The year that guerrilla warfare broke out in Spain witnessed a rebellion similar in character, though less successful, in central Europe. Austria had been defeated by Napoleon, but Andreas Hofer and the peasants of the Tyrol continued to resist the invaders. By the terms of the peace treaty, their land had ceased to be part of Austria and become Bavarian territory. The uprising of the Tyroleans has been attributed not very convincingly to a reaction against the malpractices of the Bavarian administration. The Tyrolean peasants were royalists and deeply religious and, like the Vendeans, put Church above state. Their leaders called on them to attack the infidels in the name of the Holy Trinity, a somewhat inappropriate rallying cry considering that the French and the Bavarians were also good Catholics; a shared faith in no way diminished the Tyroleans' hatred of them. It was, in essence, a struggle of the natives against foreigners, for tradition against the imposition of new, alien rulers.

The Tyrol rising lasted from April to November 1809. The first battle took place at the Isel Mountain south of Innsbruck, which was also the scene of much of the ensuing fighting.[76] Andreas Hofer and his lieutenants made good use of the difficult mountain terrain; the peasants always tried to command the heights and they invariably attacked enemy forces in narrow mountain passes. In the battles for Mount Isel, some seventeen thousand Tyroleans faced about the same number of Bavarians, with the one advantage to the peasants lying in their superior strategic position. They had no artillery other than some wooden cannon, but the Bavarians had to storm uphill and once they approached the insurgents' lines they met with deadly fire. The peasants were organized in companies of a hundred and fifty to two hundred men all hailing from the same valley; each company had some ten to twelve subsections (*Schrannen*).

Like northern Spain, the Tyrol was ideal guerrilla territory, but it was much smaller and could more easily be controlled by the invader. It would have been difficult for partisans to hide anywhere for any length of time. Besides, the Tyrolean peasants were not geared for extended guerrilla warfare; they wanted to liberate their country at one stroke. In some other respects the similarities with Spain were striking: the revolt took place after the regular Austrian armies under Chastelet and Hormayr had been defeated. It was only in the month of May that Hofer, originally the commander of a partisan unit in a little valley, became the military leader of all Tyrol.[77] Relations between the Tyroleans and the regular Austrian army and civil authorities were strained. The professional soldiers, most of them aristocrats, considered war an occupation for gentlemen and looked with contempt on the amateurish, crude and often savage warfare waged by the peasants. Atrocities were committed by both sides; the peasants threw Bavarian soldiers alive into burning houses, the Bavarians cut off the tongues and noses of captive Tyroleans. The priests told their flock that no Tyrolean above the age of twelve would go to heaven unless he (or she) killed at least three Bavarians.[78] After the first liberation of Innsbruck there were scenes of drunkenness and widespread plundering; in the evening there was a pogrom against the local Jewish community.

Patriotic historians later contended that the peasants showed moderation and magnanimity and that there were no excesses, but this is not borne out by contemporary documents.[79] Not that the peasants acted without provocation. Napoleon had sent two divisions to the Tyrol not so much to occupy the country as to spread terror and to discourage local resistance. Villages were plundered, houses burnt, and the attitude to the local population was anything

Cossacks and fifty Hussars. Bagration, who conveyed the message with regret, said that he would have given Davydov three thousand men ("I do not like to fumble things"), but the ultimate decision was, of course, with the commander in chief.

Denis Vasilevich Davydov, who was "the first to realize the significance of this terrible weapon" (Tolstoy), was a friend of Pushkin, and himself a minor poet. In his early verse he described and glorified the attraction of Hussar life. His father had also been a military man but was compelled to resign from the army and the family was impoverished. Denis Davydov's extracurricular activities were followed by the court with suspicion though he was anything but a wild revolutionary. He had to quit the army temporarily in 1804 but participated in all important campaigns from 1806 onward.

Having concentrated his checkered little army after Borodino, Davydov chose the area due west of Moscow as his field of operations. At first his men were attacked by the villagers, who thought they were French. To prevent any further such misunderstandings, Davydov decided to wear a peasant's caftan, let his beard grow, carried a large picture of St. Nicholas, and tried to talk to the peasants in their own language.[82] He attacked marauders, of whom there were a great many constantly molesting the villages, and in addition taught the villagers how to cope with the menace themselves; they should make the marauders drunk and kill them at night. Their bodies and their uniforms should be buried in the far depths of forests so as to avoid detection.

Davydov's units liberated convoys of Russian prisoners, seized French supplies, cannon and ammunition. On 12 September, near Viazma, he took two hundred and seventy prisoners including six officers; on the fifteenth he captured two hundred and sixty. Of the released Russian prisoners, two hundred and fifty joined his troops, besides which he also mobilized the peasants so that for special occasions he could count on the assistance of twenty-five hundred men. By trial and error he learned the essentials of partisan warfare; while camping, some horses had always to be saddled, guards had to be changed every two hours, and he was alert to the great value of information to be gleaned from the local population. He wrote with a certain sarcasm that some of his lieutenants had not yet mastered the art of guerrilla warfare but behaved like old generals in the tradition of Austrian military textbooks. The ideal position for a partisan was to be perpetually on the move, the enemy should never know one's whereabouts; the best slogan was *ubit-da-uiti* (kill and get away).[83] On one occasion he captured a Belgian colonel who was out hunting ("*O malheureuse passion*!" the poor man moaned), on another he seized a giant load of shoes. He also arrested and punished traitors who had collaborated with the enemy. He noted in passing that the peasants as a rule killed every enemy soldier they laid hands on; these had a chance of surviving only if one of Davydov's soldiers was present. In late October the French began to retreat and Davydov wrote: "Even if I had ten times as many Cossacks, even if everyone had ten arms, we could not have taken the tenth part of the booty."

In due course Davydov's forces united with those of Figner and Seslavin, two other partisan chiefs. Figner was a very brave man but a sadist who killed French soldiers even after they had surrendered. Such cruelty was abhorred by most Russian officers, who refused to serve in his unit.[84] The peasants, on the other hand, as already indicated, were not beset by such scruples; one Russian village elder reportedly asked the partisan leaders whether he could advise him of any new way to kill the French — all known methods had already been used on them.[85] Seslavin was also a daring officer but prone even more than the others to exaggerate the importance of his operations. He was the first to report to Russian headquarters that Napoleon was advancing to Maloyaroslavets. Commenting on his discovery, he wrote in his dispatch: "The enemy has been forstalled, the French destroyed, Russia saved, and universal peace concluded . . ." — all this as the result of his report. The bureaucrats also engaged in similar rodomontade; the local police chief of the Sytch region wrote in a report to the minister of police: "In 36 days the partisans have killed 1,720 French soldiers for the loss of only 93."[86]

Nor was Davydov free from such weaknesses, but he disarmingly admitted that in all armies the number of those killed on the other side are exaggerated.[87] He had no illusions about the limits of partisan warfare; as he put it: "Even on the retreat from Moscow, Napoleon's Old Guard went through the Cossack crowd like a warship with a hundred guns through a flotilla of fishing boats." Davydov made some interesting comments on partisan warfare both in his "diary" and in a theoretical essay published in 1836.[88] Partisan warfare, as he saw it, was neither the burning of one or two granaries nor a major frontal attack against the enemy's main forces but something in between. He referred both to the Spanish guerrilla experience and to the Suvorov tradition, to the importance of surprise and flexibility. "The partisan acts more through his skill than his strength." General staff officers and bureaucrats were quite unsuitable for this kind of warfare. The partisan needed enterprise, sagacity, strictness, unselfishness. He should not be fussy, should be ready to sleep on straw in the open. His life was a daily encounter with death.

Davydov made the important point that large modern armies had become especially vulnerable to the effects of partisan warfare. They needed ammunition and food, hospitals, clothes; their supply lines could easily be cut by light cavalry units. The partisan could also destroy bridges, collect important intelligence and, in general, demoralize the enemy and give a moral uplift to one's own side.[89] He polemicized against foreign military theorists who maintained that cavalry units should only be used as mounted infantry in the general battle. The Russian cavalry, he claimed, was exceedingly mobile and courageous, and could operate independently.

Davydov was not taken quite seriously by many of his contemporaries; he felt himself forever persecuted by enemies and not entirely without reason. The great partisan leader whose fame had spread all over Europe was only slowly and with great reluctance given promotion, and this despite the fact that General Ermolov, his cousin and protector, was one of the leading Russian commanders of the day. He was pensioned off in 1823, but at his own insistence was permitted to rejoin the army, took part in the Polish and Turkish campaign and died in 1839 of a stroke at the age of fifty-five.

Davydov's unit was active between Gzhatsk and Viasma, about a hundred miles from Moscow; Seslavin and Vadbolski raided the countryside between Mozhaisk and Moscow; Kudashev operated in the direction of Serpukhov, Figner near Zvenigorod; while there were others, like Benkendorf and Dorokhov, who also took a prominent part in partisan warfare. Napoleon later maintained that he had not lost a single courier, and not a single convoy, a claim anything but supported by his correspondence with his generals. In a letter to Berthier, for instance, Napoleon wrote that Marshal Ney was losing more soldiers in providing protection for his supply convoys than on the field of battle. Of course, if the accounts of the various partisan leaders themselves are to be believed, each of them had defeated Napoleon single-handedly. Only a few displayed modesty and a sense of proportion. One of them was Prince Volkonsky, the future Decembrist, who was to write in his memoirs: "In describing the actions of my guerrilla detachment, I shall, unlike many partisans, refrain from mystifying my readers with stories of unprecedented exploits and dangers. Thus, by scrupulous avoidance of the exaggerated accounts of other partisans, I shall inspire confidence in my notes."[90] The fantastic accounts of the partisan leaders, the quarrels between them as to who had committed more heroic acts (and who should be given the higher orders and distinctions) made the public doubt whether partisan activities had played any significant part at all. Regular army officers returning after victory over Napoleon had finally been won also denigrated the role of the partisans. It was Soviet historians who, later on, were inclined to put the emphasis on the role of the peasants in the war against France.

That the partisan leaders performed acts of great individual heroism has been fully established. Figner — "a Northerner by birth with a round face, bright eyes who knew half a dozen languages"[91] — went alone into occupied Moscow, and with the help of some urban ruffians killed French stragglers at night. Disguised as a *muzhik*, he even tried to enter the Kremlin but was stopped by the guards. Another time, dressed in a French officer's uniform, he entered an enemy camp without even knowing the password. With an air of authority he ordered the sentinels to stand to attention. Figner was killed when the Russian army crossed the Elbe in 1813. Seslavin on one of his reconnaissance raids crept up to a French sergeant, lifted him onto his saddle and bore him off to his unit. Similar tales of bravery were related about Dorokhov who appealed to the peasants to "arm yourselves and unite with us for the destruction of the enemy of the faith and the fatherland who destroys our churches."[92] Soviet historians have uncovered the valiant action of a peasant woman, Praskovia of the village of Sokolovo, who allegedly defended herself with a pitchfork against six Frenchmen including a colonel, killed three and put the others to flight. Chetvertakov, a former private in the Russian army, reportedly gathered a detachment of three hundred partisans which was joined on occasion by several thousand peasants and defeated an entire French batallion. Other peasant partisan units were headed by Samus, Stepan Eremenko, Ermolai Vasilyev.

There was some resistance in the villages and the exploits of individuals were magnified by patriotic historians. The attitude of the authorities toward such manifestations of patriotism was ambivalent. On the one hand there was the appeal of Rostopshin, the governor of Moscow, calling for a people's war; on the other, there were orders not to arm the peasants — even to disarm them. For there was always the danger that in the end they would turn against their masters.

If the war against Napoleon became a people's war, it was in part the French themselves who were responsible. During their invasion of Russia they burned villages, killed civilians and prisoners. "Is this the civilization we brought to Russia?" Caulaincourt complained in a letter to the emperor.

Napoleon was defeated, according to Russian historians, by the combined resistance of the Russian people; the war on two fronts caused great losses to the French and the attacks in the rear compelled Napoleon to withdraw forces from the front. But partisan warfare, in actual fact, only began on a substantial scale, as already pointed out, after the Battle of Borodino. Since the French order to retreat from Moscow was given a mere six weeks later, it becomes clear that partisan warfare could not possibly have been of decisive influence. Some partisan units continued to pursue the retreating French army, rounding up stragglers. But these would have been picked up anyway by the advancing regular army units. The peasants were chiefly preoccupied with protecting their own villages and later on with seizing the supply trains of the retreating French army. One of the earliest Russian appraisals of partisan warfare was perhaps the fairest: the partisans were as yet inexperienced, they did not take a single town. But their operations were useful, and caused some harm to the French; above all, their very appearance in the enemy's rear kept up morale among the Russian population. This, in the final analysis, was the main contribution of partisan warfare in Russia in 1812.

2

Small Wars and Big Armies

The century between Napoleon's fall and the outbreak of World War I saw a notable spreading of military conscription and the steady growth of large standing armies. Firepower increased enormously, while the revolutionary changes in the means of transport made for the first time for the rapid assembling and moving of vast contingents of men. Armies became highly organized and specialized bodies complete with their general staffs, their logistic support and supply departments. Military doctrine everywhere developed along parallel lines and strategy was taught as a (more or less) exact science. Militarists and antimilitarists disagreed strenuously about virtually everything, but on one point there was grim unanimity, a universal recogniton that the coming war would be a war of masses, with the outcome hinging on which side could get fastest with the greatest quantity of men and materiel to the particular area of operations. Pacifists, such as Jean de Bloch, argued that the immense improvement in the mechanism of slaughter and the unbearably high cost of modern war would lead to its abolition. The militarists put their own firm faith in the decisive value of bold leadership and maximum organization.

In the light of all this, it is hardly surprising that scant attention was paid to the possibility of guerrilla warfare; it seemed that, with the invention of the machine gun and the swift evolution in communications, the age of partisan warfare had come to an end. Small bodies of soldiers, or of civilians acting behind the enemy lines, could perhaps have a certain nuisance value, but it was unthinkable that they could effectively influence the result of a battle, let alone a campaign or a war. European armies, in short, prepared for nothing but regular warfare.

Such single-minded concentration on war between great masses of men and the unwillingness to consider any other possibilites seems nonetheless a little curious in retrospect, for between 1815 and 1914 there were only very few major wars but a great many guerrilla campaigns. Even the major wars (such as the American Civil War and the Franco-Prussian War of 1870–1871) had been accompanied by partisan operations. Most of the campaigns throughout the British Empire and in Latin America were waged against (or sometimes between) irregular forces. Guerrilla tactics figured prominently in the Polish insurrection, the Italian and Greek wars of independence, as well as in the Spanish civil wars, in the resistance of the Caucasians against the Russians and of Abd el-Kader against the French in North Africa. This list, though by no means complete, shows a discrepancy between military reality and strategic doctrine. All the stranger considering that the regular armies unprepared for irregular warfare had themselves suffered nasty surprises on more than one occasion: the Germans in 1870 were convinced that the war was over after they had defeated the regular French armies; the British in South Africa had similar illusions after the capture of Bloemfontein and Pretoria. In the event, the Germans and the British found it more difficult to cope with the irregulars, and this despite the fact that partisan warfare was almost unplanned and uncoordinated. Surely such experiences should have induced military leaders and theorists to give at least a passing thought to the possibility, however remote, that guerrilla warfare still had something of a future.

One does not have to look far for the reasons for such blindness. There was, to begin with, instinctive resistance to the employment of forces that could not be fitted into the framework of organized and disciplined armies. Guerrilla warfare was erratic, unprofessional, unpredictable; it violated all established rules. It might dovetail neatly with right- or left-wing anarchist thinking, but it was altogether alien to the makeup of the military mind. To conduct guerrilla warfare was a counsel of despair, an *ultima ratio,* to be applied by a weaker army in the case of occupation; to prepare for such an eventuality was tantamount to defeatism. It could be argued, in addition, that most nineteenth-century guerrilla wars had taken place outside Europe, between or against backward nations that were as yet incapable of conducting any other form of warfare. It seemed only natural to assume that with the spread of civilization — or, to be precise, with the spread of modern technology — guerrilla warfare would disappear even in distant and underdeveloped lands. Finally, there was the indisputable fact that, with very rare exceptions, partisans, guerrillas and *franc tireurs* had invariably been defeated in the end, however brave and well led, unless they had fought in conjunction with regular armies. If in later perspective the nineteenth-century strategists were mistaken in belittling or altogether ignoring guerrilla warfare, there were certainly weighty enough reasons at the time to bolster their attitude.

But even if guerrilla warfare was deemed a thing of the past, it was still very much in evidence, and it is to its more important manifestations in the last century that we next have to turn. Among the many guerrilla wars of the period it is impossible to find two that were identical; each had its own specific character and political context. In Italy and Spain, in Poland and Greece, regular and irregular warfare were intermingled, professionals applying unorthodox tactics, and *guerrilleros* playing regular soldiers. Guerrilla campaigns in Latin America differed in basic respects from jungle, mountain and "savage" warfare in Asia and Africa. There were some striking resemblances between Shamil's and Abd el-Kader's campaigns — and not alone because both leaders were pious Moslems — but the political, cultural and physical backgrounds had little in common; one war was conducted in the mountains, the other mainly in the desert, of necessity making the tactics used by and against the guerrillas in each case markedly dissimilar. The history of guerrilla war, in brief, varied from country to country, and sometimes even from province to province. To attempt in these circumstances to formulate a definitive theory of guerrilla warfare is a vain undertaking.

LATIN AMERICA

Latin America is the guerrilla continent *par excellence*. In the entire history of Central and South America it is difficult to point to more than a handful of full-scale, regular wars; on the other hand, there were countless external and internal guerrilla wars, too many, in fact, for enumeration. This can be laid, to a certain extent, to their own particular history and geography — the wide-open thinly populated spaces, the governments which were too poor to afford sizable regular armies. It had long been the Latin American military disposition in any event to incline more to the convention of small flexible fighting units than to large, rigidly disciplined armies. Moreover, army and politics have been traditionally closer linked in Latin America than in other parts of the world; the armies were on the whole more politically oriented, and political life more militarized than elsewhere. The dividing line between guerrilla war, banditry, the regular army and politics was, in fact, altogether blurred.

The Latin American tradition of guerrilla warfare predates the wars of independence. In the Andean regions a small white minority ruled the exploited and mistreated native Indians who periodically revolted against their masters. The most widespread risings were those of Tupac Amaru (1781–1782), who claimed to be of royal (Inca) descent, and of Pumacahua (1814–1815) — also apparently of Inca ancestry.[1] Tupac Amaru's revolt, which almost two centuries later inspired the Uruguayan urban guerrillas, began with a successful ambush in which twelve of his men captured the local *corregidor* and seized a quantity of guns. His following soon swelled to forty thousand and later to sixty thousand men who were, however, not much of a fighting force; undisciplined and poorly equipped, careless and often drunk. Tupac Amaru made a halfhearted, if somewhat unavailing, attempt to attract a few whites and the Negro slaves, and with his nonetheless still growing army, tried to defeat the enemy in frontal assaults by sheer numbers (the siege of Cuzco). But these operations failed and in the end the Indians were decisively defeated by a slow-moving seventy-year-old Spanish general commanding untrained troops, numerically far inferior to the Indians.[2] Tupac Amaru was sentenced to death and his body dismembered, he first having been compelled to watch the execution of his wife along with all his relatives. After his death the rebellion continued under Diego Tupac Amaru. This was the bloodiest period of the rising, the policy of the Indians being to kill all whites and *mestizos*. They besieged La Paz and Puno but could gain no major victory. These were curious battles accompanied apparently by more noise than actual fighting; the operations were stopped from time to time by agreements which neither side had any intention of keeping.

The same pattern recurred in the Pumacahua rebellion; he had been a commander in the Spanish army in the campaign against Tupac Amaru. His one advantage was that he had Creole support, and his force, unlike that of his predecessor, succeeded in occupying several important towns such as La Paz, Arequipa and Puno. But discipline among his troops, too, was lamentable, and once the Spanish had managed to concentrate a small force, they defeated him with the greatest of ease.[3] There were other risings but they all failed, primarily because the Indians made the same mistake as the European peasants in the Jacqueries; they were moderately successful while fighting in small groups, but the moment they tried to concentrate their forces and to imitate regular armies, they became a target that could only too easily be outmaneuvered and destroyed. The Indians fought bravely, but they needed the Creoles for military leadership and organization. In general, whenever Indians and Creoles made common cause, the prospect for victory was immeasurably greater.

What has been said about the Indians applies *a fortiori* to the Negro slave revolts of which there were several, especially toward the end of the eighteenth century, partly under the influence of the French Revolution.[4] With one famous exception they all collapsed for lack of internal solidarity, lack of weapons, insufficient military know-how and, above all, the absence of effective military leadership. The striking exception was, of course, Toussaint l'Ouverture's rising in San Domingo. It succeeded because it had a leader of genius, and because the general level of education under French rule was higher than elsewhere; no Indian leader would have been able to compose beautiful declamations in the style of the French Revolution. Toussaint also received foreign help, mainly from the British, and the French forces sent out against him were decimated by tropical disease. But insofar as guerrilla warfare is concerned, the war waged by the "Black Jacobins" offers little of interest; they did not, as a rule, conduct guerrilla warfare but defeated the French precisely because their leaders were able to establish a fairly effective regular army with all its trappings.

> The officers called themselves generals, colonels, marshals, commanders, and the leaders decorated themselves with scraps of uniforms, ribbons and orders, which they found on the plantations or took from the enemy killed in battle. . . . The insurgents had developed a method of attack based on their overwhelming numerical superiority. . . . They placed themselves in groups, choosing wooden spots in such a way as to envelop their enemy, seeking to crush him by weight of numbers.[5]

Hédouville, the French commander, was a veteran of the Vendée campaign, and if he found coping with the insurgents so difficult, it still stemmed principally, their greater numerical strength notwithstanding, from their not behaving like Chouans. After Toussaint's treacherous arrest by the French, the movement degenerated into a race war culminating in the massacre of the white inhabitants of the island.

The Latin American wars of independence involved much guerrilla fighting; "pure" *guerrilleros* were rare, but then regular warfare was also quite irregular by European standards. More often than not the campaigns consisted of a mixture of regular and guerrilla warfare. Among the more prominent guerrilla bands were the *montoneros* of the La Plata region, the Almeydas of New Granada, the guerrilla bands of Central and Upper Peru, and the units commanded by José Antonio Paez in Venezuela.[6] They varied greatly in outlook and social background; the *montoneros* were mostly *gauchos* following their local leaders, the Almeydas were the private army of a Creole clan, while the guerrillas of Central Peru consisted similarly of middle-class Creoles and Mestizos whose property and families had suffered at the hands of royalists and who sought revenge. "They were joined by delinquents, by bandit chiefs and their followers . . . who used guerrilla operations as a means of personal plunder."[7]

Paez, one of Bolivar's most able commanders, started his military career in mid-1809 when, while driving a herd along a highway in Venezuela, he encountered a slave revolt. He joined them, became their leader, and eventually had some two thousand lancers under his command. These savage bands were held together by no ardent idealism or ideology but simply by the prospect of plunder, a fact freely recognized by their leaders. Bolivar temporarily dominated the guerrilla bands by promising them land after victory, but they still went on robbing and plundering without waiting for war's end.[8] Slogans such as "independence" meant not so much national independence and unity, but independence from Spanish law and taxation; it was a rebellion against authority in general and it resulted in the transitory emergence of dozens of small, short-lived, semianarchist republics. Their revolutionary convictions were not always very deep; the great guerrilla leader Bores went over to the royalists from one day to the next without much compunction. The contribution of the *guerrilleros* to the war effort was on the whole a modest one, with the possible exception of Venezuela, where Bolívar succeeded in coordinating their operations during the crucial years. What the guerrillas lacked was not so much arms and provisions — the Spaniards were not much better equipped — but staying power, elementary discipline, cohesion, and leadership. They operated haphazardly, and whether, on balance, more for harm than for good remains debatable. Looked at in historical perspective, their chief significance lay in their setting a pattern for many years to come. A few hundred peons led by their *hacendado* or a local *caudillo* armed with machetes would rise against the local government; there were many variations on this theme, but usually those who followed him were mainly out for plunder. The political label was of importance, for it provided immunity against capital punishment.[9] European commanders quickly adjusted themselves to local customs. Garibaldi's biographer notes, quite matter-of-factly, that his hero "remained for 36 hours in Gualeguaychu [in 1845]. For the inhabitants, they were 36 hours of terror, about which their descendants still speak to day. Garibaldi's men looted the town, causing great destruction."[10] In Italy, Garibaldi would have shot soldiers found looting.

Guerrilla warfare frequently was the high road to political and economic power; yesterday's brigand could well become tomorrow's government minister, amassing a fortune on the way. The spoils were, of course, shared with his followers and the great prize was always land. This is not to say that patriotism or social protest were totally absent as motivating forces; as far as the leaders were concerned, political goals and ambitions always featured prominently. But, on the whole, political motivation played a lesser role among guerrilla movements in Latin America than in Europe.

There were some signal exceptions, above all the Mexican revolts between 1810 and 1815 led by two clergymen, Miguel Hidalgo y Costilla and José Maria Morelos.[11] They fought not only for national independence, but also for far-reaching social reforms, such as the abolition of the tribute paid by the Indians, and the establishment of a republic. While the bulk of the army — "Generalissimo" Hidalgo had an army of sixty thousand — was Indian, the leadership was Creole (native-born of European descent), and the religious element was a factor of considerable importance. (The battle cry was "Long live our Lady of Guadalupe, death to bad government.") Hidalgo and Morelos were eventually captured by the royalist forces and executed. Both were posthumously rehabilitated; Hidalgo has forever been inscribed in the history of his homeland as the "father of Mexican independence," Morelos had a province named after him, which one hundred years later was to become the scene of Zapata's operations. Great popular leaders, their genius was not in the military field, and it is doubtful whether it is really permissible to regard them as guerrilla commanders at all, as some historians have done. They would, in fact, have fared much better if they *had* applied guerrilla tactics instead of besieging (or defending) big cities. The forces under them, however, were more numerous than the enemy's and the temptation must have been great to defeat the other side in a few decisive battles rather than in protracted partisan warfare. But irregulars cannot as a rule seek conclusive battles; the failure to accept this simple truth has spelled doom time and again to popular movements in Latin America just as in other parts of the world.

Fifty years later, Mexico again became the scene of a major war. The Spanish and French armies which invaded Mexico in 1861–1862 encountered resistance from Mexican irregulars almost from the beginning (the blockade of Santa Cruz by Juarez's forces). The French contingents were led by veterans of the North African campaigns such as Bazaine, and before long counterguerrilla units were set up under the command of Stoecklin. Their anabasis to Mexico City was beset by every manner of hazard — rain, bad roads, shortage of supplies, yellow fever — but the capital was taken in June 1863, and by the end of the following year about three-quarters of the country was in their hands and they controlled all major cities. Only the state of Guerrero was still held by Juarez. Juarez's forces, weak as they were, adhered to time-honored guerrilla tactics, harassing the French lines of communications, refusing to accept battle, always retiring, biding their time.[12]

The subsequent course of events need not be retold in detail. Napoleon III decided to withdraw his forces in January 1866; poor Maximilian, a charming man but weak and vacillating, lost control and was eventually captured and executed. Napoleon's decision came not because his army had been defeated; their losses had been insubstantial. The Mexican forces improved somewhat in the course of the war and they received considerable help from the United States, while Escobedo, one of Juarez's generals, was the first to use the machine gun in battle. But the Mexican soldiers, often unpaid, lacked both training and fighting spirit, the various guerrilla units would not coordinate their efforts, and individual marksmanship was poor. Not that French morale was very high either; Porfirio Díaz, subsequently president of Mexico, had three hundred French deserters in his little army. During the last phase of the war there was a tacit understanding between the French and Profirio Díaz that the former should not be molested during the evacuation. The French, in short, suffered no military defeat and the Mexicans were lucky in that the French emperor changed his mind in midcourse about what was anyway for him no more than a minor adventure. Maximilian could have fortified his position by calling for equal treatment to be granted to the Indians and by building up a new regular army, but he failed to do either. Having outlasted the enemy, Juarez returned in triumph to his capital.

It would be tedious to enumerate the long series of armed conflicts in Latin America during the second half of the nineteenth century. There was much guerrilla activity in Venezuela in the 1860s; when Crespo entered Caracas in 1892, his forces were composed mainly of guerrilla units. Again, in the Brazilian civil war in the 1890s, guerrilla warfare was the rule rather than the exception; passing reference need only be made to Gumercindo's hit-and-run attacks, and his raid of seven hundred and fifty miles to Santa Caterina.

The Peruvian general Andrés Caceres fought a three-year guerrilla war against the victorious Chilean army in the Andean regions of Peru from 1881 to 1883 (*Campaña de la Breña*); this campaign was later studied in the Austrian war academy as an excellent illustration of successful mountain warfare.[13] But the most protracted guerrilla war took place in Cuba; it lasted from 1868 to 1878 and again flared up in January 1895, leading eventually to war between Spain and the United States. In the Cuban war more than national independence was involved; the neglect of the smaller farmers was an important issue, the question of slavery being another. Like all Cuban guerrilla wars including Castro's campaign, Oriente, the eastern province of the island, was the central scene of operations. The rebels numbered some ten to twenty thousand men and, unable to face the Spanish army in open combat, engaged for the most part in acts of sabotage. "It was less a war than a breakdown of order . . . a formalization of the violent banditry that had gone on through much of the early 19th century."[14] Two outstanding leaders emerged, the mulatto Antonio Maceo, and Máximo Gómez. This inconclusive war lasted for a decade; despite some initial successes, the rebels failed to raise the standard of revolt in prosperous western Cuba. As the war continued, dissension spread in the guerrillas' ranks, and by concentrating strong forces and conceding some of the rebels' demands, the Spanish induced them to accept an armistice.

When, two decades later, fighting broke out again, the old military leaders Máximo Gómez and Maceo were still very much to the fore. But the inspiration for the independence movement now came largely from the ideologists, from José Martí most especially in his North American exile. Nonetheless, the moment fighting started, effective control passed into the hands of the guerrilla captains. As in the 1870s, the insurgents depended to a great extent on money and arms from the United States. In contrast to the desultory fighting of 1868, the guerrilla war in 1895 was far more ferocious; the insurgents were stronger — numbering about thirty thousand men organized in some thirty bands. They threatened to burn down the plantations and to make the island uninhabitable. Eventually, some forty thousand *guerrilleros* faced eighty-five thousand Spanish troops, of which, however, only about half were battle-ready, the rest suffering from yellow fever and other diseases. (About two thousand Spaniards were to die in battle but many more of this or that tropical sickness.) The war went badly for the Spanish until General Valeriano Weyler, the toughest and most effective Spanish soldier of his time, was made commander in chief. Weyler responded to terror with counterterror, sealing off the eastern part of the country more or less effectively by means of ditches, walls and blockhouses; he used *reconcentraciones* (concentration camps) to remove civilians from the battle areas. From a purely military point of view, Weyler was winning the war by 1896, but it was by no means total victory and, as far as Spain itself was concerned, the war had gone on for too long already and become too costly. The economy of Cuba, above all the sugar crop, was in ruins under the double onslaught of guerrillas and the government troops. There was anti-Weyler campaign in Madrid, and, more damaging yet, he had an exceedingly bad press in America, preparing the ground for U.S. military intervention in Cuba. Eventually Weyler had to go and Blanco, his successor, was easily defeated by the American expeditionary force in 1898. Antonio Maceo was killed in the fighting, but Gómez lived to become a general and to enter Havana at the head of his troops. As peace came, the politicans once more took over and the guerrilla army was disbanded.[15]

Guerrilla warfare in Latin American history took many forms: wars of national liberation; struggles of landless peasants and small farmers against large landowners; fighting between local chieftains for political power. A remarkable example of the Vendean type of guerrilla warfare against political and social change with strong religious undertones is the affair of Antonio Conselheiro in Canudos.[16] Conselheiro was a primitive mystic who fell out with the Church authorities and attacked the Brazilian Republic because it had arrogated to itself jurisdiction over marriage and burial. He regarded the victory of the Republic in 1889 and the proclamation of religious tolerance as the work of Satan in the guise of Masonry, Protestantism and Positivism, about which admittedly, he had only the haziest notions. In 1893, at the age of sixty, he decided to seek refuge in a little place in the Brazilian hinterland about two hundred miles from Bahia where the police would never be able to find him. This was Canudos, formerly an old cattle ranch which had become a miserable shantytown of several thousand inhabitants virtually cut off from the outside world. Two Italian Capucine monks who visited Conselheiro and his followers in May 1895 reported that it was the hotbed of a dangerous political as well as a religious sect. The Rio de Janeiro republicans came to look on it as the center of restorationist plots which had to be destroyed.[17] In truth, Conselheiro's followers, the *jaguncos* (a term originally meaning ruffians), were not so much fanatic royalists as adherents of a messianic folk movement, caring only about their "simple-minded, visionary religion, a crude mixture of Catholic rites, African witchcraft and Indian superstition." The population was made up, according to da Cunha, of the most disparate elements, ranging from the fervent believer who had voluntarily given up all the conveniences of life elsewhere, to the solitary bandit who arrived with his blunderbuss on his shoulder in search of a new field for his exploits. "Under the spell of the place, all these elements were welded into one uniform and homogeneous community, an unconscious brute mass."[18]

The clash with the authorities began with a quarrel about cheap building material for a church which the *jaguncos* had appropriated; a small force of about one hundred soldiers sent out by the governor of Bahia was surprised by the rebels and routed. The governor asked for federal help, but a second expeditionary force of some five hundred and forty men fared even worse. They were ill prepared for a long march through unfamiliar country and faced the accurate fire of snipers whom they could not even see in the impenetrable undergrowth:

> The army has come to feel that its very strength is its weakness. Without any maneuverability, in a state of continual exhaustion, it must make its way through these desert regions under the constant dread of ambuscades and be slowly sacrificed to a dreaded enemy who does not stand and fight but flees. The conflict is an unequal one, and a military force is compelled to descend to a lower plane of

combat; it has to contend not merely with man but with the earth itself; and, when the backlands are boiling in the dry summer heat, it is not difficult to foresee which side will have the victory.[19]

The greatest disaster was the defeat of the third expeditionary force under Colonel Moreira César, a tough soldier subject to epileptic fits. His force included a squadron of cavalry and a battery of artillery. For the first time there was a clear military plan but it was a crude one — "to hurl a thousand-and-some bayonets against Canudos in double-quick time" (da Cunha). The enterprise was carried out in great haste, and all the past mistakes were repeated. Meanwhile, the people of Canudos had dug an elaborate system of trenches in and around their small capital; they were not badly armed for guerrilla warfare with their scythes, "scraping knives," muskets, shotguns and blunderbusses. Gunpowder they had bought in the neighborhood or manufactured themselves. They received reinforcements from all over the province — the "badmen" from the backlands congregated in and around Canudos. The expeditionary force managed to reach the town, but in the ensuing assault it got literally lost in the labyrinth of lanes and alleyways. César was mortally wounded in the attack and soon his soldiers were in full flight, rifles abandoned. The insurgents seized a great quantity of arms, including four Krupp field guns. The *jaguncos* took no prisoners — wounded soldiers were beheaded; later on the army retaliated with similar acts of cruelty.

This defeat caused a major crisis in Brazilian politics. "Patriotic passion was verging on insanity," is the way da Cunha put it. The fourth expedition was headed by General Artur Oscar who had been previously engaged in guerrilla warfare along the Rio Grande. But neither his own past experience nor the lessons of the previous campaigns against Canudos were heeded and on his first major encounter with the rebels he was surrounded and lost almost all his reserve supplies. Another army column saved him from complete destruction; but this force, too, had to retreat, decimated by hunger, thirst and disease. To vindicate his defeat, General Oscar claimed that the insurgents were armed with the "most modern weapons" which had allegedly been smuggled from Europe. Eventually, Canudos was destroyed but only after all the country's military resources had been mobilized and after it had been besieged from every side. Conselheiro died on 22 September 1896; on 5 October the fighting was over. "Canudos did not surrender. . . . It held out to the last man. Conquered inch by inch, in the literal meaning of the words, it fell on 5 October, towards dusk — when its last defenders fell, dying every man of them. There were only four of them left: an old man, two other full-grown men and a child, facing a furiously raging army of five thousand soldiers."[20]

EUROPE

Guerrilla warfare in post-Napoleonic Europe was limited mainly to the south and east of the continent; more often than not it occurred in the wider context of wars of national liberation or civil wars. This applies, for instance, to the first and second Carlist wars in Spain (1833–1840 and 1872). These dynastic wars were at one and the same time conflicts between a backward countryside resentful of change and the modern town, and between the clergy and the free-thinkers. The Basques' desire to maintain their traditional privileges against central state power was also a factor of some importance. The Carlists, broadly speaking, were fighting for tradition and the old Spain, and the Cristinos (with the help of a British legion and other foreign volunteers) for change and modernism of a very moderate variety. The larger cities usually supported the Liberals, while much of the countryside sympathized with the Carlists. It may be recalled that the guerrilla leaders of the war against Napoleon found themselves here in opposite camps, Merino, the priest, fighting with the Carlists, Mina throwing in his lot with the government forces. But these were no longer the prime leaders; the military command was by now in the hands of men of another generation. Two of them in particular distinguished themselves — Colonel Tomás Zumalacarreguy, who had begun his military career under a minor guerrilla leader in 1810 and led the Carlists until his death in 1835, and Ramón Cabrera (the "tiger of Maeztrazgo"), who then took over; having married an English woman, Cabrera was to spend his declining years as a liberal country gentleman near Virginia Water.[21]

Zumalacarreguy, who had at the start no more than a mere few hundred ill-armed adherents, forged them gradually into an effective fighting force. He subdivided his little army into battalions, which would occasionally meet for some major action, but most of the time acted independently. By 1835 he was strong enough to engage in regular warfare and force a decision, but Carlos, envious of his general's popularity, ordered him to seize Bilbao rather than march on Madrid. Zumalacarreguy was wounded in the fighting for Bilbao and died soon after from his wounds. With the demise of its most gifted soldier, the Carlist cause, already undermined by internal intrigues, suffered a lasting blow. Cabrera, his successor, had been trained as a priest, but even his closest friends would not claim that Christian charity was his outstanding virtue. Under his leadership, acts of atrocity became ever more recurrent in a war which had been cruel from its inception. Prisoners were frequently shot, and after the Cristinos had killed Cabrera's mother, he no longer showed any restraint whatsoever. His skill as a guerrilla leader was undoubtedly considerable. If he suffered a reverse, he would send his troops to rest for a fortnight "to change their shirts. Soon afterwards they would reassemble and fight again."[22] His aides, Batanero (yet another priest) and Miguel Gomez, would engage in long penetration raids from Biscay to Old Castile or even to Gibraltar and back. But apart from showing the flag and engaging in brigandage, these ventures were militarily without value and they clearly pointed to the limitations of guerrilla warfare. Defeated by the Cristinos, Cabrera crossed in 1840 into France with the remnants of his forces. In 1848 fighting in the mountains of Catalonia, he made a short-lived and ineffectual comeback. Once more he had to leave his native country and toward the end of his life, much to the disgust of the diehard Carlists, he made his peace with the Spanish government.

The Carlist wars were demonstrable proof that the guerrilla tradition had become deeply rooted in Spain; it was no mere coincidence that its principal bases of operations were yet again in the northern regions such as Navarre, Catalonia and the Basque mountains. But despite superior leadership, guerrilla tactics were in the last resort less than effective in this prolonged conflict. In the Greek War of Independence they proved on occasion more successful, and this for all the absence of good commanders. Applying guerrilla methods, the Greeks contrived to liberate part of their country in the early phase of the war (1821–1822). Later on they tried to transform their bands into a regular army with the help of some European well-wishers. The results were disastrous; they suffered an almost unending series of defeats. The discipline, the drill, the organizational effort were not to the liking of the Greeks who were saved in the end from almost certain defeat only by the intervention of the European powers.[23] But nor is it certain that they would have fared any better had they stuck to their initial tactics. They succeeded in the first stage of the war because their attacks took the Turks by almost complete surprise. Once the element of surprise was gone and the Turks had dispatched new forces to the field, the Greeks simply had no answer. Happily for them, their war differed in one essential respect from other such campaigns and this helped to some extent to restore the balance. The distinctive element lay in sea power; whatever their weaknesses on land, the Greeks proved more than a match for the Turks at sea.

If the Carlist wars were scarcely marked by compassion, the Greek War of Independence was almost genocidal in character, the Greeks' premise being that if they exterminated the Turkish communities in their midst, they would eventually be masters in their own home. As in the Iberian peninsula, the clergy took a prominent part both in the fighting and the atrocities. The Greeks had the enthusiastic support of most of Christian Europe, and there was a steady flow of volunteers and money. Later, many of their erstwhile supporters turned against them, some even to become their worst enemies. The Greek intellectuals living in exile who had first lifted the banner of independence were not military men, and the leadership in the war passed to Klepht chieftains like Kolokotronis who were no budding Napoleons either. That brigands were potentially excellent guerrilla leaders is a well-known fact, but an analysis of the battles of the Greek War of Independence (such as Kaki Skala, Elaphos and Trete) makes plain that the Klephts never really had any coherent policy on what kind of war they intended to fight. They had no plan or general strategic concept, nor were they very good at improvising. Their experience was limited to the command of smaller bands; they were simply not accustomed to cooperating within a larger framework. Much of their time was spent quarreling, both with the government and with each other. Many of them had been reluctant to join the rising in the first place, preferring the certainty of an arrangement with the Turkish authorities to the doubtful proposition of a civil war.

Kolokotronis, who was the most prominent of the Klepht leaders, came from a family which had engaged in officially licensed brigandage for several generations. A historian of the Greek Revolution wrote of him that he could never distinguish very clearly right from wrong, and that he had an instinctive aversion to order and law. "His patriotism was selfish and his occasional acts of magnanimity cannot efface the memory of his egotistical ambition and sordid avarice during the period of his greatest power."[24] At the head of a band of some three hundred warriors, Kolokotronis enlisted the local peasants at Karitena and eventually had some six thousand men under his command. But this formidable force could still not resist the onslaught of five hundred Turkish horsemen and Kolokotronis even lost his rifle in the affray. These and similar such encounters left him with a great deal of contempt for the military qualities of the peasantry, ascribing all the successes in the war to the prowess and the fighting experience of the brigands and *armatoli*. But this judgment has not been generally accepted. "A careful study of the history of the Revolution has established the fact that the perseverance and self-devotion of the peasantry really brought the contest to a successful termination. When the Klephts shrank back, and the *armatoli* were defeated, the peasantry prolonged their resistance, and renewed the struggle after every defeat with indomitable obstinacy."[25] The Greek War of Independence was essentially a series of uncoordinated operations carried out by irregular troops. The Peloponnese (Morea), where most military actions took place in the early phase of the war, had been classic brigand territory since time immemorial; hilly northern Greece offered even better protection to large guerrilla units. It was in the north that under Karaiskakis, a former officer of Ali Pasha of Janina, some of the major guerrilla operations took place in the later years of the war. But, whereas in the early days the Klephts had the sympathy of the rural population, the depredations of the bands antagonized so the peasants that when the Turks returned to Central Greece in 1824 they were frequently welcomed as liberators.

Attention has been drawn more than once to the often decisive importance of intervention by outside powers: what happened if other powers would not, or could not intervene is well illustrated by the Polish example. The three Polish insurrections (1793, 1831 and 1863) were a blend of regular and guerrilla warfare. In some measure they were a people's war, but the support of the peasants waned in the course of time. Many thousands of peasant scythemen in their white cloaks fought under Kosciusko in 1793, but peasant participation in 1831 was lukewarm at best, and in 1863 only the cities, broadly speaking, responded to the revolt. This erosion of peasant support was the result of the reluctance of the "Whites," the Polish aristocratic party, to carry out any agrarian reform. In Galicia, the Austrian authorities effectively thwarted a rebellion by inciting the peasants to kill the landowners and to turn against the middle-class revolutionaries. There was not much guerrilla fighting in the 1793 rising save for some sniping from Warsaw windows and rooftops. Kosciusko, the military leader, based his strategy on the experience of the revolutionary war in France — massed attacks and bayonet charges.[26] Politically the Poles were isolated; the French stayed clear of any active help, Austria took a benevolent attitude but left it at that. The international constellation in 1831 and 1863 was, if possible, even worse. Britain and France made perfunctory representations to the Russian capital, but the Prussians, alert to the direct and decidedly undesirable repercussions a Polish victory could have in their eastern districts, closed the borders to the rebels. The Russians always had numerical superiority; there had been some hundred thousand Russians against the Poles' sixty thousand in 1793. Diebitsch, in 1831, had 127,000 men at his disposal, and in 1863 the Poles were much of the time outnumbered by as many as ten to one. But the Russians had to keep their forces dispersed over the entire country for fear of the revolt spreading; there were local uprisings in 1863 in distant Polish Lithuania and Livonia. In 1831 the Russian forces were reduced to half their strength by hunger and disease. Polish leadership was bad in 1831 and indifferent in 1863; furthermore, there were unending internal squabbles among the insurrectionist leaders. Many Polish officers serving with the Russian army refused to join the rebellion in the first place because they saw no possible chance of their country winning independence in an armed struggle against Russia.

Large parts of Poland are quite flat and provide little effective cover for guerrilla operations. Only in the east and the north were conditions more favorable and it was there that small Polish units caused considerable damage to the Russians in 1831 (Worcell near Lutsk, Puschet and Selon in the forests of Augustowo, the partisan bands in the Bialowicza forest on the road to Brest). In 1863 Augustowo again became an important theater of guerrilla warfare, but there were also sizable operations in the Radom district (under the command of Langiewicz who had fought with Garibaldi in Italy) and near Wengrow.[27] The insurgents were on the whole meagerly armed, "raw and undisciplined levies, no more conversant with war than are English yeomen and shopkeepers."[28] Only few had muskets, most of them having to make do with pikes, scythes and sticks. The Poles would launch a surprise attack against the Russian units from the forests, then disperse and return to their hideouts. These small skirmishes were often successful, whereas the major battles were always costly and usually ended in a Polish defeat. A hostile observer noted that it was one of the primary mistakes of the Poles that they did not stick to small-war tactics but tried to act like a regular army. In the process, he went on, good partisans became bad soldiers who fled whenever they suffered a setback. Following the Polish concentration of their forces, the Russians were able to withdraw their units from various parts of the country and to crush the insurrection by delivering a massive blow to the Polish force.[29] The Polish leaders were ambivalent in their attitude to guerrilla tactics. Mieroslawski, one of the leaders of the "Red" party, who for a short while in 1863 served as "dictator" of Poland, wrote that it was dangerous to stick too rigidly to partisan warfare, that it should always be closely coordinated and should never clash with the general, overall strategy.[30] He "hated" partisan warfare, he wrote, but nonetheless did not deny that it could be very useful, given political control and good leadership. After the failure of their insurrections the Polish veterans saw action in revolutionary wars all over Europe — Mieroslawski in Baden and Sicily, Dembinski and Bem against Austria in 1848, the poet Mickiewicz in the short-lived Roman republic. Usually they were on the losing side and were employed as general military experts rather than as specialists in guerrilla warfare which they regarded as of marginal importance only. Their inclination was to apply Napoleonic tactics in wars of national liberation, and for this reason, if for no others, they stood no hope of winning.

The Italians, in contrast to the Poles, had a base for their military operations — Piedmont. Victor Emanuel II and Cavour had grave reservations about Garibaldi's exploits; they never gave him all he demanded to launch his spectacular, if not always well-conceived campaigns. When Garibaldi returned to Italy in 1848 from a long stay in South America, he already had the reputation of a great guerrilla leader as well as a daring naval commander. He certainly was a brave man and a born leader; whether his guerrilla reputation is entirely merited is a moot point. For the kind of cavalry charges he had specialized in, with the lance as the favorite weapon, was not really in true guerrilla tradition. During the next two decades he emerged as Europe's most dashing and admired revolutionary hero, but it was only on rare occasions that he engaged in battle against forces greatly superior to his own, applying guerrilla tactics. The exceptions were the retreat from Rome in 1848 and the battle for Palermo when his forces were greatly outnumbered. Commenting on Garibaldi's exploits in 1848 in Upper Italy (from a base in Switzerland), a critic wrote that he was not really a good guerrilla leader, because he exhausted his men by long, pointless marches, made inadequate provisions for feeding them, and when he found a good, defensive position, waited there for the enemy to attack, instead of attacking the enemy.[31] It is only fair to add that, like the Polish insurgent leaders in their combats, Garibaldi found little patriotic enthusiasm among the rural population. The peasants were reluctant to cooperate or even to provide food. Nor was the quality of his soldiers outstanding; more than once he was to complain that his Italians were not as good as the Latin American *gauchos*. The first Italian legion of 1849 was composed chiefly of artisans, shop assistants and a great many students. There were also a few convicts for — shades of Fanon! — "to fight for Italy would cure all moral diseases."[32] The composition of the "thousand" with whom he conquered Sicily and Naples was similar. There were a hundred and fifty lawyers, a hundred doctors, a hundred merchants, half were workingmen, but there was not a single peasant.[33] The *Garibaldini* about to enter Palermo looked like scarecrows, "resembling in appearance a Boer commando towards the close of war" (Trevelyan), limping, and their clothes in tatters. They had bad weapons (smooth-bore muskests which were just about accurate at fifty yards) and antique artillery. But they were enthusiastic young men, they had a leader who had learned from previous mistakes, and the morale of the Neapolitan soldiers facing them was very low indeed. Twenty thousand Neapolitans evacuated Palermo, unable to resist the onslaught of a far smaller attacking force. Garibaldi had the support of some local *squadri*, but they were only of scant use because they lacked the *sang froid* to participate in a bayonet charge and in any case preferred to return home after a few days.

Garibaldi's military career, after the March of the Thousand, came as an anticlimax. In Mentana in 1867 the better equipment of the French forces was telling, and at Dijon in January 1871 he and his sons fought the Germans without conspicuous success, even though Victor Hugo wrote that Garibaldi was the only French general (he was holding a French command at the time) not to be defeated in the Franco-Prussian war. His soldiers were not considered guerrillas by the Germans, who drew a distinction between *franc tireurs* on the one hand, and the Garibaldini on the other. The former when captured were to be shot, the latter were treated as prisoners of war.[34]

The element of political propaganda and indoctrination in Garibaldi's campaigns foreshadowed guerrilla wars of a later age. But Garibaldi's inclination to give battle and to attack frontally rather than to harass the enemy in less forthright ways was hardly in the guerrilla tradition. Garibaldi's conspicuous white poncho and the red shirts of his soldiers would have been shunned by a true guerrilla. Needs of necessity vary with the circumstances, and customary Spanish guerrilla methods would not have been feasible in Italy; the Spanish *guerrilleros* of 1809 had their cardinal support in the countryside,with the urban population on the whole anything but enthusiastic. In southern Italy, to the contrary, the "reactionary" peasants were slaying the "liberal" landlords just as sixty years earlier they had attacked urban republicans and democrats. In addition, the Italian clergy was deeply hostile to the insurgents, while the Spanish guerrillas had the solid encouragement of the priests. In general, then, Garibaldi could not expect much assistance in or from the villages. He was, however, in a more fortunate position than the Poles who received only a trickle of supplies via Cracow, whereas his forces had the direct support of the Piedmontese, and indirect aid from Britain and other European powers. The political situation, in brief, was more auspicious than in Poland in 1831 or 1863, and in view of the different social character of the movement, predicated a strategy different from the guerrilla war in Spain.

EMPIRES VERSUS GUERRILLA

During the period of imperialist expansion the European colonial powers faced resistance frequently in the form of guerrilla warfare by native tribes or peoples. Russia, expanding in an eastward direction, and the United States, opening up the West, both fought wars on their frontiers. It usually proved easier, however, to conquer new colonies than to hold them against a hostile population; the occupiers were few, the natives many, climatic conditions were adverse and the Europeans had little immunity against indigenous diseases. In retrospect, it is surprising that the imperialist powers suffered in the event only temporary setbacks. But then, more often than not, they faced disunited tribes, lacking modern arms and reliable supply lines. Guerrilla warfare waged by them was usually of the most primitive kind, deficient in leadership, direction and endurance; it was only seldom that an inspired leader would emerge in Asia or Africa to offer effective defiance. Our knowledge of these wars is mostly based on accounts by the invaders, which does not necessarily mean that it is one-sided; Shamil and the Boers were folk heroes all over Europe, and the French had considerable respect for Abd el-Kader.

The two longest and in many ways most interesting guerrilla wars were those waged in North Africa and the Caucasus. France had had its eye on Algeria for a long time and in 1827 an expeditionary force of twenty-seven thousand was sent to the country by Charles X. The French behaved with scant regard for local customs and mores, occupied land, seized property, and soon found themselves under attack by the tribes of western Algeria led by Abd el-Kader, the newly elected twenty-four-year-old emir of Mascara. The North Africans would lead the invaders on wild-goose chases into mountainous country or the desert; the French would never even spot the enemy, and thirst and exhaustion would claim countless victims. They were not mobile enough and had no system for controlling the country they had seized. Abd el-Kader's columns would appear suddenly and hit at them. Small French detachments were surprised, escorts carried off, depleted garrisons destroyed, provisions were cut and there were no regular communications even between the principal towns. Victory bulletins were dispatched to Paris, but at the same time there were constant requests for further reinforcements. Thus, the Algerian war proved to be far more costly than the French had bargained for. A static fortifications system a hundred and twenty miles long and consisting of a hundred and sixty blockhouses and ditches was built, but it proved to be ineffective. The situation changed only with the arrival of General (later Marshal) Bugeaud in 1836. In his first address to his officers he told them bluntly, "*Messieurs, vous aurez beaucoup à oublier.*" He was new to Africa, but it was immediately obvious to him that their methods of pursuing the Arabs were wholly unsatisfactory. He had campaigned in Spain in 1812 and found many analogies between the war there and in Algeria. The French columns would have to be broken up, disencumbered of artillery and heavy baggage. In sum, the French troops would have to be free in their movements.[35] There was some muttering among his lieutenants — their men would lose confidence without artillery — but Bugeaud made short shrift of these objections.

He requested mules rather than horses for desert warfare and divided his army into eighteen flying columns, each consisting of two battalions of *chasseurs*, a battalion of Zouaves, one or two squadrons of native levies (*Chasseurs d'Afrique*) and two pieces of small mountain artillery. Bugeaud taught his soldiers to travel light; instead of the old heavy campaigning bag, they should pack their few belongings in a piece of canvas (which, joined to similar pieces, would form a tent), dress in loose clothing, not take along spare shoes. The columns would start well before daybreak and make a halt every hour.[36]

A French officer provided a vivid account of guard duty while on a *razzia:*

> Passing the night on guard, to one who knows not by experience what war is, especially partisan war, awakens only the idea of a certain number of men sleeping at 200 or 300 paces distance, with a small band in advance, one of whom walks up and down with a musket on his shoulder. It is thus that we are represented in the theatres at Paris; but in Africa the night guards are as unlike this picture as possible. No one sleeps, everyone watches. If the rain falls, if the north-wind blows ice in your face, there must be no fire to warm the limbs fatigued by the day's march. A fire may betray the post. Everyone must be on the alert constantly, close to his arms; and those who are on sentry, crouching like wild beasts among the bushes spying out the slightest movement, listening to catch the slightest sound, are glad to do all this to keep their eyes, heavy with sleep, from closing. The safety of all may depend on their wakefulness. Further, should the enemy attack, no firing; the bayonet is for defence; no false alarms; the sleep of the bivouac must on no account be disturbed. Such is the point of honour.[37]

Bugeaud clearly recognized that strategies appropriate for European theaters of war were not suitable in North Africa. It was pointless to seize the centers of population, of trade and industry in Algeria, because there were none. The right approach, as he explained in a speech in the French parliament, was to keep a flying column of seven thousand well-led soldiers operating *razzias* near the desert. This was sufficient to beat the largest possible collection of Arabs who were nothing but a "tumultuous gathering," a multitude of very brave individuals without the capacity for united action. He would give orders to his commanders not to pursue the fleeing tribesmen — which was useless — but to prevent them from sowing, reaping the harvest and pasturing their cattle. "The Arabs can fly from your columns into the desert, but they cannot remain there, they must capitulate."[38] Bugeaud's predecessors had been in a constant state of alarm, whereas the new commander, in the words of an admiring junior officer, Saint Arnaud (who was to command the French forces in the Crimean war), *"se bat quand il veut, il cherche, il poursuit l'ennemi, l'inquiète, il se fait craindre."* Abd el-Kader's tactics, in a nutshell, were turned against him, although it still took quite a while before the new approach was to show results. Another junior officer serving at the time under Bugeaud, Trochu (the future commander of Paris in 1870–1871), wrote that "this campaign has not been the most fruitful in dangerous and brilliant combats, but the most extended, the most active and the most effectual. . . . Marches and counter-marches, crushing fatigue, unheard-of efforts, were exacted from all; but no one had any serious fighting with the enemy, for not having any organization they remained invisible and could not be caught." In the event it was by sheer accident that one of the flying columns consisting of six hundred soldiers stumbled on Abd el-Kader with five thousand of his men at Temda and inflicted a crushing defeat on him from which he was not to recover. Abd el-Kader crossed into Morocco, which provided a temporary sanctuary. But then Bugeaud routed the sultan of Morocco's army at Isly. (He was subsequently made duc d'Isly.) Again on the run, Abd el-Kader eluded the French for three more years, but he was no longer a serious threat.

There was less cruelty in the campaigns of North Africa than in most colonial wars. The French esteemed Abd el-Kader as a fighting man; Bugeaud once met him during a short truce and reported to the Minister of Foreign Affairs that the rebel chief was "pale and a good deal resembles the portrait often given of Jesus Christ."[39] When he finally surrendered in December 1847, he was exiled to Damascus, given a pension and toward the end of life made his peace with the French, saving many Christians at the time of the Damascus riots of 1860. The unexpected resistance the French encountered in North Africa was motivated by Muslim fundamentalism, a rudimentary form of patriotism, and hatred of the foreigners who had appropriated the best parts of the land. Brigandage was also a factor of some importance. Some native tribes supported Abd el-Kader but others opposed him, there was no close cooperation between Arabs and the Kabyles, this lack of unity contributing its share to his ultimate defeat.

If it took the French some fifteen years to "pacify" Algeria, the Russians had to fight twice as long to subdue the Caucasian tribes from the time of Kazi Mulla's first appeal for a holy war in 1829 to Shamil's final capitulation in 1859.[40] The conquest of the Caucasian mountains began with the arrival of General Ermolov (a cousin of Denis Davydov) in 1816, but Russian military occupation was limited at first to the main strongpoints. General Paskevich, like Bugeaud's predecessors in Algeria, had proposed a network of forts and blockhouses to control the area, but since the fighting tribesmen were not confined in their operations to the main roads (of which there were very few), it was clear that this system was not well suited to local conditions. The Russians were fully conscious of the soldiering qualities of the enemy. "The mounted natives," wrote General Velyaminov, "are very superior in many ways both to our regular cavalry and to the Cossacks, they can ride between dawn and sunset one hundred miles. They are born on horseback, their weapons, carefully selected, were private property and kept in excellent state." Ermolov's tactics were ruthless; when he commanded the Russian army, many mountain villages (*auls*) were destroyed and the inhabitants killed. These outrages precipitated a general rising of the mountain people against the Russians; Ermolov "conquered the mountains but the forests defied him."[41]

The Shamil rebellion coincided — and was to a large extent connected — with the rise, in the Daghestan mountains, of Muridism, a revivalist movement derived from Sufi'ism, an Islamic religious trend. Shamil, who became the leader of the movement, strictly enforced the *Shari'at*, the law of Islam. His reputation both among his fellow tribesmen and some of the Russians was that of a superman. He could make himself invisible, and on several occasions is said to have jumped with ease over a ditch twenty-seven feet wide, which, if true, would have stood as the world broad jump record for more than a century. In 1839, after the surrender of Akhoulgo, a costly battle for both sides, the Russians thought the war was over, but Shamil escaped from the besieged fortress and most of the fighting was still to come. There were years in which the Russian forces lost up to twenty thousand men in the struggle against this invisible enemy.[42] Exact figures do not exist; Allen and Muratoff maintain that the Russians lost some five thousand men in 1840–1842, but they counted only battlefield casualties, not those who succumbed to disease.[43] Ermolov had been the first to use flying columns, but this strategy was given up because the Russian soldier, accustomed to fighting in the plains, proved to be inept in mountain warfare. He needed to see his neighbor and was short on initiative if left without explicit orders from his superior officer. Marksmanship, too, was bad — the Russians all too often fired without even bothering to take aim. It was only in the later years of the war that General Voronzov again reinstated the mobile columns with greater effect. By that time the Russian army had adjusted itself to the technique of mountain warfare, it was better equipped and it had even greater numerical superiority (a hundred and fifty thousand). Voronzov and Evdokimov, unlike their forerunners, realized that the intelligent course was to exhaust Shamil gradually, rather than seek to destroy him with one numbing coup. The new Russian strategy frustrated the Caucasians. "When time after time they found that in fact they could never come to blows, their weapons fell from their hands. Beaten they would have gathered again on the morrow. Circumvented and forced to disperse without fighting, while their villages were occupied without opposition, they came in next day and offered their submission."[44]

The Shamil movement has remained a bone of contention to Russian and Soviet historians to this day. Some praised him as an opponent of feudalism and imperialism, a revolutionary democrat and a fighter for national liberation. After 1947, there was a reversal in the party line and Shamil was condemned as a religious obscurantist, a reactionary and a hireling of foreign imperialism. In the post-Stalin period a compromise was reached: Muridism is still considered essentially reactionary, but while the Caucasian aristocracy is said to have been opposed to everything Russian, the masses had great love for the Russian people; that, despite the involvement of foreign intrigues, their struggle was anti-Tsarist, not anti-Russian in character. As a compromise formula the new version had much to recommend it; whether it corresponds to historical truth is a different question altogether.

In 1846, at the height of his power, Shamil had some twenty thousand men under him, subdivided into units of either a thousand or five hundred warriors. All of them were horsemen, who could be assembled and dispersed in a very short time. There were no baggage trains, every mountaineer carried with him what he needed. The men wore yellow robes and green turbans, the officers black robes with cartridge cases of silver sewn across their chests. The Circassians' rifles were of better quality than those of the Russians, and the Caucasian *shashka* was superior to the Russian sabre. In 1847 Shamil acquired some artillery, most of it captured from the Russians, but it was of no great service to him; it may, in fact, have hampered his movements. He was at his best in surprise attacks against small Russian depots or forts, in disrupting their lines of communications, and denying them local supplies until the Russians, in the words of a historian, "might as well have been in the middle of the Sahara. Shamil had taken or destroyed everything eatable by human beings for miles round."[45] He was a superb commander of five hundred raiders; five thousand he found unwieldy to handle and coordinate. With his primitive religious fanaticism he showed surprising sophistication in conducting political warfare, encouraging desertion from the Russian ranks;

deserters were well received, joined his small army, and those who had fled because they were ill treated by Russian officers became the staunchest and most implacable of fighters. From time to time Shamil ventured into the plains in an effort to raise the Kabardins and other tribes, without, however, any pronounced success, the "lowlanders" being too exposed to Russian reprisals to dare join the revolt. During the Crimean War Shamil was Turkey's ally and indirectly also that of Britain and France, if the help he either obtained or himself gave here was all but negligible. On the other hand, he was never entirely cut off from the outside world, and arms and supplies continued to reach him all along. Shamil surrendered in August 1859, lived in fairly comfortable exile, first in Kaluga, later in Kiev, and died in 1871 while on a pilgrimage to Medina. Some of his followers emigrated to Palestine and Jordan and settled there. Shamil's regime was harsh, even despotic; perhaps it was the only possible way to spur the mountain people into battle, though in the long run it certainly did not make for unity in his ranks, or for solidarity between the mountain tribes. But Shamil's warriors, however brave, would have been defeated in any case because the odds were too heavily against them. Russia was so much stronger both in numbers and materiel and it could not possibly be deflected from expansion. The wonder is not that Shamil was vanquished, but that he held out for so long.

Compared with the battles for the Caucasus, the Russian expansion in Central Asia was a walkover. True, the Turcomans occasionally put up stiff resistance. A correspondent of the *New York Herald* has given the following account of the kind of skirmish General Kaufmann encountered on his march to Khiva:

> "Gotovo, charge," shouts the Prince and we are down on them like an avalanche. A cloud of dust, the panting of horses, the rattle of harness, a flash of sabres and we are there.
>
> But the Turcomans are not. Three hundred yards further on we see them, they are going in a gentle canter, not seeming to be in the slightest hurry, and evidently not in the least apprehensive of our overtaking them. We continue the chase a short distance with no result. It is exasperating. We may as well charge a flock of wild geese and we give things up.[46]

The Russians were exasperated, but not for very long. The Turcomen were excellent horsemen, brave in individual combat. But like the Red Indians, indeed like all primitive people, they were incapable of fighting in large units, and since guerrilla operations, no less than regular warfare, called for an overall strategy and careful organization, they never had a chance.

During most of the nineteenth century Great Britain engaged in what some contemporaries termed, crudely but not altogether inaccurately, "nigger bashing" — small, and not so small colonial wars in various parts of the empire. Each of them came into its own due share of publicity at the time; there are many streets in London named Magdala, Gwalior, Cawnpore and Khartoum after the sites of notorious battles fought there, and as many after the generals who led the British troops. (There are four Outram streets in the British capital, five were named after Kitchener, three in honor of Brackenbury.) Up to about 1860–1865, the majority of these wars took place in Asia, subsequently the scene shifted primarily to Africa. Most of them were not really guerrilla in character; the Afghans whom the British fought thrice had a regular army, so had the Egyptians, and even the Zulus, and the Mahdi in the Sudan. The native armies were ill equipped and their leadership was usually not very competent, but they nonetheless practiced regular, not guerrilla warfare; the British commanders were frequently surprised by the lack of enterprise they displayed. In the Abyssinian campaign, as an example, the British expeditionary forces under Sir Robert Napier eventually reached Magdala, a seemingly unassailable fortress, Napier himself writing later that if old women had been at the top and, hiding behind the brow, had thrown down stones, they would have caused any force a serious loss.[47] The British were still, however, defeated on more than one occasion — in the Afghan wars, for instance, the Zulu wars, and against the Boers at Majuba Hill — this usually as a result of underrating the opposition's numerical strength or fighting qualities. During the Indian Mutiny there were incidents of guerrilla warfare, such as Tantia Topi's raids, especially in the later stages of the rebellion. The *talukhdars* attacked British convoys, surprised small detachments and engaged in general brigandage. But this was the exception, not the rule, and many observers, including Marx and Engels, expressed astonishment that guerrilla warfare had not been more widely applied by the insurgents; in view of the wide spaces of India and the small number of British troops, guerrilla warfare would have been infinitely more effective than sieges and open field battles. But the mutiny lacked a broad popular base, and the precondition for a successful people's war did not exist.[48]

The forces facing the British in India were numerically quite strong — perhaps sixty thousand men in the second Sikh war; the

British had to fight fifty thousand dervishes near Omdurman, and forty thousand Zulus. But invariably the natives were subdued in the end, again for the standard reasons of poor leadership and the needed discipline for fighting in large platoons. They were incapable of carrying out any complicated maneuvers once the battle had started. Whenever, on the other hand, a colonial army had to cope with guerrilla tactics, the war was likely to be undramatic, costly and prolonged. The Pathans went on fighting in the Northwest Frontier region for many decades, while the French needed a long time to "pacify" Indochina until at last Gallieni and Lyautey hit on the right tactics in the 1890s — surprise raids by converging columns along with simultaneous doses of political warfare, aimed at depriving the rebels of the support of the local population. Often the colonial armies were hampered by adverse conditions. Wolsely, for one, found the going rough in the steaming jungles of Ashantiland, what with the Ashantis harrying the British lines of communications, and the Black Watch prone in the confusion to mistake the Welch Fusilliers for the enemy.

The Maori wars in the north of New Zealand went on for twelve years from 1860 to 1872 and, more perhaps than any other, bore the characteristics of a guerrilla war. It had begun, strictly speaking, even earlier, with the Wairam incident in 1843 when white surveyors were killed as the result of an incident which they had unnecessarily provoked. The white settlers bitterly complained about the "brutal tortures of the cruel Maoris," but the war had certainly not been started by the Maoris. It was a conflict over the ownership of land, and the local whites were far more avid to attack the Maoris than were the military authorities and the British government. The Christian missionaries, too, sympathized with the Maoris. The British troops soon found that regular army tactics such as fixed bayonet charges were of little avail in this war. The Maoris skillfully used the high grass as cover and built ingenious fighting positions in the form of trenches (*pas*), access to which was barred by felled trees and other obstacles, and designed not as fortresses but to impede the advance of the British troops and to inflict losses on them.[49] The Maoris were led by a gifted chieftain, Tito Kowaru, "the De Wet of the Maoris," who still could not prevent his followers from being gradually pushed back.

British soldiers saw combat in the jungle, the bush and in mountain passes; the Shundur Pass which was the scene of much fighting in the Clitral campaign (1895) is situated twelve thousand feet above sea level. The Dutch encountered armed resistance in Sumatra (Achim), the French in Tonkin (1882–1895), in Madagascar (1884 and 1895), and in Tunis, the Germans in Southwest Africa (the Herero revolt), the Spanish in Morocco (1892–1895), the Americans in the Philippines (1899–1902). There were many other armed conflicts, some involving small detachments, others, thousands of men. Sooner or later these insurrections were suppressed, sometimes by brute force, on other occasions pacification coming about as a result of combined military and political action.

These exotic wars frequently caught the contemporary imagination, stirred by what seemed like nothing so much as glorious adventure, laced even with a certain romanticism, at least from a distance. The same applies, *a fortiori*, to the Indian wars in the United States. But they were only seldom guerrilla wars, and sweeping statements such as "the Apaches were in fact guerrillas,"[50] are of not much use toward an understanding of the specific character of these wars. It is perfectly true that the Indian braves showed great resources of courage, that they engaged in hit-and-run attacks, that they were past masters in woodcraft, and at ambushing. But in other essential respects they were anything but *guerrilleros*. They hardly ever operated in the enemy's rear, certainly not in any systematic way; whenever they could, they refrained from night attacks; their frontal assaults, in wave after wave, were often suicidal owing to the far greater firepower of the enemy. With a few exceptions, Tecumseh paramount among them, the Indian chiefs were quite incapable of leading sizable contingents in their campaigns. The Red Indian tactics, as Fletcher Pratt has noted, were those of the squad — they could not combine their operations and were unable to think in larger terms. The U.S. Army, like the British in Asia and Africa, met with the occasional defeat, as on the banks of the Little Bighorn in 1876, when Custer made the fatal mistake of sending out units that were too small and when he declined to take his Gatling guns which might have made all the difference. In the early decades of the nineteenth century, when the Indians were not yet outnumbered by the white man, such reverses were less uncommon. Major Dade's whole force was massacred by Seminole Indians in the swamps of Florida in December 1835. But the forces involved at the time were small; there were altogether only 536 U.S. soldiers in the entire state of Florida.[51] The Seminole wars are of interest, be it solely because they cost in their seven-year span more U.S. soldiers' lives than all the other Indian wars of the nineteenth century put together, plus the staggering sum of forty million dollars.[52] In the end some eight thousand men and naval support ships had to be concentrated to subdue the Seminoles. Patrolling, "blind raids," and bloodhounds had proved unproductive; the campaign succeeded only after the troops were ordered to persist in their efforts throughout the summer as well, the great heat notwithstanding, and after their blows were directed no longer against a forever elusive enemy, but against their villages, crops and food supplies. Deprived at last of their sustenance, the Seminoles had no alternative but to surrender.

With the opening of the West to white settlement and the spread of a railway network, the Indians were still further depleted and their remnants herded into reservations. Dissension had always been rife in their ranks; there was not, after all, one cohesive Indian nation, only a disparate collection of many tribes, frequently on the warpath against each other. Their arms were inferior in quality and they had no powerful outside ally. Their foe gradually became as adept as they were themselves at woodcraft and scouting.

THE SMALL WAR IN THE BIG WAR

In three of the four major wars of the nineteenth century, guerrilla warfare played a certain, albeit minor, role — the American Civil War, the Franco-Prussian War, and the Boer War. Differing as greatly in almost every respect as these wars did, the guerrilla activities in each warrant individual examination. Considerable claims have been made about the impact of the Confederate guerrillas; according to Virgil Carrington Jones, the war would have ended in 1864, eight to nine months earlier than it did, but for the operations of the raiders who prevented Sheridan from clearing up the Shenandoah valley so that Grant could pursue his campaign.[53] But even if this claim is accepted, the guerrilla activities merely prolonged the agony of the South. Leading Confederate generals had misgivings about the irregulars from start to finish. "I regard the whole [partisan] system as an unmixed evil," was General Lee's own unvarnished way of putting it. Perhaps it was simply the resentment of the regular soldier against an unconventional manner of warfare, and snobbism vis-à-vis its practitioners who were not always gentlemen. Not that it was snobs alone who objected to William Quantrill, the gang leader who captured Independence, Missouri, and plundered Lawrence (1862–1863). He was a vicious murderer, and his men consisted chiefly of cutthroats like Jesse James and his brothers.[54] They developed a strange ritual (a black flag and a black oath), but their prime object, as far as can be ascertained, was to pillage and to destroy, with no great pains taken to distinguish between friend or foe.

True, it would be unjust to regard Quantrill and his men as the typical Confederate guerrilla; Mosby, Morgan, Johnson, Ashby, Stuart Sheridan and their raiders were men of quite different caliber. To single out only the two most famous and effective among them, Morgan was a businessman in Lexington, Kentucky, thirty-six when the war broke out — a daring cavalryman and lover of thoroughbred horses who had seen action as a captain in the Mexican campaigns. Mosby was a twenty-eight-year-old lawyer, who while a student at the University of Virginia had killed a fellow student in an unprovoked duel, and had read law while in prison.[55] Morgan's raiders came from the best families of the South — deep-rooted gentry as for instance the Bullitts, Colemans, Breckinridges — and were themselves lawyers, physicians, merchants, journalists for the most part, drawn at the time to what one Richmond newspaper called "the most attractive of the services for all young men of a daring and adventurous nature." In equipping his troops, Morgan discarded the saber in favor of pistols and carbines; everyone dressed according to his taste — broad-brimmed hats, the pants stuck into high boots, a pair of pistols buckled around the waist. Mosby later denied that his men habitually wore blue overcoats to mislead the enemy; they did so only when they could get no others.[56] The question was of more than academic importance, for the Northerners were not at all sure whether captured partisans should be shot or treated as prisoners of war — and those in civilian clothes, it was argued, had forfeited all rights. Mortan was the first to stage long-distance raids into the enemy rear in the cold winter of 1861–1862 after the South had sustained some unexpected defeats. When about to enter battle, the raiders dismounted and fought as if they were an infantry unit.

Of Morgan's four major raids which took him to Indiana, Ohio and Tennessee, the first and third were the most successful. In the first raid he covered a distance of a thousand miles in twenty-four days and for the loss of a hundred of the eight hundred men with him, did much damage to the Union forces and captured (and paroled) twelve hundred prisoners. In the third raid, this time with four thousand men, he took over eighteen hundred prisoners for the loss of only two men, and destroyed two million dollars' worth of property.[57] The fourth (Ohio) raid in July 1863 was the most spectacular but also the most pointless. With some twenty-five hundred men he ventured far afield from his base; on one occasion in Indiana he covered ninety miles in thirty-five hours.[58] But eventually his force was surrounded and destroyed and Morgan had to surrender. A new raid into Tennessee ended in total disaster; at Greeneville he was tracked down by Union troops and killed.

The composition of his band had changed greatly toward the end. The Southern aristocracy was replaced by Southern riffraff; bank robberies as well as petty thieving became quite common. Morgan was a courageous, impulsive, and impatient commander, conspicuously lacking some of the essential qualities of a guerrilla leader. His great achievement was to tie up to thirty thousand Union soldiers with a force which never extended beyond a few thousand men, but usually was far smaller. He excelled in public relations; newspapermen were always welcome at his camp and he had a good press, except, of course, in the North. But his victories, however dashing, never influenced the issue of a campaign, let alone the war.[59]

Mosby's rangers, who numbered only two hundred, began their actions on a small scale, cutting telegraph wires; later on they hung around enemy camps, shot at sentinels and pickets, intercepted couriers and supply wagons and forced the Union army to move only in large bodies. On one occasion Mosby captured a Union general in his bed (General Stoughton, not one of the outstanding leaders of the North); on another he seized a payroll of $173,000. His activities were given wide publicity and many Southern soldiers of the line asked for a transfer to this unit or to another raider command.

Among the generals and colonels in charge of these units, Nathan Bedford Forrest was considered by many to be the greatest. According to some, he was the most accomplished fighter to emerge in the war, and this despite his having had no previous military training.[60] A wealthy businessman from Tennessee, he had raised a battalion at his own expense in 1861. In a major raid with a thousand men in July 1862, he overpowered an enemy brigade, destroyed railway bridges and captured a load of supplies. His second raid in December 1862, into Kentucky with twenty-five hundred men, was less successful, but in the Atlanta campaign, during the second half of 1864, he played a considerable role, diverting major enemy units for a dismayingly long time with a force of between two to five thousand men. Sherman was brought to the point of saying that "the devil Forrest must be hunted down and killed if it costs ten thousand lives and bankrupts the treasury."[61] His operations resembled those of Denis Davydov in the War of 1812; deep-penetration raids into the enemy's rear rather than guerrilla warfare.

By mid-September 1862, partisan rangers had been organized into six regiments and nine battalions in half a dozen Southern states, and units of company size existed in Florida and Mississippi. The North, too, had a few rangers, if their activity was mainly confined to the Leesburg area. But the strong opposition partisans and their operations had aroused from the beginning tended to increase rather than diminish in both North and South and cannot readily be explained away, exist though it may well have done at times, as simple professional jealousy. General Heth, commanding the West Virginia military district, wrote that the partisans were just organized bands of robbers, that they were more ready to plunder friends than enemies (because it was less dangerous), that their leaders were unable to enforce discipline and that their interpretation of fighting was roaming over the country, taking what they wanted — and doing nothing.[62] Seddon, the Confederate secretary of war, wrote President Davis in 1863 that they (the partisans) had not infrequently caused more odium and done more damage with friends than enemies.

The treatment of guerrillas by the North varied from state to state. Assistant Secretary of War P. H. Watson wrote that they were the "common enemies of mankind" and should be shot without challenge. Generals McClellan and Halleck issued orders instituting the death penalty for insurgent rebels apprehended in the act of destroying bridges, railway and telegraph lines. When Northern forces threatened to execute two Confederate officers, Lee warned McClellan in a letter that he would retaliate and the Union Command did not carry out its threat (on a previous occasion, after the execution of six of Mosby's men by Custer, Mosby had retaliated by killing six of Custer's soldiers). Grant thought that guerrillas without uniform should not be treated as prisoners of war. Washington consulted Francis Lieber, professor of law at Columbia, about the status of the guerrilla in international law. His treatise, learned and fair as it was, did little to elucidate the issue. Partisans (he wrote) were entitled to the privileges of the law of war provided they opposed the invader openly and in respectable numbers and operated in the yet uninvaded portions of the hostile country; on the other hand, no army and society could allow unpunished assassination, robbery and devastation.[63] But this left many questions open. What was "open resistance"? What were "respectable numbers"? And what if the front line was not clearly delineated?

As resistance against guerrillaism grew, the Confederacy in early

1864 repealed the act which had authorized the formation of partisan units. Some nevertheless continued to operate, and a few continued to fight on even after the South had surrendered. The Union generals responded to the raids during the last year of the war with the systematic destruction of farms, crops and livestock and the carrying off of all men under the age of fifty; in this way, Grant had told Sheridan, "you will get many of Mosby's men."

The subsequent fate of some of the leading partisan commanders is not without interest. Mosby tried his luck in politics, made his peace with the North, and was helped by President Grant, much to the disgust of his fellow Southerners. He became consul general in Hong Kong and eventually acted as an attorney for railroad companies. Duke became a congressman and published a number of books about his experiences in the war. Forrest, who had been a millionaire before the war and lost his property, became a planter and apparently also the Grand Wizard of the Ku Klux Klan. "Flying Joe Wheeler" also became a congressman, and a general in the U.S. Army who was to see action in the Philippines against another partisan, Aguinaldo, three decades later. General Adam R. Johnson outlived them all; blinded in battle, he settled in a new town which he had founded on the shores of the Colorado River in Texas; he died in 1922.

The question of the effectiveness of partisans and rangers in this war has remained, as noted initially, a matter of dispute to this day. That the guerrillas were a "hornet's nest" and that they caused damage to the North is beyond doubt. But Union retaliation could be telling; if the rangers attacked their supply lines, they would live off the land, much to the detriment of the South. The rangers were at their most potent when operating among a friendly population, their deep-penetration raids frequently taking them into hostile territory where they could not count on the valuable goodwill of the civilians. Some of the large-scale raids involving thousands of men were quite successful, but they were also far more risky and liable to end in disaster. The smaller units, such as Mosby's, were more elusive and therefore, for that very reason perhaps, on balance more effective.

PARTISANS IN THE FRANCO-PRUSSIAN WAR

The war between France and Prussia began in mid-July 1870; two months later the French armies were beaten, the emperor had abdicated, Paris was under siege and Moltke, the German commander in chief, was reasonably certain that he would be back on his farm in Silesia in October in time for the hunting season. But outside Paris a government of national defense had taken over with the ringing slogan, *guerre à l'outrance*, and a huge new army was mobilized by Gambetta and his military deputy, Freycinet. The underlying concept was that this new army should act as a vast guerrilla force, harassing the enemy rather than engaging in frontal attack. Memories of the Vendée and of Spain were conjured up; some of the partisan units were commanded by officers who were descendants of leading Vendée rebels such as La Rochejacquelin. The guerrilla concept did not lack plausibility; the Germans had an efficiently organized fighting machine, but they might well find themselves hard pressed to adjust to an unfamiliar type of war. It was thought, furthermore, that guerrilla warfare would give the French soldier a chance to exhibit his real prowess. In the first phase of the war the soldiers had fought well, whereas the higher command had failed. Guerrilla warfare, however, demanded no staff experience and planning; it could be carried on by zealous citizens even if they had no profound knowledge of strategic theory or its practice.

The new government nevertheless decided on 29 September 1870 to put the *franc tireur* units under the general command of the army. In all, some fifty-seven thousand officers and men enlisted in the free corps (*corps francs*); some small units of perhaps two to three thousand men had been in existence even before then.[64] But the original enthusiasm for a vast *chouannerie* did not last, for several reasons. Above all, Gambetta realized that it would take time to organize a people's war, longer time than was available to lessen the pressure on the besieged capital. Hence, priority had of necessity to be given to the establishing of a new regular army which could be readied more quickly to help relieve Paris. Secondly, there was passive resistance on the part both of army officers and the civil administration. There were reports that the *franc tireurs* were misbehaving, scandalizing the population by their brigandage, and that they were none too eager to engage the enemy.[65] New measures were proclaimed to intensify control over the free corps; each such unit was to be directly responsible to the local military command, every officer had to report twice weekly on the activities of his unit. On 14 January 1871 it was announced that no new free corps would be established.

The opposition to *franc tireur* operations stemmed partly from the innate conservatism of unimaginative army officers who feared that with the spread of partisan units the line between soldier and civilian would be blurred.[66] But their aversion was not entirely unjustified, for the *franc tireurs* indeed lacked discipline, they were incapable of carrying out sustained military operations, joining or absenting themselves from their units as it suited them. Lastly, patriotic enthusiasm was strongest in the towns and weakest in the countryside. The peasants did not receive the Germans with open arms, and more often than not they refused to collaborate, but neither was there any great willingness on their part to leave home and farm to join the *franc tireurs*. There was a general feeling of apathy, and since the Chouans lacked enthusiasm, there could be no *chouannerie*.

The *franc tireurs* were badly equipped, their leadership was indifferent and they missed countless opportunities. Even their two most spectacular operations were of no military significance. By the time they mined the viaduct of Fontenoy (22 January 1871), the railway line was no longer of vital importance for the Germans.[67] And the capture of the village of Le Bourget, north of Paris (the site long since of a famous airport), by Parisian *franc tireurs* on 27 October 1870 provided a psychological boost but little more; the Germans took it back four days later.

And yet, uncoordinated and badly executed as partisan warfare was, it produced some startling results. As the war progressed — and as it emerged once it was over — the Germans had to deploy some hundred and twenty thousand men, a quarter of their total force, to protect their lines of communication, mainly the railways. The people's war between the Seine and Loire caught the Germans altogether unprepared, both politically and militarily. Politically, because Bismarck feared that the longer the war lasted, the more likely the diplomatic intervention of the other European powers, which would deprive the Germans of at least some of the fruits of victory. On that count alone, Bismarck had every incentive to bring the war to a speedy end. Militarily, the Prussians were superbly prepared to fight against a regular army, but an elusive enemy was not that easy to destroy. France is a big country, the German armies combined numbered fewer than half a million soldiers and the farther they advanced, the more thinned out they became, for garrisons had to be left behind in every town and strongpoint that was occupied; not too small garrisons either since an attack or an insurrection could never be ruled out. Altogether the Germans lost more than a thousand men in *franc tireur* warfare, a not insubstantial figure in terms of casualties in general. Of more import was the pervasive atmosphere of insecurity generated by *franc tireur* operations, and German commentators freely admitted after the war that the irregulars had caused them serious problems.[68] A people's war conjured up the specter of a revolution. The Germans would have greatly preferred to make peace with the emperor; instead, they had to deal with a republican government, and there was the danger of further radicalization.

The war was conducted cruelly on both sides; the *franc tireurs* committed acts of individual terror, the Germans retaliated by executing hostages and burning villages. French publicists, including Victor Hugo, called blatantly for total war, the extermination of every last German; Frau Bismarck was not alone in suggesting to her husband that all Frenchmen should be shot and stabbed to death down to the smallest infant. But Bismarck and the old emperor, despite occasional expressions of violent anger, were sober and farsighted enough to reject such advice. They rightly feared the incalculable consequences for the future relations between France and Germany if these atrocities should spread and become common practice.

The *franc tireur* war consisted of innumerable small actions such as destroying railway lines and bridges; the French irregulars also tried to blow up tunnels but lacked the know-how and sufficient quantities of explosives. On several occasions they succeeded in freeing transports of French prisoners of war. Thus, on the road from Soissons to Château Thierry, between three to four hundred prisoners escaped during a *franc tireur* attack.[69] Telegraph lines were cut and supply columns attacked. The scope of *franc tireurs* activities would have been wider but for the lack of cavalry which restricted their movements on the whole to forests and other inaccessible regions. They engaged in night attacks on small German garrisons, as at Chatillon in November 1870.[70] In this instance the Germans lost 192 officers and men. Many of these were taken prisoner and the French threatened that they would be executed unless the Germans treated captured irregulars as prisoners of war. Auxon, near Troyes, had to be evacuated temporarily under *franc tireur* pressure, and a first German attempt to enter the city of St. Quentin and to arrest the local prefect was beaten back. The Bavarians and the troops of the Grand Duke of Mecklenburg ran into difficulties near Orléans, on the road to Chartres and in the Dijon area, and almost invariably in hilly or wooded country.

The *franc tireur* units, hastily established, were a mixed bag and it is almost impossible to generalize about their composition, political orientation and military efficiency. Some bands had only a handful of members, others several hundred. Most were set up on a local basis, with the men fighting in the vicinity of their homes, but there were also partisan units from Bretagne, from Nice and from North Africa, not to mention Garibaldi's irregulars. They wore every kind of fantasy uniform, and some wore no uniform at all. Some were radical left wing in inspiration, a number had a conservative and monarchist bias. Some were relatively well organized and disciplined and operated to all intents and purposes as small military units would have done. Others, wandering aimlessly from village to village, showed greater proclivity for marauding than fighting the German enemy.[71]

After all the initial enthusiasm for a people's war, French resistance collapsed during the early months of 1871. France was not the Vendée or Spain; the great majority of Frenchmen, much as they hated the Germans, lacked the fanaticism and the stamina for the *guerre à l'outrance* which had been so loudly proclaimed at the start. The psychological shock of the defeat had been immense; for two centuries, if not longer, Frenchmen had believed their country to be militarily superior to all other European powers, and the surrender of their armies had destroyed their self-confidence — it was not just a crisis but a national disaster, the collapse of a whole world. To prolong resistance now, was the despondent attitude, would be but to devastate their towns and villages further, to no conceivably different end. It is idle to speculate what might have happened if resistance had continued for six more months or even a year. Moltke and the war party were only too eager to carry on the campaign, but domestic pressure on them to end it was growing. It was not only Bismarck's apprehensions about diplomatic intervention, but the war was becoming increasingly expensive, and daily more unpopular at home. Even the front-line troops were weary and the war minister, Roon, wrote that it might take years to occupy the whole of France. But the years were not called upon: French resistance faded and died away first. Guerrilla warfare, as the average Frenchman saw it, would never bring about the liberation of his country, whereas peace would perhaps open new perspectives and possibilities for a national recovery. Which all points up the more strongly that the operations of the *franc tireurs* neither could nor did change anything insofar as the military results of the war or — even less — the conditions of the peace treaty were concerned. German demands certainly did not become more moderate as the people's war continued. It was political considerations at the last, however, quite unconnected with events on the battlefield, that fairly narrowly circumscribed the terms that Germany could finally impose on her defeated neighbor.

COMMANDO

In May 1900 there was every reason to assume that the end of the Boer War was in sight. Cronje had surrendered, the siege of Mafeking had been raised, Bloemfontein, the capital of the Orange Free State, had been taken and in early June the British troops entered Pretoria. Field Marshal Roberts, commanding the British forces, predicted the impending collapse of Boer resistance; the Burghers had become, in the words of one of their commanders, a "disorderly crowd of terrified men fleeing before the enemy." He let his men go home, for "I cannot catch a hare with unwilling dogs."[72] When President Kruger announced that the war would begin only now, the British generals were inclined to dismiss this as idle talk on the part of an old man, a civilian who lacked understanding of military realities.

At the start of the war, in October 1899, the Boers had at first beaten the British. But as massive reinforcements streamed into South Africa, the tide began to turn despite poor generalship on the British side. The Boers were excellent horsemen and crack shots. They knew the terrain and used it very well in their operations. Above all, they were fighting for their homes and national independence. According to Boer common law, every Burgher between sixteen and sixty had to be prepared to fight for his country at any moment; he had to have a riding horse, saddle, bridle, a rifle, thirty cartridges and food for eight days. They went to war in their working clothes. But they lacked military experience and discipline and were unaccustomed to receive and carry out orders. It was truly a citizens' army; on more than one occasion their elected generals were outvoted by the corporals and the privates. Their forces were subdivided into commandos of between three hundred and three thousand men. Joubert and Cronje, who led them in the early phase of the war, were old and overcautious men; having scored a victory, they failed to press it home and make it complete. They lacked any overall strategic concept; they were capable of carrying out daring raids and surprise attacks, but there was little coordination between the commandos, and no general plan. This is not to say that they could have won the war with better leadership and a well-trained regular army; the contest was too unequal. Thus they would score some remarkable victories, such as Magersfontein, Colenso and Spionkop, but in the end the British would get fresh reinforcements, whereas all the Boers got were messages of sympathy from various parts of the world.

The British were at first exceedingly bad at reconnoitering and in general at adjusting themselves to local conditions. But as the war continued, they improved; they had guns in plenty, while the Boers had only a few and did not make good use of their artillery. It was during the second phase of the war, which lasted from, roughly, March 1901 to May 1902, that guerrilla tactics were more and more often adopted.[73] In the beginning it had been the "last of the Gentleman's Wars"; the Boers usually released their prisoners after a day or a week, if only because they had no facilities for keeping them, and the British, too, acted with restraint. But in the guerrilla phase of the war the British began to burn farms on a massive scale; the women and children who had lived on these farms were evacuated to refugee camps, known also as concentration camps. Similar practices had been employed by General Weyler in Cuba. These camps had nothing but the name in common with the concentration camps of the Nazi era, but by the standards of a more civilized period these measures were considered barbarous in the extreme; there was an outcry in Britain, the Boer resistance became even stiffer. Patriotic feeling, however, was running high in Britain, and for all the sense of outrage at this inhumanity toward civilians, not everyone by any means was happy with the relatively lenient treatment meted out to members of the Boer commandos who were exiled to St. Helena or Bermuda. Maguire, the guerrilla theoretician, wrote that if it became generally known that guerillas or irregulars would be treated like the guerrillas or irregulars in South Africa were treated, "there will be plenty of guerrillas and irregulars in every future war. It will be the most prosperous career possible."

The second guerrilla phase of the Boer War was highlighted by the commando raids of de Wet, Smuts, de la Rey, Botha and Viljoen. At the beginning of the war, the British garrison consisted of a mere 12,000 men, by the end of December 1901 there were 388,000, and when the war ended 448,000. About 46,000 of them were killed or wounded during the war, or died of disease. There had been some 60,000 soldiers in the Boer army at the beginning of the war, but their number shrank to fewer than 20,000 in the guerrilla stage.

Despite the numerical superiority, the British commanders faced the

> silent disability of a regular army in contest with a horde of guerrillas manoeuvring about their own country. Seldom in the course of the whole campaign in South Africa was it possible for the British Commander-in-Chief or any of his lieutenants, to select their own sites for battle or ground for manoeuvre. Well-nigh invariably these spots were dictated by the enemy, insignificant numbers of whom led great armies whither they would.[74]

Boer tactics were not, of course, decisive, but it was "exceedingly humiliating to be thus bandied about at the will of handfuls of evasive freebooters."[75] Even by early 1902 full control had not been reestablished in the Orange River Colony:

> There was not a convoy whose safe arrival could be counted on, not a garrison that did not stand continually to arms, not a column which even whilst it marched against the enemy had not to move with the strictest precautions of the defensive.[76]

Hardly a day elapsed, according to another well-known chronicler of the Boer War, that the railway line was not cut at some point.[77]

With the beginning of the guerrilla phase a rough division of labor was decided upon by the Boer leaders; de Wet and Hertzog transferred their activities to the Free State, Botha to eastern Transvaal and on to Natal, de la Rey and Smuts to western Transvaal. Operations in the Free State became very difficult indeed, because the territory had been laid waste by the British. Smuts's raid into Cape Colony, in the course of which he covered two thousand miles, was one of the most successful operations of the whole war. He set out with three hundred and sixty men; evenutally his force swelled to almost four thousand. He did not succeed in stirring up a general rising in Cape Colony as he had hoped, but he kept tens of thousands of British soldiers busy for a long time.[78]

The Boer columns caused greater damage to the British in this period of the war than ever before. Their incessant maneuvers and frequent attacks exhausted the British forces; their horses died by the thousands. There were daring attacks on such strategic targets as the Bloemfontein waterworks. De Wet wrote that it was painful for him to see any railway line and not be able to damage it. At first the commandos used a primitive land mine. The barrel and lock of a gun connected to a dynamite cartridge were placed under a sleeper; when a passing engine pressed the rail to this machine, it exploded.[79] Later on the system was perfected; the gravel was hollowed out, the machine was placed under a sleeper and covered up again. The British trebled their guards but there were still explosions, and no trains could run at night.

To isolate the commandos and to prevent a breakthrough, the British built a network of blockhouses at a distance of between eighty and eight hundred yards, which were connected by barbed wire entanglements, trenches and stone walls. To man them and to maintain other garrisons, some hundred thousand troops were needed. De Wet was scornful about these "white elephants" and claimed that he always fought his way through, that not a single soldier was captured as a result of the "policy of the blockhead."[80] He found it more difficult to cope with British night attacks and, like other commando leaders, was paralyzed and eventually defeated by the British scorched-earth policy. Kitchener's flying columns sweeping the area beyond the blockhouses were not effective; if the Boers lost heart, it was not as the result of these drives but because of the strategy of steady attrition. It was a race against time for both sides; opposition to the war in Britain was rising and Kitchener was by no means optimistic. "The dark days are on us again," he wrote in March 1902. Four months earlier Smuts's column had entered the western regions; at first they were "hunted like outlaws," but, "today," Smuts wrote, "we practically held the whole area from the Olifants to the Orange river 400 miles away, save for small garrison towns here and there."[81] But there were some hundred thousand Boer women and children in the concentration camps, and the number of Boer prisoners of war (thirty-two thousand) who had been exiled was by early 1902 considerably in excess of the number of those still in the field (eighteen thousand). Thus, after heated internal discussions at Vereeniging, with Steyn and de Wet demanding a fight to the bitter end, the Boer leaders capitulated. What clinched the matter was apparently an aside by Kitchener in conversation with Smuts — that in all probability two years hence a Liberal government would come to power granting a constitution to South Africa which would meet the Boers' demands for national autonomy.[82] The peace terms were not too harsh; within the next five years Britain was to pay the Boers some ten million pounds in compensation for the property that had been destroyed, and in 1910 the Act of Union came into force, the first major step on the road to an independent South Africa.

The Boer War had certain unique features distinguishing it from all other wars of the period. Was it a guerrilla war? De Wet did not think so. "I was always at a loss to understand by what right the British designated us guerrillas," he wrote. In his view, the only case in which the term could be used was when one civilized nation had so completely vanquished another that not only the capital was taken but the whole country from border to border occupied, and this clearly was not so in South Africa.[83] But de Wet labored under the misapprehension that brigandage and guerrilla warfare were more or less synonyms; a deeply religious man, like most Boer leaders, it was for him a "war of religion." "My people will perhaps say 'our generals see only the religious side of the question.' They will be right."[84] They had begun the war "strong in the belief in God," because they thought it was the right thing to do, and the possibility of defeat had not entered their minds.

It was a "gentleman's war" to the extent that nongentlemen — that is, black people — were not mobilized by either side. The Boers underrated the enemy in the light of their past experience with the British (their victory at Majuba Hill in 1882), and the victories in the early months of the war seemed to justify their optimism. But they underestimated the resources of the British and, once the war became less gentlemanly (the scorched-earth tactics and the concentration camps), the Boers found themselves not only without food and in rags, but having great trouble getting weapons and ammunition. Reitz relates that he had exactly four bullets left when he joined Smuts's raid into Cape Colony, and others were no better off. Had there been two or three million Boers, they could have held out almost indefinitely against the British, but even their fighting spirit and commando tactics were not adequate enough substitute for their paucity of numbers.

The war had long been at an end when the leaders of the Boer commandos were to find themselves on opposite sides of the barricades — like the Spanish *guerrilleros* in the Carlist wars, and like many other guerrillas before and since once their wars were over. When World War I broke out, de Wet and de la Rey felt the time had come to shake off the British yoke. Hertzog was wavering, but Smuts and Botha suppressed the rising by force of arms. While de la Rey was killed in the fighting, Smuts lived to become a British field marshal and member of the British war cabinet. He commanded the British troops in German East Africa in World War I, while Botha fought the Germans in South West Africa. Reitz commanded the Royal Scots Fusiliers, one of the oldest British regiments, on the Western Front and eventually became a South African cabinet minister. The Boer War generation dominated the South African political scene for many decades; when the Nationalists at long last broke Unionist rule in the elections of 1948, the former were still led by General Hertzog, the latter by Smuts. And when the South African Republic was proclaimed and in May 1961 left the Commonwealth, there were still some of those alive who had fought for independence sixty years earlier.

BRIGANDAGE AND GUERRILLA WARFARE

After his return from a visit to the distant provinces of northern China, Eric Teichman, a British diplomat, wrote in 1917 that the north of Shensi Province was at the time of his visit in the hands of organized troops of brigands of a semipolitical character, "robbers one day, rebels the next, and perhaps successful revolutionaries the next." It was in this very area that the Chinese Communists established their main base after the Long March. But the phenomenon was by no means specifically Chinese — in Latin America throughout the eighteenth century, and elsewhere up to and including the present time there have been similar phenomena.[85] There is in fact frequently no clear dividing line between guerrilla warfare, terror and brigandage. No one, to be sure, thought of nineteenth-century Russian and French anarchists as guerrillas; their attacks were directed against leading figures of the political establishment, and sometimes indiscriminately against the public at large. But in later years the border line between guerrilla warfare and assassination became hazy; the activities of the Irish rebels and the Macedonian IMRO serve as an illustration. The demarcation between guerrilla and banditry had all along been less than clear. For ages past, the world over, bandits operating from hideouts in inaccessible regions such as mountains or forests used the technique of the hit-and-run raid and the ambush, they had to be good shots and good horsemen to succeed in their chosen profession. There were, of course, important differences, not least on the tactical level; the marauders usually operated in very small groups and, even where brigandry was endemic, cooperation was rare between one band and the other. (Outside Europe, however, bandits sometimes operated in units of many hundreds.) Above all, they lacked a political incentive, robbery being in the main and chief objective; the richer the victims, the better. Nonetheless, as with every rule, the exceptions did sometimes exist and a political element would enter into brigandage — as, looking back, with the Haiduks in the Balkans, while the activity of *dacoits* in Burma can also not be ignored in this context. After the defeat of King Thibaw in the third Burma war (1885), armed gangs of patriotic robbers continued to harass the British for several years. But they also fought their own countrymen, and, of course, each other.

Certain early anarchist and socialist ideologists such as Bakunin and Weitling set great store by the bandit, the "genuine and sole revolutionary — a revolutionary without fine phrases, without learned rhetoric, irreconcilable, indefatigable and indomitable, a popular and social revolutionary" (Bakunin). Such expectations seemed perhaps only logical, up to a point; the bandits were a subversive force, undermining existing society — like the early Fascists, they were the outcasts of all classes. But, unlike the Fascists, they were highly individualistic people, they had no intention whatsoever of establishing a mass movement and of overthrowing the entrenched order. They were not even interested in expanding their ranks beyond a certain limit. Bakunin's fantasy had a revival in the theories (and practices) of some twentieth-century revolutionaries in Europe and the Americas, as we shall see in turn.

Banditry has been inherent in all known societies since the beginning of time, but robbers have differed from each other not less than sociologists and philosophers. There was, at one extreme, the sadistic outlaw who found fulfillment in murder for murder's sake, and on the other the noble bandit (or *bandolero*) who robbed the rich and distributed some or even most of his loot among the poor. In times of general political and social disturbance robbers would become guerrillas, some because they found it convenient to pursue their old exploits under a new respectable cloak, others because they were patriotically inclined and capable of disinterested action. Furthermore, most guerrilla movements included members of semirespectable professions such as smuggling and poaching; in peacetime these activities were strictly illegal, but smugglers and poachers were hardly regarded by society as major criminals and in time of war moral standards were invariably lowered. Smugglers and poachers knew the countryside better than anyone else, they had a lifetime of experience of being on the run and were often of considerable help to guerrilla strategists.

There were other affinities between guerrilla and bandit, inasmuch as the former, "living off the land," had to appropriate horses, food and other supplies from the local population, usually without paying for them. He did this in the name of a cause, whereas the robbers did it for less elevated reasons, but for those deprived of their belongings, the effect was all one. In addition to the official requisitions ordered by the guerrilla chiefs, there was invariably a good deal of private enterprise marauding. It would be hard to point to a single guerrilla campaign in which looting did not occur, if only because strict discipline was difficult or impossible to enforce among dispersed irregulars. It was rare in some guerrilla movements led by men of integrity such as Garibaldi or the Boer generals, but more often than not the guerrillas took a share of the spoils as their due, simply because this was the only kind of payment they had any hope of getting. This was all but standard operating practice in nineteenth-century Latin America, but it also happened in the Vendée, in the Spanish rising against Napoleon, in the American Civil War, among Abd el-Kader's followers and in countless other instances. The more farsighted guerrilla leaders did what they could to prevent systematic and too-frequent looting because they knew that, in the long run, they depended on the goodwill and the collaboration of the local inhabitants — mention has been made of the extermination of robber bands by Mina in the Spanish wars —but even they had their hands full trying to impose a guerrilla order. Mention has also been made of the almost imperceptible transition from brigandage (or *bandolerismo*) to partisan warfare in Cuba in the 1870s and the same applies, in some degree, to Mexico — Pancho Villa was perhaps the most famous case of a bandit turned guerrilla, but there were many others.[86] And this leaving out, nearer to our own times, Algeria, Vietnam, and the Cuban war in the 1950s. In some countries *bandolerismo* was a concomitant of the social struggle — this certainly goes for Mexico and for Andalusia at the end of the eighteenth century — elsewhere, as in Greece, the *armatoli* were on the contrary a conservative force forming part of the established social system.

Pancho Villa turned bandit, so he claimed, only because he wanted to defend the honor of his mother.[87] Su San, the female gang leader who earned a unique place in the history of the Taiping revolution, had become an outlaw (and the head of a major bandit gang) after the death of her husband. She organized a posse to hunt down his murderer, killed him with her own hands and in time became the chief of a band which had the reputation of robbing the rich to help the poor. Eventually she joined the Taiping army with two thousand men, winning immortality in the poems of contemporary Chinese literati.[88]

During its guerrilla phase, Communism in China drew not a few of its recruits from the ranks of robber bands. Writing about the composition of the Red Army, Mao declared it was not true to say (as the Hunan Provincial Committee had done) that all the soldiers were *éléments déclassés,* meaning deserters, robbers, beggars and prostitutes, but he admitted that the majority consisted of such men and women.[89] In principle it was quite true that they should be replaced by peasants and workers but in practice it was impossible to find replacements. Hence the necessity to intensify political training "so as to effect a qualitative change in these elements."[90] The *éléments déclassés* (*yumin*) were especially good fighters, they were courageous, and under the right leadership they could become a revolutionary force. It was surely no mere accident that the Communist guerrillas appeared precisely in those parts of China such as the Hua Yin region in which banditry on a mass scale had been endemic for a long time. In the early phase the Communist guerrillas had much in common with other armed bands such as the *t'u-fei* (bandits), and Mao for one displayed great interest in the *t'u-fei* tradition.[91] In later years the Chinese Red Armies only tolerated them in regions in which Communist rule had not yet been firmly established.

The ecology of guerrilla war and banditry is identical to all intents and purposes. This applies to all the more recent major guerrilla wars — China, Cuba, Algeria, Vietnam, Greece, the Philippines, Malaya, and so on. Naturally both guerrillas and bandits looked for hideouts in difficult terrain. But there was also usually a regional tradition of "young people taking to the hills."

If some bandits would turn to the left, others would join right-wing forces; the story of the resistance against Napoleon in southern Italy is enlightening in this respect. As in Spain, the French army of occupation could maintain itself only in the large towns; the countryside was in the hands of men who hated the foreign invaders, had lost their jobs or did not want to be conscripted. They preferred to pursue guerrilla warfare: "too weak for such an operation, they were still strong enough to turn brigands."[92] But they did not think of themselves as robbers, almost to a man these brigands died courageously when apprehended by the French. "*I ladri siete voi*" (You are the robbers), a Calabrian peasant proudly declared when facing the tribunal at Monteleone. "I carried my rifle and knife for King Ferdinand whom my God restore."[93] Militarily these brigands' gangs were by no means insignificant; the occupation army could transverse the country only in large units, small detachments were almost certainly bound to be destroyed. Murat's lines of communication were constantly disrupted, several battalions always had to be ready to fight the bandits. In the end, ten thousand soldiers were spread over two provinces. General Championnet once admitted that Fra Diavolo's band gave him more difficulty than any division of the royalist army.[94] The French position improved after they had enlisted the help of Andrea Orlando, himself an ex-bandit who knew most of the hideouts of his former comrades, and made him head of a counterguerrilla detachment. The bandit guerrillas committed innumerable atrocities, to which the French responded with "extraordinary measures"; since they found it impossible to chase the brigands, they turned against the villages, compelling them (in the words of a French officer) "to extirpate the brigands of themselves under penalty of being regarded as their complices and abetters." (*Lettres sur les Calabres, par un officier français,* Paris, n.d.)

Fra Diavolo, the most notorious of these robbers, was in fact born Michele Pezza, his better-known sobriquet indicating the cunning of a priest and the malice of the devil. It is reported that French officers who fell into his hands were burned at the stake while the villagers danced around this auto-da-fé. Like other prominent robbers such as Gasparone, he became the protagonist of an opera (by Auber, with the libretto by Scribe).[95] He had been made a colonel by King Ferdinand and after his death — he was hanged by the French in November 1806 — his family received a royal pension. Mammone, Fra Diavolo's almost equally famous colleague, had the reputation, perhaps apocryphical, of a cannibal: "The inhabitants [of his native village] assert that he hung about the butchers' stalls for an opportunity to put his mouth to the gashed throats of bullocks and swine."[96]

In this guerrilla war in the southern Italian provinces, the patriotic forces consisted of an alliance between the Bourbonists, the nobility, the clergy and the brigands, while the liberals, the republicans and the French were the enemy. The reactionary forces were headed by Cardinal Ruffo, who coined the slogan "*Fernando e la Santa Fede*"; the brigands, among them many released convicts, were fighting in the name of the holy faith. Ruffo promised the citizens faithful to the king exemption from taxes for six years and celestial delights for all eternity:

> The rabble took up its line of march as a disorderly religious procession. They tore down the trees of liberty, set up crosses in their place,

entered villages and visited churches with the most sacred forms and ceremonies of the Roman Church, the Cardinal in his purple blessing the people and their arms.[97]

When the Bourbon forces entered Naples, the republican sympathizers were lynched and their property looted. It was later bruited about that the British had first suggested that the government of Sicily should be relieved of the great burden of maintaining so many convicts and transport them to Calabria, making them useful to the public cause. Ruffo, according to this version, made the best of a difficult situation by reeducating the cutthroats with the help of his chaplains who were acting as political commissars: "He turned this unpromising human clay into brigadier generals and saints."[98] Ruffo's efforts, as subsequent events were to demonstrate, were not altogether successful.

Why certain bandits defended the existing order while others fought for the revolution depended partly on the general character of the gang, partly on the political and social background, but also on the character of the gang leader. In many cases this was a matter of sheer accident. Su San, herself turned bandit and then a leader of the Taiping through unordained circumstances, is reported by some to have later married Lo Ta-kang, another prominent ex-bandit. Together with seven other leaders of river pirates, Lo had shown up in Chin-t'ien, the main base of the Taiping, in 1851 or 1852. He became one of the most distinguished generals of the movement, and after his death in battle he had bestowed on him by the Heavenly King the noble title of *Fen wang* (Endeavour King).[99] The other chiefs of the water pirates deserted the Taiping and went over to the emperor's army. There is no known way of sociological or psychological analysis to explain why Lo became a pillar of the Taiping whereas the other pirates rejoined the antirevolutionary forces, why Mina fought with the Cristinos and Merino with the Carlists, Fra Diavolo with the Bourbons and Andrea Orlando with the Napoleonic forces. In a war against a foreign invader the choice for patriotically inclined bandits was more obvious than in a civil war.

THE OUTLOOK FOR GUERRILLA WAR: 1901

The days of guerrilla wars seemed to be over as the nineteenth century drew to its close. One of the few who dissented was that remarkable Russo-Polish-Jewish businessman and strategist Jean de Bloch, author of *The War of the Future*, who claimed that the modern rifle favoring individual action and sharpshooting and requiring the abandonment of close formations, was primarily a guerrilla weapon and tended to put the civilian on a level with the regular soldier. A guerrilla war, he declared on more than one occasion, would inevitably follow regular resistance in the future.[100] A German officer, writing in *Deutsche Revue*, sharply disagreed; it was not true that, as Bloch had argued, a guerrilla war in Europe would make a decisive result impossible. The action of the *franc tireurs* in 1870 had been quite ineffectual. Futhermore, a highly civilized nation could not carry on a guerrilla war, it would not have the patient capacity to endure the burdens, privations and sacrifices of such a war and the longing for peace would become overwhelming.[101] Bloch retorted that the case of the *franc tireurs* proved nothing, and that anyway a protracted war with large standing armies such as foreseen by the leading strategic thinkers of the day could only lead to social cataclysms and violent revolutions. Both Bloch and his Prussian critic were to be proved right by the events of the next two decades. With a few exceptions, the nations of Europe were indeed too civilized for guerrilla warfare, but the social cataclysms and the revolutions came with a vengeance.

3

The Origins of Guerrilla Doctrine

In the nineteenth century "small war," partisan and guerrilla warfare fell into disregard in Europe's more developed countries. When irregular warfare was rediscovered towards the middle of the twentieth century its antecedents had been forgotten. It was generally assumed that the history of guerrilla warfare began with the Spanish insurrection against Napoleon — as if there had been no wars of liberation and wars of opinion throughout history. But since they had not been "revolutionary wars" in the fashionable twentieth-century sense they were thought to be of little interest. It was widely believed, even among experts, that previous to Mao Tse-tung no military thinker had ever systematically studied guerrilla warfare — with the possible exception of T. E. Lawrence, an amateur of genius, but no military philosopher. In actual fact the problem had preoccupied eighteenth- and nineteenth-century students of war in many lands; it is of some interest to establish why exactly their writings have been forgotten.[1]

Small-war strategists of the last two centuries have anticipated most present day guerrilla tactics. Furthermore, already in the 1820s and 1830s some of them were perfectly aware of the political potential of guerrilla warfare. To retrace the genealogy of guerrilla warfare and doctrine is not a purely academic exercise; the assumption that guerrilla warfare in the post-1945 era is an essentially new phenomenon is not only historically incorrect, it is bound to give rise to misconceptions about the origins, the character, and the future course of "revolutionary war."

The theory of small warfare (*petite guerre*) has its origins in the seventeenth century. It was mainly based on the experiences of the Thirty Years' War (which, perhaps more than any other, had been a "war without fronts"), of the Spanish War of Succession, and the wars of Frederick the Great. These early reflections deal with the amount of scope to be given to the activities of relatively small detachments. The American War of Independence provided more examples of the many uses to which small, highly mobile units could be put. However, the literature published before 1810 did not accord an independent role to these units and was exclusively concerned with the operations of professional soldiers acting in close cooperation with the main body of the army. This single-minded approach was modified only in the light of the experience in the Vendée, Spain, Tyrol and Russia. The concept of a national war, "the most formidable of all," as Jomini put it, emerged (or re-emerged) only in the Napoleonic age; a war that had to be fought against a united people (or at any rate against its great majority), determined to preserve their independence. Every step in such a war was contested, an invading army only held its camping ground, supplies could only be obtained at the point of the sword, convoys everywhere were threatened or captured. But Jomini also thought that popular uprisings without the support of a disciplined and regular army would always be suppressed, though the suppression could be protracted such as in the case of the Vendée.[2]

Eighteenth-century military thinkers were very much preoccupied with surprise attacks, ambushes and other operations which, by necessity, had to be carried out by relatively small units.[3] A great many ruses and strategems were listed, such as, to give but one illustration, the despatch of officers and soldiers, pretending to be deserters, to the enemy camp before an attack was launched against it. But there was not much in these proposals that had not been known to the Romans and even before them, and it was perhaps indicative that the *Strategemata* of Sextus Julius Frontinus (30–104), a collection of such anecdotes, was republished and studied throughout the eighteenth and even early nineteenth centuries.[4] Specific works on partisan warfare only began to appear in the mid-eighteenth century and they drew their main inspiration from the activities of the small, light, highly mobile (and semiprivate) units which were employed in the Austrian army since the seventeenth century. These units, composed of Pandurs and Croats, had amassed considerable combat experience in the areas bordering Turkey. Later on, albeit on a smaller scale, such detachments became part of the French army.

In a book published in 1752, de la Croix, a French officer of English origin, defined the function of "parties" (*partis, partidas*) as that of moving ahead of the regular army and gathering information about the enemy's movements.[5] The author described the many ruses of these "free corps," and defended them against their detractors. In particular, he stressed the need for caution and discipline in all their operations. He had himself served as commander of such a unit and was able to provide many vivid illustrations from his own experience. The book was published posthumously by his son who had served under him. Not all his readers were equally impressed; the Prince de Ligne noted that "every one of our Croats knows as much about war ruses as de la Croix, and if he only could write, would teach him a few more."[6] For a considerable period of time the Croats and Pandurs specialized in attacking isolated enemy outposts, and cutting off supplies. But they did not bother unduly to differentiate between friend and foe, and though no one doubted their courage, their reputation was a bad one; they lacked discipline and hardly ever passed over an opportunity to rob and plunder. A fair example was Isolano (Schiller's Isolani) who was involved in the conspiracy that led to Wallenstein's murder.

Among the most influential early authors on small warfare was Grandmaison, a lieutenant-colonel in a Flanders volunteer corps, whose book Frederick the Great recommended to his officers. Grandmaison referred to Hannibal's Numidian cavalry, to the Albanian *stachiots,* to Louis le Grand's light units as forerunners whose exploits were of enduring significance. He dealt in some detail with the qualities necessary for soldiers engaged in partisan warfare: they ought to be robust; not too tall ("better five feet than five and a half"); young but not so young as to be unable to endure fatigue and various privations.[7] Similar observations about the suitability of the "shortest sized men" were made by the Count de Saxe:

> It has frequently been proved that a horse which will carry a man thirty leagues a day, whose weight does not exceed eight or nine stones, which is usually about that of a man of five feet, two inches high, will hardly be able to carry one of from ten to twelve stones, half the same distance.[8]

Grandmaison recommended muskets, pistols and sabres; the sword (*epée*) was only good for the parade ground. Hungarian and Ardenne horses were the most suitable for this kind of operation. He was a great believer in night attacks; ambushes at night were almost always successful, causing confusion out of all proportion to the effort required from the attackers.[9]

According to de Jeney, one of the leading early small-war theorists, the successful partisan needed an almost impossible combination of talents: a fertile imagination, a penetrating and intrepid spirit, a firm countenance, a good memory, alertness, the gift to size up a situation quickly, and overt self-confidence.[10] De Jeney had served as a captain with the French army on the Rhine. His book on partisan tactics, which also appeared in German (Vienna, 1785) and English (it was in Thomas Jefferson's library) was apparently the first of its kind to include maps, sketches, and even advice on first aid. He complained that partisan warfare was the least respected of all the military professions despite the fact that it was the most dangerous and fatiguing.

In the 1760s and 1770s many more books appeared on the subject, mainly in French and German; one was attributed to Frederick the Great, another to Prince de Ligne, a third to General von Kleist. Some of these books were translated into the main European languages, and a few remained in circulation for almost a century. They contained practical advice on the movement of patrols by night, the capture of prisoners, and gave instructions on how to act if a small detachment was cut off from the main body of the army.[11] Some of the authors were not paragons of precision and brevity: Ray de Saint Genie needed six volumes to develop his views on *L'Officier partisan* (Paris, 1769) and not all the advice was either original or very helpful. Baron de Wüst, a colonel of the Hussars who had served in Southeast Asia, suggested that cats living in haystacks in enemy territory should be caught, soaked in alcohol and set on fire — on the assumption that they would run back to their haystacks and so set fire to enemy supplies.[12] But de Wüst also made a great many sensible proposals: the partisan commander should lead an abstemious life with regard to both spirits and women (*"il doit se méfier de sexe en général"*), he should know several languages, should pay for food (but not punish his soldiers when they forgot to do so), he should never eat dinner where he had eaten his lunch and, at night, he should not sleep at either place. Above all, he emphasized the importance of good intelligence; ideally, the partisan commander should visit a theater of war three months before an outbreak of hostilities. The optimal size of a unit was a thousand horsemen and five hundred infantry, and the commander should have at least the rank of colonel. Among other studies on small war mention should be made of Colonel de la

Roche Aymon's *Essay sur la petite guerre* (Paris, 1770), and Scharnhorst's *Militärisches Taschenbuch* (Hanover, 1792). The most interesting works, however, were books by Ewald, Emmerich, and Valentini's *Abhandlungen über den kleinen Krieg*. Andreas Emmerich (1737–1809) and Johann von Ewald (1744–1813) were both born in Hesse, took part in the Seven Years' War and later fought on the English side in the American War of Independence. Ewald joined the Danish army in 1788, eventually becoming a lieutenant-general; in 1809 he fought against Major Schill, the Prussian officer, who started a rebellion against Napoleon. In the same year Colonel Emmerich, aged seventy-two, played a leading part in an anti-French rising in Marburg, and was executed by the French after its suppression. His book, *The Partisan in War or the Use of a Corps of Light Troops to an Army*, was first published in English in London in 1789.[13]

Emmerich insisted that a wartime army could not exist without light troops. He was fairly dogmatic about their number. A unit should consist of not less than a thousand, and not more than seventeen hundred soldiers, all of whom should be volunteers. They should constitute the avant-garde of an army on the march, covering its flanks, and harassing the enemy rear guard. On the other hand, when the main body of their own army was on the retreat, the light troops should cover the rear. Emmerich noted that the commanding officer of such a corps ought to be sober and reliable, a man of initiative and great endurance. Defeat in battle could never be excluded but his unit should never, under any circumstances whatsoever, be taken by surprise. The main danger that faced a detachment of raiders was their own negligence and lack of caution. As an example of unforgivable negligence he cited the American attack in the War of Independence at Trenton, Delaware, in which Colonel Rall, a reliable officer with an unblemished record, was off guard for only a few moments, with fatal consequences for himself and his troops.[14]

Emmerich analyzed in considerable detail various situations the partisan was likely to face, such as night marches and attacks. Infantry, cavalry, or mixed units could be employed. Drawing on his personal experience, he described the two months he spent with a detachment of infantry behind enemy lines in the depths of winter, seizing couriers and enemy officers, destroying supplies, generally causing damage and spreading confusion. He had to cross frozen rivers and use snowbound roads.[15] Emmerich, like other contemporaneous authors, freely offered practical advice: at night, horsemen, whether on patrol or at rest, should play with the reins of their horses lest they neigh and betray their presence. It was essential to prevent the raiders from falling asleep while in the saddle, and to cover wooden bridges with straw so as to minimize noise when crossing them at night. There were strict injunctions that patrols never dismount or remove the saddles from their horses; dogs were not to be kept, for their barking was always a source of grave danger. Guides from among the civilian population should be employed only if they volunteered. Again and again Emmerich emphasized that a partisan officer needed special qualities, particularly the ability to act independently of his commanding general, who could not possibly give him orders covering all eventualities.[16] He stressed the importance of changing camps as frequently as possible; he warned against the delusion that bad weather offered a guarantee against enemy attack, again quoting several incidents from the American War of Independence (General Matthew's surprise attack on Young's House in 1780). Spies should be well paid and taught to be punctual; their identity should be known only to the commanding officer. If a partisan discovered an enemy spy in his own ranks it was always advisable to "turn him around."[17] He recommended employing women, and officers who had been dishonorably discharged and who badly needed additional sources of income. Emmerich repeated his commanding general's useful advice before one of his first major raids, two hundred miles into the enemy rear: never offend or mistreat civilians; do not permit plunder; and treat prisoners of war decently.[18]

Ewald's contribution to the theory of partisan warfare was at least as important as Emmerich's; he was frequently quoted by Clausewitz and subsequent authors, but a detailed review seems unnecessary in this context because his views overlap with Emmerich's in most essential aspects. His books survey and analyze the lessons of recently fought wars, in particular the Seven Years' War and the American War of Independence. In his later works he also dwelt upon the experiences gained in the war against revolutionary France in the 1790s.[19] One of the important points made by Ewald was that officers — especially young officers — all too often lacked even the rudiments of theoretical knowledge. Any lawyer, doctor or forester would read some professional literature, but a young officer all too frequently believed that he could acquire the essentials of his trade at the gambling table, an inn, or perhaps while he was asleep.[20] Ewald described the "ideal officer" *pace* de Jeney: a combination of manly virtues, modesty, courage, humaneness and intellectual curiosity. Like Emmerich, he stressed that there was no excuse whatsoever for being taken by surprise — but as officers are only human, he devoted several chapters to the techniques of surprise attacks and ambuscades. Certain of his suggestions have become part and parcel of guerrilla practice in succeeding ages: for example, that there are hardly any regions from which surprise attacks cannot be successfully launched, and that some have succeeded precisely because they were carried out where least expected — in open terrain, from behind fruit trees, isolated houses, etc. Not only should dogs and horses not be employed on an ambush, soldiers suffering from a cold and likely to cough or sneeze should not be a party either.[21] Given the necessity of relying on the population's goodwill, it was of the utmost importance to punish marauders severely — either putting them to death or, at least, giving them a sound beating. It was important to study the psychology of the enemy leader; if he lacked experience and was impulsive, he would tend to be hyperactive and therefore a likely victim for an ambuscade. Ewald's experience in America taught him that however brave the English soldiers, they were not really suitable for a small war, because they lacked sufficient patience for the difficulties and arduousness this kind of warfare entailed.[22]

George Wilhelm von Valentini (subsequently a Prussian lieutenant-general) was only twenty-four years of age when his book was published, by which time he was already a veteran of the wars of 1792–1796.[23] He based his book partly upon his own experience in the field of partisan warfare and also upon that of others, such as Ewald, although their views were often contradictory.* Valentini believed that a small war could be decisive in the last resort. The French *tirailleurs* had harassed the Austrians like a pack of hounds in the winter campaign of 1793, compelling them to retreat though not a single major encounter had taken place. (Almost a century earlier, Maurice de Saxe had compared an army without light cavalry units to a knight in heavy armor who has been forced to retreat in disgrace by a crowd of schoolboys throwing sticks and stones.) According to Valentini, the early campaigns of the revolutionary

* When Valentini's *Abhandlungen* first appeared in 1799 they were the most intelligent and comprehensive guide available on partisan warfare. Valentini's subsequent military record was not unblemished. He was York's and Bülow's chief of staff but they were not quite satisfied with his performance. He quarreled bitterly with Gneisenau. Valentini had the reputation of being a highly educated, easygoing commander, a good diplomat. In 1828 he was made chief of the Prussian army education service, an appointment closely corresponding to his talents and inclinations.

wars were an excellent school for small warfare. The French had employed small-war tactics on a massive scale, conducting a war of destruction against the enemy forces. On the other hand Valentini argued not very convincingly that the fighting in the Vendée and in the Tyrol was not small warfare.[24] He thought that mountaineers and hunters were the most likely candidates for partisan warfare; those without such natural training would need to be highly educated and young. With regard to the essential qualities needed by a commander of a partisan unit, Valentini agreed with other authors on the topic that much would depend on his ability to make quick decisions; a brave officer was able to challenge fortune against overwhelming odds, indeed against all the laws of probability. Above all, he needed moral force, the charisma to inspire officers and men in a decisive moment. Of great interest are Valentini's remarks on surprise attacks. He noted that the Croats were once past masters of this art but seemed to have lost their talent. He believed that if surprise were complete the enemy would offer little resistance even though his forces might be numerically superior. However, well-disciplined troops were needed for a surprise attack; infantry units should carry out the attack by night, holding a cavalry detachment in reserve to pursue the enemy troops in their flight.

Of the leading strategic thinkers of the period, Napoleon paid little attention to the problem of small warfare. Other French military leaders had thought highly of partisans; Maurice de Saxe had said that a brave *parti* with three hundred or four hundred men could create utter chaos amongst an entire army. Napoleon's comments on partisan warfare were totally negative. The affair in the Vendée had been a simple case of *brigandage*, very much in contrast to the Spartacus revolt which had been led by a great man, aspiring to the human ideal of liberty. The Vendéeans just wanted to rob and destroy, they had established a veritable *république des anarchistes*, while professing to fight for the Church and monarchy. True, the rising had been popular and spontaneous to begin with but the insurgents later became the tools of the British. The rebellion would have been nipped in the bud had the local administration been any good, and Kléber and Marceau been given a free hand from the very beginning.

In Napoleon's eyes, the Cossacks were cowards and wretches, the "scum of the earth"; if he had been defeated in Russia, it was through a series of accidents, by General Morozov (frost) rather than by Marshal Kutuzov. The German free corps (such as Lützow's) were "hideous militias"; what would the great Frederick have said about such soldiers? Napoleon wrote of the Spanish soldiers in 1808 that it was impossible to find worse troops anywhere in the world; they could not fight in the mountains or on the plains. They were ignorant, cowardly and cruel; the monks and the inquisition had destroyed this nation. But in retrospect Napoleon modified his views: he wrote in exile that the Spanish *en masse* had acquitted themselves like honorable men. His own subsequent misfortunes had their origins in Spain without exception; he had had to divide his forces, therewith exacerbating all his problems and detrimentally affecting his image throughout Europe. But his Spanish experience did not affect his views about partisan and people's wars. The Prussian *Landsturm* was barbaric in inspiration; *levée en masse* would always cause terrible trouble.[25]

Guibet, an erratic and unorthodox military writer, who opposed big armies and thought that the role of artillery and fortification had been much inflated, nevertheless did not gravitate towards *petite guerre*.[26] On the contrary, he regarded light mobile units as superfluous and an artificial innovation. Since armies had become so big and unwieldy, something had to be created to give employment to those who were underemployed, but it was a mistake to copy the Austrian example of employing small, semi-independent, light cavalry units. It was a gross exaggeration to claim that these units were one of the most useful corps in an army; small war could never be decisive.

Dietrich von Bülow, like Guibet, was an eccentric thinker, with occasional flashes of genius. He published his comments on the "modern system of warfare" citing Carnot as his main hero and the open formation of *tirailleurs* as the key to success. He was among the first to realize that the old linear drill of Frederick the Great was not only incompatible with elementary human dignity, it was also militarily ineffective. He foresaw that all infantry would become light infantry, and expected skirmishes rather than major battles to become the main feature of modern war.[27] But again, like Guibet, he saw no room for partisan warfare in the modern system. Bülow advocated "human tactics" against the outmoded clockwork practices of the eighteenth century. The old-style soldier had been treated and had behaved like a slave; modern war put a premium on individual courage, intellect and initiative. But at the same time he saw the mobilization of the masses and numerical superiority as the only decisive factor in a future war; hence his disregard of small warfare.

The experience of the American War of Independence, as he saw it, had been unhelpful. He wrote scathingly about the lack of discipline among the North American militias whose soldiers came and went to war as they saw fit.[28] That the Americans prevailed in the end was mainly due to Washington's political (not military) genius and also to the many mistakes committed by the British who singularly lacked initiative.

Mention has already been made of Jomini's observations on national wars. He had been at the receiving end in Spain on one of the first occasions when guerrilla warfare was waged on a large scale in modern times. He wrote:

> The spectacle of a spontaneous uprising of a nation is rarely seen; and, though there be in it something grand and noble which commands our admiration the consequences are so terrible that, for the sake of humanity, we ought to hope never to see it.[29]

The commander of an occupying force in enemy territory could all too easily find himself in the position of Don Quixote attacking windmills, whereas his opponent knew the most minute, obscure paths, and had friends and relations everywhere to help him. He quoted illustrations from the war in Spain — entire companies had disappeared completely without trace. All the gold in Mexico could not procure reliable information for the French, who received only false trails that led them into snares:

> No army, however disciplined, can contend successfully against such a system applied by a great nation, unless it be strong enough to hold all the essential points of the country, cover its communications and at the same time furnish an active force sufficient to defeat the enemy wherever he may present himself.

It should be recalled, however, that according to Jomini, guerrilla units would not succeed in holding out for any length of time without the support of regular troops. His belief was shared by other authors of the period who specialized in the study of the art of small wars such as the Austrian Captain Schels who wrote a multivolume opus on the subject.[30] Schels and San Juan offered much useful advice but their very attempt to provide a system in minute detail limits the value of their books. Schels was aware that small, highly mobile units could be as valuable as a whole army, but like many subsequent authors he always regarded them as a part of the regular army and did not consider their guerrilla potential.[31]

CLAUSEWITZ

Major Clausewitz became a professor at the Prussian war academy in 1810 and, on 15 October of that year, gave the first of 156 lectures, which were spread over nine months, on the subject "small war" (*Kleiner Krieg*). His copious notes — almost four hundred printed pages — were first published in 1966.[32] They provide a detailed analysis of the methods and scope of operations that had been undertaken by units of between twenty and four hundred men. For his illustrations Clausewitz drew heavily on contemporaneous studies of the subject (Ewald, Emmerich, Valentini and, above all, Scharnhorst's *Militärisches Taschenbuch*). Like these writers he was not concerned with the general political context of the "small war," which he regarded as just an extension of a "big war" by other means. Though he was familiar with events in the Vendée and in Spain, he referred only to regular army units in his lectures, never to insurgents. He was, after all, addressing the lieutenants and captains of the Prussian army, not Chouans or *guerrilleros*. Nevertheless, from a military point of view his observations are relevant to the development of partisan warfare.

According to Clausewitz, the small war differed from a big battle in that it involved not only greater courage and temerity but also demanded the utmost caution. Since in a small war the partisans almost always faced superior forces they had to avoid danger whenever possible; otherwise their units would not last long.[33] Ideally, such warfare should be carried out by infantry in close collaboration with mounted units. Whenever possible they should march at night and should camp in small detachments in a forest during the day. They should move forward on concealed roads, should obtain food from the most remote villages, and never reveal the total strength of their detachment. If the presence of the unit were discovered by civilians, they should be detained until such time as the soldiers left the area. However, as the raiders depended in so many ways upon the goodwill of the local population it was imperative to treat them in a friendly way. In some cases money would buy goodwill, but it might become necessary to threaten murder and arson. If the raiders could not pay for the food they needed, they should at least give receipts; messengers should always be given gifts.[34] Following Ewald (and Sun Tzu) Clausewitz emphasized that secrecy was of paramount importance; few people should know about the intention (and direction) of a raid. Clausewitz provided the most detailed instructions regarding patrols and advance guard duties. The duty of a scout was to observe, not to fight. At night he should not draw his cap over his ears so that he could note every noise. Only one scout at a time should water his horse, and half the number of horses should be saddled at any given moment. Basing himself on the Tyrolean experience, he suggested that in mountainous territory the detachment should take up positions on the heights, not in a narrow pass which could be outflanked.[35]

Surprise attacks were best carried out at night, or at midday when the enemy would be cooking in their camps and least prepared to face an onslaught. Causing false alarms in the enemy camp the night before an attack was always advisable, since this would result in less vigilance the following day. Clausewitz was skeptical about the use of artillery in small wars unless the raiders intended to hold the position they had gained. It had become less easy to lay ambushes because advance parties usually moved in front of the main body of a big unit. Nevertheless in difficult terrain such as forests or mountains, there would always be some scope for ambushes — more likely in one's own territory than in the enemy's. What if a small detachment found itself encircled? In most cases small groups of soldiers could fight their way through the enemy lines more easily than bigger units, hence it was advisable to split up the detachment in such an emergency. But a new meeting place should be fixed and every soldier informed accordingly.

In dealing with specific assignments for the partisans (*Partheygänger*), Clausewitz singled out the following: to collect intelligence; to arrest enemy couriers; to kidnap enemy generals or other important persons; to destroy bridges and arms stores; to make roads impassable; to seize enemy funds and supplies.[36] He thought that it would be exceedingly difficult to kidnap enemy generals in an orthodox war between regular armies, as an operation of this kind called for skill and courage. It was, therefore, more likely to occur in civil and people's wars, and if such an operation were indeed successfully carried out would spread despondency in the enemy camp.*

* In Clausewitz's *magnum opus* one brief chapter (the twenty-sixth) deals with problems arising out of a people's war. He stressed the moral-political importance of people's war. A war of this kind, Clausewitz wrote, ought to be protracted to be successful. He stressed the importance of the terrain; the popular units should disperse, not concentrate and their blows should not be directed against the main enemy force. But Clausewitz envisaged mainly a militia-type (*Landsturm*) resistance rather than typical guerrilla warfare. *Vom Kriege*, 16th ed. (Bonn, 1952), 697–704.

Clausewitz strongly advised against a worst-case-expectation strategy. The enemy could not possibly do everything at the same time, he was bound to be more aware of his own weaknesses than his attackers. To take into account all (theoretically) possible dangers was tantamount to magnifying them. Clausewitz doubted whether once the commander had taken a decision, his junior officers should then be consulted since their opinions might only make him waver. When facing capitulation the commander should weigh up the alternatives of a last counterattack or an order to disperse, thus giving the officers and soldiers a chance to escape.[37]

SCHARNHORST AND GNEISENAU

The Spanish experience inspired Scharnhorst's and Gneisenau's drafts and memoranda proposing the establishment of a national militia (*Landsturm*) in Prussia (1808 and 1811). The aim of the Prussian military reformers was to hinder the enemy's advance and bar his retreat, to keep him continually on the move, to capture his ammunition, food, supplies, couriers and recruits, to seize his hospitals, and to attack him by night, in short, harassing, tormenting, tiring and destroying him either individually or in his units wherever possible.[38] To that end the militia was to be trained in guerrilla tactics and to use them so that the main body of the enemy army would be virtually cut off, without information, unable to send out small detachments to obtain fresh supplies or patrol the vicinity. They were to pursue a scorched-earth policy — villages were to be destroyed, food and drink burned or spoiled, horses and cattle removed. Such revolutionary measures, especially in Gneisenau's radical version, were anathema to a bureaucracy which regarded order and obedience as its supreme values; the enemy should be fought, to be sure, but not at the price of utter chaos. When, after Napoleon's retreat from Russia, Prussia joined the war against France, the *Landsturm* was ordered to assist the regular army in its operations and was given no opportunity to make use of the tactics proposed by the reformers.

SMALL WAR DOCTRINE AFTER NAPOLEON

The study of small warfare led a modest existence throughout the second half of the nineteenth century and up to the First World War.

Though not entirely forgotten it was on the whole neglected, surprisingly so in view of the fact that so many small wars were being fought at that time all over the world. The Napoleonic wars themselves had generated sufficient interest for several important studies, the more noteworthy of which were the works of Le Mière de Corvey, Decker, Stolzman and Chrzanowski (French, Prussian and two Polish officers, respectively).[39]

Jean Frederic Auguste Le Mière de Corvey, who was born in Rennes in 1770 and died of cholera in Paris in 1832, was better known in his lifetime as a composer than as a strategist. He wrote several operas (including *La Blonde et la Brune*, [1795]) and symphonies (including *La Bataille de Jena*), he arranged Rossini's *The Lady of the Lake*, based on Sir Walter Scott's poem, and *Tancred;* on one occasion he even put a newspaper article on Custine's defense of Mainz into music. But he also had a military career; significantly he had started as a sub-lieutenant in the Vendée, continued as aide-de-camp to General Thibaud in Belgium, Germany and Spain, and ended his career as lieutenant-colonel at the Battle of Waterloo. His work on partisan warfare, published in 1823, which heavily relied on his experience in the Vendée and Spain, was in some respects the first truly modern work on the guerrilla.[40]

Le Mière was struck by the tactics of the Vendeans which, in contrast to most of his fellow officers, he did not regard as primitive and atavistic but as essentially novel. He tended, if anything, to exaggerate the amount of damage inflicted on the French, estimating that they lost half a million people at the hands of the *guerrilleros* in Spain. He noted the similarity between warfare in the Vendée and Spain and argued that far from belittling these lessons (as almost all other military writers were doing), they should be carefully studied and drawn upon by others in the event of a foreign invasion.[41] Civilians normally would not take up arms against regular troops: it was difficult to imagine, for instance, the merchants of Paris constituting themselves into a fighting force. But this situation might suddenly alter if the house of a civilian was destroyed and his wife or children killed. Thus, there was something to be learned by all European nations from the guerrilla experience. True, some areas were better suited than others, especially mountainous regions such as the Pyrenees, the Alps or the Vosges in France, Scotland, the Tyrol or Greece. In his work, Le Mière traced partisan warfare throughout the ages and noted that while partisans were frequently used as a corollary to regular armies, they assumed far greater importance once the national armies had been destroyed. He concluded from a detailed analysis of the Spanish experience that it was not sufficient for an invader to seize the major towns since his lines of communication would still remain open to attack. Traditional military doctrine was of little use in combating the partisans. Who were the leaders of the Spanish guerrillas, who defeated the brave French generals in Spain? A miller, a doctor, a shepherd, a curate and some deserters . . .[42] He noted, furthermore, that for obvious reasons the local populace would always be the most adept at defending their native regions. He dealt in considerable detail with the organization of guerrilla units, their tactics, their weapons, and even their uniforms. He was less doctrinaire than other authors, recognizing that guerrilla units followed tactics essentially different from those of light units attached to regular armies. And in contrast to other authors, Le Mière put great, perhaps decisive, emphasis on psychological factors. That a guerrilla had to be courageous went without saying — once he was attacked he could not look back. Above all, guerrilla warfare *faut un peu de fanatisme*, for this was a war of extermination; the enemy armies would use reprisals and treat the partisans as mere brigands.[43] Though the author very much regretted this —for *guerres d'opinion* (ideological wars) had terrible consequences — he accepted this change in the character of war as an unalterable historical fact.

Major (subsequently General) von Decker, the least politically minded of the four, presented a useful and systematic summary of the topic which, of necessity, repeated much of the advice proffered by earlier authors; for instance, he advised guerrillas to change their quarters rapidly, after both victory and defeat, never to take their safety for granted, to march at night and to camp in the most remote of villages during a raid. He pointed out the great importance of maintaining good relations with the local population; the partisan should be welcome everywhere, he should be considered a liberator, not a pirate or filibuster. To be thus considered entailed strict discipline, and the giving of payment for supplies received. Decker's ideal "party" was smaller than those advocated by the eighteenth-century theorists; at most it should number a hundred to a hundred and fifty men. Except for high-ranking enemy officers, no prisoners should be taken whose presence would only slow down the movement of the raiders. He counseled extreme prudence when enlisting new soldiers, and stressed the importance of having spies in all classes. The partisan's appearance should inspire confidence, and it was particularly important for him to be on good terms with priests and women ("no one will be able to obtain secrets which neither priests nor women can penetrate").

On some occasions he also emphasized the importance of psychological warfare: for instance, during the Napoleonic wars when the French general Vandamme offered a prize of a thousand thaler for the head of the rebel leader Bork, the latter countered by offering two francs for the head of Vandamme. Partisan warfare, as Decker saw it, was more difficult than *la grande guerre:* even a mediocre talent could make a useful contribution in regular warfare, whereas partisan warfare called for very special qualities.[44]

In a subsequent study, Decker commented on the war in Algeria; this was one of the very first attempts to analyze the new military problems beyond the confines of Europe which faced the colonial powers.[45] Decker did not rate French chances of success very high, thinking that the French would secure a few towns near the coast at best. A European army with its baggage trains and other encumbrances was unsuited to fight in such unfamiliar conditions. It could not come to grips with the enemy, there were no centers of power to be attacked and the absence of good roads impeded movement. It was quite pointless in these circumstances to aim at inconclusive victories; the side which would last longest would ultimately emerge victorious irrespective of how often it had been defeated.[46] Decker correctly analyzed the problems which would one day face the colonial powers but his misgivings were premature as regards the immediate future. The French, under Marshal Bugeaud, developed highly effective "counter-guerrilla" tactics. Abd el-Kader surrendered only three years after Decker's book appeared, and for a century French North Africa remained relatively quiet.

Karol (Charles) Bogumir Stolzman (1793–1854) was an artillery captain who had taken part in the Polish rising of 1830–1831; later he lived as an emigrant in France and Switzerland, and he represented "Young Poland" in Mazzini's *Jeune Europe.** His remarkable treatise was the forerunner of a whole twentieth-century "do-it-yourself" literature; it was reprinted in Warsaw in 1959. It gave practical advice on how to produce explosives in a kitchen or garden shed, it provided exact figures on how much powder was needed to produce land mines, and the required size of a mine for

* Stolzman had fought at Leipzig and other battles of the Napoleonic wars. In the emigration he was a leading member of the West European radical-democratic underground. The last time he saw action was during the revolution of 1848 when at the head of a battalion of Polish volunteers he unsuccessfully tried to reach Frankfurt. Eventually Stolzman made his home in England and married an Englishwoman; he is buried in the village of Millon in Cumberland.

blowing up a wall or a bridge.[47] Like other authors he referred to the historical predecessors of modern partisan warfare (Skanderbeg, Spain, the Caucasus), stressing the popular character and national inspiration in a modern small war. Surely twenty-two million Poles were as brave as twelve million Spaniards had been. They too would make life intolerable for the occupation forces of their enslaved country. Among the problems which preoccupied him was the question of maintenance of discipline in an irregular unit, and the advisability of awarding decorations for actions of special valor. He suggested that after termination of hostilities a roll of honor should be published, listing those who had distinguished themselves.[48] But much of his book was devoted to advice on eminently practical issues: how to cross a river; how to defend a house (or church); how to prepare a code for secret correspondence; how to use scythes in battle if more effective weapons were not available.

Like Stolzman General Wojciech Chrzanowski (1793–1861) had participated in the Polish rising of 1830.* His observations on partisan warfare are of great interest because they contain, in a nutshell, most of the basic ideas of twentieth-century guerrilla warfare. Sometimes there is an almost textual overlapping with Mao's doctrine: namely the importance of guerrilla bases, the idea of protracted warfare and even the gradual transition from guerrilla to mobile warfare.[49] Chrzanowski noted that guerrilla warfare could be successful only if the enemy army was not large enough to occupy the whole territory, but this, he added, was seldom likely to happen. To be effective partisan warfare had to be protracted; the longer it continued, the better the chances for victory, for while the guerrillas grew stronger, the enemy units became weaker and more demoralized. Guerrilla war, as envisaged by Chrzanowski, would at first be conducted against individual enemy soldiers, then against small units, and eventually against larger bodies. He stressed, however, the importance of attacking the enemy only from a position of marked superiority. Attacks should be launched if possible from the flanks.

* As a young lieutenant he participated in Napoleon's invasion of Russia and was the first soldier of the *Grande Armée* to enter Smolensk. He participated in the Russian-Turkish war in 1828 and was chief of staff of the rebel Polish army in 1831. In 1833 he became an adviser to the British government on Ottoman affairs, visited Turkey several times and organized an Arab cavalry regiment in Baghdad. In 1849 he was commander in chief of the Sardinian army. Chrzanowski died in Paris in 1861. There is an Italian biography: G. Roberti, *Il generale Chrzanowski* (Rome, 1901).

Swiss writers on guerrilla warfare developed the idea of the popular-patriotic partisan war in the same decade. Thus J. M. Rudolph: "Without the support of local inhabitants even the most gifted partisans will be unable to succeed."[50] Gingens-La Sarraz emphasized in his study the great importance of the moral factor — the success of an invader, however well he was organized, would be ephemeral against an insurrectional war conducted with energy and intelligence. Counterinsurgency would be of no avail against a total war which denied the enemy supplies and in which no quarter was given.[51] A widely quoted Swiss manual on guerrilla warfare by Major von Dach, published almost one hundred years later, rests on the very same basic concept; only the technical details have changed.[52] Partisan warfare was studied by military men in most European countries in the nineteenth century, including Spain and even Serbia.[53] The nineteenth-century authors quite often plagiarized each other, but what matters in this context is simply the fact that the subject was never entirely neglected.

Stolzman's work was one of the last books on partisan warfare of the pre-railway and pre-telegraph age; two decades later another important book on the topic was published, the contents of which reflected the far-reaching technical changes which had taken place in the intervening years. Its author, Wilhelm Rüstow, a former German officer, was a radical democrat who had taken part in the revolution of 1848. Having settled in Switzerland he was a respected figure in emigré circles, and was to see military service again in Garibaldi's little army. Rüstow (1821–1878) was a well-known and prolific writer on military affairs and his book covered a range of issues from high strategy to minute details on equipment.* He was opposed to high boots, and came out strongly in favor of short loose shirts or tunics ("Garibaldi shirts" or *Schifferhemden*) with dark, not shiny, buttons. He also dealt with the shape of saddles and the size of rifles and carbines, preferring the small-caliber guns. He made the sensible point that the partisan should not have to carry more than twenty-three to twenty-four pounds on his march.[54] Rüstow favored the employment of very small tactical units for a variety of reasons — one of which was to mislead the enemy about their real strength. He thought that technical develop-

* Rüstow was a gifted, prolific and controversial author on military topics. His best-known work was a history of infantry; it induced the young Hans Delbrück to become a military historian. In 1877 Rüstow became the first holder of a new chair for military science at the Zürich Federal Polytechnic. But his contract was not renewed and he committed suicide the year after.

ment had made night attacks very risky, but on the whole he believed in ambushes and, generally speaking, in bold, dare-devil action, even in the age of the railway and telegraph line, which, needless to say, figured highly on his list of targets.[55]

The Prussian wars of 1864 and 1866, and particularly the war with France in 1870, gave fresh impetus to the study of irregular warfare. A. von Boguslawski, a German lieutenant-colonel (and subsequently a general) was perhaps the first to discuss the legal questions involved. He thought it highly unlikely that there would ever be a generally accepted international convention with regard to partisan warfare. It had been suggested at a conference in Brussels that the civil population would no longer be entitled to continue a war once the country was occupied by enemy forces.[56] Bluntschli, the famous Swiss expert on international law, had stated in his *Modern Law of War* that a popular rising in the rear of the enemy was illegal, and those who took part in it should be treated as rebels according to martial law. Bluntschli added, however, that if the insurrection was on a large scale and if it was at least partially successful this was bound to change the situation and also the status of those who took part in the insurrection. What, Boguslawski asked, constituted the effective occupation of a village? Was it the presence of three soldiers or five hundred? He did not regard an attack by civilians against regular army units as unethical even if the former were not in uniform. But what if a peasant killed a passing officer, threw his rifle into the nearest hedge, and continued to plough his field? A rising in the enemy's rear could be perfectly justified, but could it still be regarded as legitimate warfare if a soldier was killed in his sleep by the owner in whose house he was billeted? Boguslawski argued that there could never be a clear dividing line between the defensible and the indefensible and that, in the last resort, each civilized nation would react to partisan warfare as it saw fit; it certainly would not be bound by international conventions. The author also maintained, somewhat surprisingly for a Prussian officer, that it was very mistaken to believe that a soldier needed no knowledge of politics whatsoever.[57]

Again, in advance of his time, Boguslawski considered the possibilities of political warfare in combination with military operations. He foresaw that an invading army might want to incite a popular rising; the Piedmontese had done so in Lombardy in 1848, and again in 1859. The question of whether soldiers could be prepared in peace for small warfare intrigued him very much. He thought that such training did not basically differ from all military training apart, perhaps, from teaching the well-known fact that small war depended to a much greater degree on the initiative of the individual officer and soldier. He suggested that it would be helpful if officers, even when on holiday, scanned their surroundings with a professional eye to explore the possibilities for a small war. The idea was not entirely original; Clausewitz's letters from the Silesian spas in summer 1811 to his friend Gneisenau showed that he was heavily preoccupied with such problems and it is unlikely that he was the first strategist whose mind continued to function on professional lines even while on leave.

Boguslawski thought a total war, involving the active participation of women and children such as had occurred in Spain and the Tyrol, unlikely to recur because the mass of the population was not composed of heroes. Nevertheless, as wars had become national ones and were no longer contests between professional soldiers, such an eventuality had to be taken into account, especially when geographical conditions favored it. He analyzed in some detail the attacks of the *franc tireurs* in the autumn and winter of 1870, a topic which also preoccupied a French officer, Captain Devaureix.

Devaureix's detailed study of partisan warfare was published in 1880; it opened with the melancholy observation that this kind of war, which was so "eminently French" had fallen into disfavor after our "recent disasters."[58] But had not the lessons of 1870–1871 amply demonstrated the uses of partisan warfare? The Germans themselves admitted that they had suffered more losses in the second phase of the war than in the first, despite the fact that the French army had been routed. Devaureix claimed that the Germans were not prepared for the privations and fatigues of a lengthy war and he quoted with approval an observation of Marshal Bugeaud that the way to defeat the Germans was to cut their lines of supply, keep them on the march and deprive them of their sleep. Devaureix admitted that the behavior of some of the *franc tireurs* had been scandalous, and he made it clear that he opposed freewheeling, independent guerrilla activities. Partisans would have to operate under the command of the regular army. Like other authors he saw the main task of partisan warfare as disrupting the enemy's lines of communications and spreading confusion; again like others, he referred to the experiences of Frederick the Great in Bohemia in 1740–1741, and of Napoleon in Russia and Germany after 1812, when these military leaders lost contact with sections of their armies because their couriers were unable to get through to them. The basic effect of partisan warfare was on morale, but had not the great Napoleon said that in war three-quarters of the outcome depended upon morale? He also quoted General Duhesme, author of the leading infantry manual of the Napoleonic age, that partisan warfare had an "immense psychological effect." Much of Devaureix's study is devoted to an excessively detailed historical outline, establishing that real partisan warfare developed only after the end of the seventeenth century, when Turenne's and Montecucoli's last campaigns put warfare on a truly modern and organized basis. Before that time armies had lived off the land, virtually independent of lines of communication and supply. With changes in the art of warfare, armies had become far more vulnerable precisely because their dependence on supplies had become so much greater. While regular warfare became subject to rules which were rigidly observed, partisan warfare had no rules and was by its very essence adventurous and independent. Maurice de Saxe was the first French military leader to realize the importance of a free corps and to establish such a unit (the *Légion de Grassin*). This, however, was not really a partisan detachment in the modern sense but simply a light cavalry unit operating within the framework of the regular army.

After the end of the Seven Years' War, under the impact of the victories of Frederick II, independent light units were abolished almost everywhere only to re-emerge a hundred years later. Devaureix referred to the popular resistance encountered by the French in the Tyrol (1796) as well as in Verona, Venice and the Naples region and, of course, in Spain. He attributed — wrongly no doubt — the disunity of Napoleon's marshals as the main reason for France's failure. He cited Napoleon to the effect that the guerrillas came into being one year after his departure from Spain because of the pillage and other abuses committed against his strictest orders by several of his generals, above all Soult, who should have been shot.[59]

The effects of partisan warfare in Russia were again felt mainly on morale: the French had nothing to eat and could not get any sleep. The actual military value of the Russian partisans was very nearly nil. Devaureix quoted the 29th Bulletin of the *Grande Armée*, which compared the Cossacks to Arabs in the desert, always evading any serious battle: "this contemptible cavalry which makes nothing but noise and is incapable of even breaking through a company of light infantry. . . ." Contemptible or not, the Cossacks were quite effective and this was all that finally mattered. Referring to the lessons of the American Civil War — "as yet insufficiently studied" — and, of course, the Franco-Prussian War, Devaureix reached the conclusion that partisan warfare still had a future — and an important one at that.

This view was shared by another French student of partisan warfare, Captain Charenton, who, writing in the year 1900, noted that in the next war the *franc tireurs* would have to fight against two formidable opponents — the telegraph and the railways.[60] The former made it difficult for them to hide their presence for any length of time, the latter helped the enemy to concentrate his forces against the partisans. Hence the importance of directing the first blow against these new inventions. Charenton like Devaureix was concerned with the operations of semiregular units which were authorized by the military command and operated under its direction. He quoted as an illustration of a successful *franc tireur* raid the destruction on 22 January 1871 of the viaduct at Fontenoy over which all trains from Paris to Orléans had to pass. The French had used various ruses: scouts disguised as peasants were sent out to reconnoiter; the raiders wore German *Landwehr* caps when they attacked the German unit guarding the viaduct; they made a bayonet charge, not firing a single shot. Like other authors Charenton stressed the importance of keeping iron discipline, for otherwise the semiregular units would degenerate into gangs of robbers "as in the American Civil War." If the free corps were to be constituted of elite troops, however, and they realized that they would always have to be on the offensive, such a corps could well play an important role in any future war. But these unorthodox views were not favored by the general staff of the main European armies, and it was not during the First World War but only in its aftermath that the free corps came into their own again in Central and Eastern Europe.

BRITAIN AND THE COLONIAL EXPERIENCE

Guerrilla warfare as a legitimate subject of study received a new, albeit short lease of life in Britain in the wake of the Boer War. Before that war, as an observer noted, it had not been part of the curriculum of military officers, although the British army had been engaged in the reign "of our late Gracious Majesty Queen Victoria" in no fewer than eighty-two campaigns, most of them small, irregular wars in the bush and desert between "armies" of a few thousand men; wars in which artillery had played no leading role. There was, in January 1900, according to this witness, not a single work on guerrilla warfare available in any London bookshop: "some of our statesmen were amazed that any nation should be so foolish and absurd as to continue any warfare after the regular armies of the country were defeated or after the capital of the country were taken. . . ."[61] The author, a barrister of Inner Temple, was certainly correct in noting the general trend in military thought of divorcing doctrine and practice.[62] Strictly speaking his above-quoted strictures were exaggerated; there were available Captain Johnson's survey of night attacks (from Gideon's battles to Napoleon's attack on Mantua), and Colonel Malleson's book on *Ambushes and Surprises* ("being a description of some of the most famous instances of the leading into ambush and the surprises of armies, from the time of Hannibal to the period of the Indian Mutiny").[63] In addition there were the writings of officers of the Indian army who had frequently encountered guerrilla warfare and who early on had realized that the objective of a war in Asia was basically different from that which was being taught in the European military academies. Usually there were no enemy capitals to be captured, protection of one's lines of communication was exceedingly difficult, if not impossible, and it was pointless to fire volleys at an enemy who usually did not launch concentrated attacks.[64]

Admittedly these studies dealt with the operations of regular armies. But there was also Callwell's *Small Wars*, which presented what was certainly the fullest account of all unorthodox campaigns in nineteenth-century experience. Captain Callwell saw these campaigns as an inevitable consequence of keeping order throughout the confines of the British Empire. Far from romanticizing them, with the Light Brigade charging howling dervishes in the desert sunset, or the Bengali lancers fighting a treacherous enemy at the Khyber Pass, he noted time and time again the general rule that the "quelling of a rebellion in distant colonies means a protracted, thankless, invertebrate war." He warned that "guerrilla warfare, regular armies always have to dread, and when this is directed by a leader with a genius for war, an effective campaign becomes well nigh impossible."[65]

Callwell's study and Maguire's later work are eminently pragmatic books devoid of any ambition to develop a general theory of guerrilla warfare. But from time to time the authors pause for reflection, providing not just practical advice for counterinsurgency but speculating on future developments. That guerrilla warfare ought in fact to be met with an "abnormal system of strategy and tactics" goes without saying — the rules of the game as played in Europe would not work: "the inner line principle is not so effective against invaders as it has been in France, Bohemia and the United States, as the savage has no idea of strategy. . . ."[66] The savage, in other words, did not know when he was beaten. Callwell and Maguire were fully aware of the fact that the more irregular and dispersed an enemy force, the more difficult it was to pursue once it had been defeated. The American rangers often could not find the camps of the Indians, just as the French in Algeria had been unable to locate the Kabyles and the British had lost tract of the Zulus for a time. On the other hand, the enemy always seemed to know the movements of the regular army. Hence the conclusion that it was always better to fight the irregulars than to maneuver against them — provided, of course, contact could be established at all. Callwell noted that generalizations about effective counteraction to guerrilla warfare were always dangerous: in the Maori wars, the British faced an enemy who was poorly armed and, on the whole, not very spirited (not all observers agreed with this view), whereas the Austrians in Bosnia and the Turks in Montenegro fought opponents who were well armed and eager. But, Maguire asked, was this not the shape of things to come? The native ("the natural man") had better eyesight and woodcraft, could manage with less food than the city dweller, was hardly affected by heat or cold, and was seldom ill — in short, he was tougher than his civilized brother, an ideal recruit, superior to him in everything except discipline and armament.[67] Acquisition of the latest weapons was merely a question of money. In future the European powers would face in Africa and Asia opponents "individually superior to the vast majority of our men in all the qualities that go to make a good soldier" and who no longer wielded swords and spears, but rifles:

> If fuzzy-wuzzy be, as he often is, as good a man as Tommy Atkins, or Fritz, or Jacques, and is even approximately as well armed, numerical superiority, knowledge of the country, and better health will go a long way to redress the balance in our favour, which experience and discipline in these days of loosened fighting may produce. Both sides — nature and civilization — being once more on an equality, the scale must be turned by better generalship in the future, as it has been in the past.[68]

In the writings of these British authors no clear distinction was drawn between guerrilla and small warfare; perhaps experience had taught them that there was no clear dividing line. Callwell and Maguire agreed that a guerrilla war was something to be avoided; it was desultory, claimed more victims from disease and exhaustion than from gunshot, it was demoralizing because of the futile marches involved. The way to deal with guerrilla warfare was to adapt one's own methods to that of the enemy, to use flying columns, like Hoche in the Vendée and Bugeaud in Algeria had used.[69] It was of paramount importance always to maintain the initiative. Callwell argued that the strength of the flying columns ought to depend on circumstances — in Burma three hundred men with one or two guns proved sufficient. In the fight against Abd el-Kader, Marshal Bugeaud had employed as many as three to four battalions with cavalry. Infantry alone could be used in the bush; on the prairie and steppes, however, only mounted men would stand a chance, their mobility compensating for lack of cover. Again it depended entirely on the circumstances how severely mutineers should be handled. Hoche succeeded where his predecessors had failed precisely because he did not advocate a policy of devastation. In Burma the rural population supported the British against the *dacoits* and villages which were merely victims of dacoitry had to be recognized and spared. Elsewhere the maxim *les représailles sont toujours inutiles* would not apply, because "fanatics and savages would misinterpret leniency for weakness."[70]

Callwell and Maguire thought that it was dangerous to surround the enemy completely; it almost always involved heavy losses for the attacker, since a "savage" would fight to the end. It was as effective, and less costly, to leave the enemy a line of retreat and then engage in vigorous pursuit. Both authors had misgivings about night attacks, quoting the Duke of Wellington to the effect that night attacks against good troops were seldom successful, and citing Napoleon, who regarded success or failure in a night attack as dependent on entirely unpredictable circumstances such as the barking of a dog.

Callwell defined the essential element of success in guerrilla warfare as surprise, followed by immediate retreat, before the opponent could recover. Operations were necessarily on a small scale ("petty annoyance — not operations of a dramatic kind") because surprise would be difficult to achieve with large bodies of men.[71] Maguire, and in far greater detail Callwell, dealt with many aspects of the technique of guerrilla and partisan warfare, such as attack tactics, weapons, the blockhouse system for counterguerrilla operations, blowing up of railways, bridges and viaducts, mountain and jungle warfare. It is difficult to think of any major omission with one important exception: political aspects were hardly ever mentioned. Callwell apparently believed that guerrilla warfare was a transient phenomenon that was encountered by imperial powers in distant countries. Maguire, who wrote his book a few years later (and who had the experience of the Boer War to guide him), did not exclude the possibility that with the change in the character of war since the eighteenth century, irregular or guerrilla warfare might increasingly be applied to Europe and America.

Lieutenant Frankland of the Royal Dublin Fusiliers, one of the very few other British commentators on guerrilla warfare, dissented: guerrillas were doomed in a civilized country, the loss of the capital and other main towns would paralyze all further action. But in other parts of the world given effective leadership the guerrillas were virtually insurmountable: "only by actually capturing or killing each individual can the prospective conqueror, so long as the patriotism of the inhabitants remains firm, hope to terminate the struggle." Frankland was fully aware of the cardinal principle of protracted war:

> Guerrilla warfare has as its object the exhaustion of the invader, for the primary aim of driving him away can only be brought about in this way; unable to bar his progress to any part of their country, or to prevent his occupation of what territory he chooses, the guerrilla can at least dog his steps, delay his progress and sap his strength until exhaustion or intervention causes the invader to withdraw.[72]

The guerrillas could recede like the tide, they had no organization, untrammeled by detailed orders they could move hither and thither until their presence was reported in several places at one and the same time — to the despair of even the most competent intelligence officer. But despite all these advantages, Frankland thought that the guerrillas were bound to lose sooner or later, provided the conqueror applied the correct methods and, unlike Napoleon in Spain, had the patience and the resources to carry them out.

PARTISAN WARFARE AND EAST EUROPEAN MILITARY THOUGHT

The Russian tradition in partisan warfare dates back to the eighteenth century: a biographer of Barclay de Tolly noted that his hero was "initiated into the practice of partisan warfare by that well known Caucasian, Count Tsitsianov."[73] But the real hero was the poet-warrior Denis Davydov whose notable contribution to the theory of partisan warfare is discussed in some detail elsewhere in the present study. Russian military doctrine did not entirely neglect partisan warfare, though much of its effort was directed towards a precise theoretical definition of the subject — an enterprise of doubtful promise. According to the Russian Military Encyclopedia, there was a substantial difference between "small war" and "partisan warfare" — the latter being conducted by a detachment cut off from the main army. Partisan warfare, according to this definition, only took place when the rear of the enemy was vulnerable, and the more vulnerable it was, the more promising the outlook.[74] But there was also a difference between partisan and popular (i.e., guerrilla) warfare; the latter was carried on at their own risk by groups of men tied to their native soil.

The same trend towards systematization can be found in much of the Russian literature on the subject; furthermore the stress was always on big units operating in close cooperation with the regular army. The very title of an article by Count Golitsyn first published in 1857 — the most noteworthy contribution since Davydov — reflects this tendency perfectly: "on partisan operations on a large scale brought into a regular system." General Golitsyn (1809–1892), incidentally, was the only infantry officer among Russian writers on the subject; he is mainly remembered as the author of a fifteen volume military history and the editor of a well-known journal, *Russki Invalid*.

Russian advocates of partisan warfare faced a real dilemma, in that unorthodox practices had to be accommodated within the policies of a Tsarist autocracy. Partisan warfare put a premium on personal initiative and independent action unlikely to be adopted by a political system which regarded such qualities with disfavor and suspicion. While the Russian army had considerable experience in combating partisans and guerrillas of sorts (sometimes by adopting their tactics) in Poland, the Caucasus and Central Asia, Russian military authors ignored the lessons of these campaigns on the whole, referring almost exclusively to examples from wars elsewhere in Europe, or America, or of course to the campaign of 1812. Perhaps they thought in retrospect that their colonial campaigns had little to teach them and that, anyway, such wars were a thing of the past. This applies equally to the works of Novitski and Vuich who wrote about small warfare in general, as to the more specific studies of partisan warfare by Gershelman and Klembovski. Colonel Vuich, in a textbook written for the students at the Imperial War Academy, dismissed partisan warfare in one short chapter and popular risings in one paragraph.[75] In his definition small wars were all operations carried out by small detachments; they were obviously actions of secondary importance which, unaided, could not possibly achieve the main aim, namely the defeat of the enemy in open battle. But they could contribute to the attainment of this goal, and since in every war there would be some elements of small warfare, it was a legitimate subject of study.

Some three decades later Fyodor Gershelman, a colonel on the general staff and commander of the Orenburg Cossack officers' academy, criticized Vuich for not having made it sufficiently clear that there was a basic difference between a partisan unit and a light detachment. The assignment of partisans was not to act as scouts and patrols, nor was it correct to argue, as some French authors (such as Thibault) had done, that a unit should consist as a norm of two hundred to three hundred riders; in fact it could consist of several thousand men and deploy field artillery.[76] A partisan unit, according to Gershelman, was one that had no lines of supply and communications, its task (and here he followed Decker) was to harass the enemy, without risking too much, particularly in places where large units could not operate freely. Success depended largely on surprise: this meant that their movements had to be unobserved and quick and, to this end, the partisan units ought to be constituted mainly of cavalry detachments. While a small war has a tactical connection with big operations, partisan actions have purely strategic significance. What the author somewhat clumsily and schematically wanted to stress was that since the partisans operated completely independently, their contribution to the warfare was, generally speaking, to weaken the enemy without making a specific contribution to any major battle. While a people's war (guerrilla warfare) in the rear of an enemy uses the same means as partisan warfare, the two are quite dissimilar in their scope and character.[77] Gershelman, like almost all Russian authors, did not deal with a war of this kind, only with partisan units comprised of regular army officers and soldiers. A small partisan unit consisted of a thousand horsemen, big ones of twelve thousand or more. Refuting the arguments of the opponents of partisan warfare, Gershelman claimed that despite the different topographical character of Central and Western Europe and the relative density of population, partisan warfare could be conducted there too; it could even be conducted in enemy territory, against a hostile population.[78] He stressed that since partisans could be made combat-ready immediately they could play an important role at the very beginning of a war; regular armies were still taking some six to twelve days to mobilize. German military observers were aware of this danger and one of them suggested planting big blackthorn hedges on the border of East Prussia, putting up barbed wire entanglements and arming the local population against an eventuality of this kind. (It was also proposed that partisan Cossacks should be denied the status of prisoner of war.)[79] Gershelman, who also discussed antipartisan measures, much regretted that the theory and practice of partisan warfare were not taught in Russia; similar laments by British, French and German authors have already been noted.

Victor Napoleonovich (*sic*) Klembovski's work on partisan operations was published in 1894; he subsequently became a general and was wounded in the war against Japan.[80] Like Gershelman, he was mainly interested in the activities of big, flying columns and most of his illustrations were drawn from the American Civil War and the operations of the French *franc tireurs* in 1870–1871. One of his main heroes was the Russian general Geismar, whose exploits in France in 1814 tended to fortify the thesis that partisan warfare was indeed possible in enemy country. He discounted the argument that partisans could succeed only if they faced young, inexperienced soldiers. When they attacked an army's rear, the men who covered these long lines of communication were as likely to be as experienced as anyone in the front line. He believed, like Gershelman, that partisan warfare was perfectly possible, and indeed likely, in a coming European war.

Russian comments on partisan warfare were closely followed in Vienna. The Russian cavalry is trained to conduct partisan activities *par excellence*, an Austrian military observer noted in 1885; was it not a matter of elementary caution to watch these preparations?[81] The Austrians had pioneered old-fashioned partisan warfare in the seventeenth and eighteenth centuries; J. B. Schels, of whom mention has already been made, was one of their chief theorists. Another notable contribution was made by Wlodimir Stanislaus Ritter von Wilczynski, a Pole serving in the Austrian army, who based himself to a considerable extent on the experience gained in the Polish insurrections. The partisan units, as envisaged by him, would consist of several units of "scythe men" (*kossiniere*), and some light cannon. The various partisan units in a given province would be under the overall authority of a district commander. Each unit should not be too large but constitute a "family," obeying its head "like a father."[82] The unit commander could appoint (or depose) his officers, and was entitled to a pension and all the other privileges of a regular army officer. Unlike the Russian theorists, Wilczynski put as much emphasis on infantry as on cavalry units within the general framework of partisan warfare, and he even made provision for the presence of a surgeon and a padre.

As the nineteenth century drew to its close, Austrian strategists, like those of other European countries, reached the conclusion that the small war had lost much of its importance — new inventions such as the railways, the telegraph ("and in future also the balloons") would no doubt shorten a future war; a mass army of half a million or more soldiers concentrated in a small space could sleep peacefully, pistol shots no longer would disquieten them.[83] Some of the Austrian writers nevertheless thought that partisan warfare still had a limited future in view of the mountainous terrain of Austria's border regions, in the Tyrol, the Carpathian Mountains, and above all in Bosnia and Herzegovina, where the Austrian forces had encountered guerrilla warfare on a small scale in 1878/79. Hence the conclusion that it was premature to regard the small war as a mere game.

Partisan units could be of particular use when the main body of the army suffered a setback and needed time to recover. Hron's emphasis on ambushes and surprise attacks offered little that was new, except perhaps in his comments on the lessons of the war in Bosnia. In this mountainous territory, which sixty years later became once again the scene of a major guerrilla war, horses were of little or no use. The partisans had to follow the smallest and most tortuous mountain paths and employ artillery only in exceptional circumstances. Hron thought that the ideal size of a partisan detachment ought to be between eight hundred and a thousand men — if it were larger it would lose mobility, if smaller, the unit would be aware of its insufficient strength which could adversely affect its fighting spirit.[84] The lot of the partisan officer was an enviable one, provided he had "a streak of genius." That his men would have to be tough and fearless went without saying; it was unrealistic to expect that such men would have the character of a saint. The "Southern Slav character," as Hron saw it, had always proved itself in partisan warfare, provided that the command was in the right hands.[85]

The Austrian army, as Hron and others had predicted, did have to fight enemy guerrilla units during the First World War, especially in Serbia. Their activities became fairly intensive in 1917. The chief organizer of the bands was Kosta Vojnovic, a Serbian army captain, later reinforced by Captain Pecanac who had been parachuted by air from Allied Headquarters in Saloniki. The Austrians coped with the problem by establishing small flying columns of about forty men and by organizing Turkish and Albanian counterguerrilla units. The detachments used by both sides were smaller than had been anticipated by the theorists, and horses — against expectations — were widely used. Allied Headquarters prepared a general rising behind the Austrian and Bulgarian lines in March 1917 which was to coincide with an Allied offensive. But the enterprise failed, partly because the secret was not well kept and partly because the insurgents were not sufficiently well armed.[86]

SOCIALISM AND INSURRECTION

The idea of insurrection played a central role in European nineteenth-century revolutionary doctrine while the technique of insurrection was very much neglected. Insurgency was in the air from Babeuf's conspiracy to the Carbonari and the revolutionaries of the 1830s and 1840s. Philipe Buonarroti provides the link between the old generation of extreme Jacobins and the young French, Belgian and Italian revolutionaries of the 1820s and 1830s. But Buonarroti's legacy did not include any clear, systematic doctrine of how to make a revolution, except that conspiracy was needed and that after the victory of the revolution there would have to be a dictatorship for a transitional period.[87] The technique of insurrection became a little more tangible and was discussed in detail in the works of three nineteenth-century revolutionaries — Blanqui, Mazzini and Carlo Bianco, Conte di Saint Jorioz; all of whom had known Buonarroti and were, to a varying extent, influenced by him.

Carlo Bianco (1795–1843) is the least known of the three but it is precisely in his writing that the link between guerrilla warfare and radical politics was first established. It was probably no accident that modern political guerrilla doctrine appeared first among Italian radicals. The resistance against Napoleon in Spain had been a people's war but the Spanish partisan leaders were guerrillas by instinct; they lacked the intellectual equipment to draw generalizations from their experience and to develop a system or a doctrine. The French revolutionaries based themselves on the support of the urban middle class and the workers; this ruled out a guerrilla campaign. In Italy, on the other hand, the radicals confronted both foreign rulers and domestic tyrants. Furthermore, Italy was much less urbanized than France.

Carlo Bianco was the son of a recently ennobled Turin lawyer; he studied law but later joined the army and served in Spain. A radical democrat, he became a member of the Italian nationalist underground; his last years were spent in great poverty in French, Swiss and Belgian exile. His main work on partisan warfare was written in Malta in 1828/29.[88] It was based on a comprehensive study of the existing military literature, and on the experience of partisan warfare in many parts of the world, in particular the guerrilla war against Napoleon and his own experiences in Spain. Bianco began with the assumption that modern (i.e., Napoleonic) warfare was quite unsuitable for the liberation of Italy — the insurgents would be unable to collect the money, obtain the weapons and mobilize the mass armies they needed. On the other hand, two million Italians could easily be mobilized for a people's war, which a group of conspirators could organize. Such a war would be most cruel, even terrorist in character; in this context Bianco referred to the forty-year struggle waged by Pasquale Paoli in Corsica. It would be a war in which the sacred end would justify all means, including a scorched-earth policy, and the evacuation of large parts of the population to the mountains. While putting great faith in the ardent patriotism of his fellow Italians, Bianco was realist enough to understand that, given "the present state of the world," one could not ignore such "ignoble pretensions as the love of money" — hence the necessity to distribute booty, or at least some of it, among the freedom fighters.

A neo-Jacobin, Bianco believed not just in national independence but equally in a free, republican Italy. Hence the necessity of a transitional period of revolutionary terror; once a certain area was liberated, the internal enemy, too, would have to be purged and even exterminated. It would be a war to the death. Only in exceptional circumstances should prisoners not be killed, for in a war of constant movement there would be no facilities to detain them. Every month bayonets would have to be checked and volunteers whose weapons were not covered with enemy blood would be publicly disgraced.

Bianco proposed a system of "democratic centralism." For purposes of organizing the conspiracy and conducting the war, Italy would be divided into four major provinces, every province into five cantons, and each canton into ten sub-districts. There would be elections on a regional basis but the leadership would be appointed. The central junta (*Consulta Suprema*) would be responsible to the supreme commander (*Condottiero Supremo*) and not to the nation. At the same time, while the war continued, the leaders of the guerrilla units would have maximum freedom of action. Guerrilla units should comprise only ten to fifty fighters in the early phase of the struggle, for units larger than these would be exposed to unnecessary danger and could easily be infiltrated by enemy agents.[89] Mobility was the essence of partisan warfare; sudden surprise attacks followed by quick retreat. In time a true people's war would evolve; women, children and the elderly, too, would play an active part, preparing ammunition, food and medical supplies. The peasants would assist in the transport of arms and supplies, and also help to spread panic through acts of individual terror: one suggestion was to overpower enemy soldiers after having made them drunk.

Bianco's "infamous" *Trattato* was followed three years later by another book in which he partly repeated what he had put forward in his previous manual.[90] There were a few new ideas: volunteers from foreign countries would join in the war of liberation; during the later stages of the war, flying columns would be formed and eventually a regular army would come into being.

ITALIAN-POLISH INTERLUDE

Throughout the first part of the nineteenth century, guerrilla doctrine was more widely discussed in Italy and Poland than in any other country, and the reasons are manifestly plain. The writings of Le Mière and Decker were read all over Europe at the time, and other writers were to borrow heavily from them for decades to come. But in the last resort these were technical manuals, devoid of direct political implications. In Poland and Italy, on the other hand, the questions of national independence and unification were the burning issues of the day: a search was on for an answer as to which was the most effective military-political approach to liberate a country from foreign occupation. In contrast to Decker, the Polish and Italian guerrilla strategists were deeply preoccupied with the political aims and context of a war of national liberation. It is for this reason that, however unsuccessful in practice, they anticipated many of the twentieth-century discussions on partisan warfare.

Carlo Bianco has been singled out for attention as a pioneer in the field. There had been individual publications even before, such as one by an anonymous Neapolitan author, who stressed the advantages of operations carried out by very small guerrilla bands of no more than ten to twenty men acting independently of each other. He also elaborated on the political and psychological differences between professional soldiers and partisans: the military man is guided by his honor, the volunteer-guerrilla has no other guide than the good of the fatherland. The professional army officer is to brave danger, the duty of the partisan to inflict maximum damage to the enemy with the least risk to himself.[91] But if Carlo Bianco was not the first to deal with the topic, no one in Italy had previously provided a systematic and detailed analysis of the problems of guerrilla warfare in Italy, and his writings certainly influenced his contemporaries, from Mazzini onwards, whereas the notes of the anonymous Neapolitan writer remained unnoticed.

Not long after the appearance of the *Trattato*, an article was published in the first issue of *L'amico del pòpolo italiano* in which the author drew the attention of his compatriots to the "*grandiosi risultamenti*" a mere twenty Corsican partisans (*voltigeurs*) had achieved: how much more could be attained by a brave band of men following similar tactics in Italy?[92] Even more optimistic in vein were the writings of General Guglielmo Pepe: everything in Italy pointed to the success of an insurrection once it had been started. All social classes would join it, for the local rulers were universally detested, and the clergy no longer exercised the influence that had been theirs a few decades previously under the French occupation. In Pepe's view it was essential for the success of the popular rising that a "liberated zone" be established in Calabria early on (and, if possible, a second, in the center of Sicily near Castrogiovanni). The enemy would no doubt counter by dispatching a major army against these southern *foci*, but three-quarters of his forces would have to remain stationed in northern Italy; and the rest, far away from their bases, would suffer defeat at the hands of the insurgents fighting according to the rules of *guerra alla spicciolata*, i.e., guerrilla warfare.[93] Pepe frequently drew attention to the lessons of the Spanish experience from which the Italians had much to learn. Cesare Balbo devoted a whole book to the subject, written in 1817 but only published thirty years later.[94] He had visited Madrid in 1815 accompanying his father, the ambassador.

Balbo also envisaged an invasion by the Austrian army in the event of a popular rising but, like Pepe, he emphasized that Austria would have to keep back many units to forestall similar risings in Hungary and Bohemia. The Italian irregulars would fight in the cities, the fields and the mountains and, supported by regular forces, would wear out the Austrian expeditionary corps. Yet another manual on the techniques of guerrilla warfare was written by Enrico Gentilini, an early utopian socialist thinker; it was published in 1848, the year of the revolution. The author was a self-made man, and his book is mostly derived from the writings of other military men. For the student of political theory there is more interest in Gentilini's essays in which he envisaged far-reaching social changes, implicitly regarding guerrilla warfare as a prelude to a people's war.[95] He thought that such a war would be a protracted one, and in his scheme the guerrillas would operate without support from regular forces.

Also socialist in inspiration was the work published in London in 1843 of Giuseppe Budini, a printer. He addressed himself not only to the spiritual motivation of the partisans but to the question of popular support, the "mass basis." He considered that since the uprising would confer material benefits upon the popular classes, it would receive their support. Ideally the revolution should break out simultaneously in the Kingdom of Naples and Piedmont; again the emphasis was on operations carried out by *"bande nazionale."*[96]

After the defeat of 1848, the general tendency among revolutionary writers of the 1850s was to de-emphasize the partisan element in the forthcoming military struggle, though some of them depreciated old-style regular army tactics even more. This refers, for instance, to General Allemandi who advocated small-war tactics carried out by militia battalions on the Swiss model.[97] La Masa had in mind a national Italian army of 600,000 men only one-quarter of which, however, ought to be a voluntary militia of partisans fighting in mobile units in the hilly regions of the Tyrol and Friuli,[98] while the regular army would fight in the plains of Lombardy and Veneto. Pisacane, one of the most interesting figures of the Risorgimento, commented on the "chimerical idea" that all that was needed were a few groups of bold young patriots (*giovani arditi*) who, fighting in the mountains, would successfully ward off a superior enemy. Pisacane wrote that attack, not defense, was essential in a revolutionary war. On another occasion he drew attention to the fact that in Spain, despite the topographical conditions favoring partisan warfare, it was the presence of a British expeditionary corps which proved decisive. Don Carlos's defeat by Cristina was additional proof that partisan war in isolation could not succeed. Pisacane's central concept was that of a nation in arms (*Nazione Armata*); within this framework there would be room for partisans but they could not possibly replace the regular army.[99] Many basic ideas of twentieth-century guerrilla doctrine can be found in these writings: the assumption that guerrilla warfare is the first stage in a people's war which would culminate in the establishment of a revolutionary army (Bianco and Mazzini); the idea that an armed struggle in Italy would prelude a general European revolution; the concept of *foci* and liberated zones; the importance of the social struggle and political indoctrination (Mazzini, Gentilini, Budini); the suggestion that the armed struggle would have to start in regions remote from the enemy's main concentrations (Pepe). The idea that moral purification and a re-education of asocial elements could be attained through an armed struggle first appeared in Bianco's and Mazzini's writings; it was taken up by Garibaldi. These thinkers did not, of course, see eye to eye on all the details of the strategy to be pursued. Pisacane thought that it was a fallacy to believe (as did Bianco and Mazzini) that the townspeople were ideally suited to mountain warfare; he advocated guerrilla warfare in combination with an urban insurrection, and the earliest possible formation of a regular force of half a million soldiers. Bianco's emphasis on terrorism ("cold terrorism of the brain not the heart" on behalf of all suffering mankind) was not shared with equal enthusiasm by the others.

From a purely technical point of view, the contribution of the Italian school of guerrilla warfare was small. But they pioneered certain political-strategic concepts and they constituted the link between traditional partisan warfare and modern radical politics, predating Blanqui whose only concern was with urban insurrection.

Karol Stolzman has been singled out in the preceding pages as a representative writer of the Polish school of guerrilla warfare. But unlike Bianco he was not the first military author in his country to address himself to the subject, nor was his position as pre-eminent. In the year 1800 a book had been published in Paris which questioned whether the Poles were prepared to fight for their independence.[100] The author was Tadeusz Kosciuszko; his little book was based on the experience of the insurrection of 1794, and it advocated a revolutionary people's war against the occupying forces.

But Kosciuszko contemplated a mass army applying guerrilla tactics; up to a million Poles were to be enlisted against 450,000 Russians, Prussians and Austrians. While he invoked the examples of Switzerland, the Netherlands and the United States it was not quite clear how a people of serfs could possibly emulate a society of free men. Thus the unleashing of a "people's war" really depended on the degree of patriotism shown by the landowning gentry, who still dominated the peasantry.

The Polish insurrection of 1830/31 produced a spate of partisan literature by men fascinated like Kosciuszko with the idea of a mass army or at least a giant militia. Stolzman thought that Poland could mobilize four million men; more modestly Bem and Kamienski believed that a million could be gathered.[101] But opinions differed fairly sharply on the role of partisan warfare within a general framework of national insurrection. A few writers such as Major Ludwik Bystrzonowski thought that guerrillas fighting in small formations would be able to liberate their country single-handed without the help of regular forces.[102] He assigned to Polesie (the area of the Pripet marshes) the role of an ideal battleground for an "unending" guerrilla war. It is interesting to note that the author of the thesis of "pure" and "eternal" partisan war was a monarchist whereas his opponents, particularly Stolzman, Mieroslawski and Kamienski, belonged to the democratic wing of the Polish national movement. They thought like Jelowicki that partisan warfare was merely the first phase in an insurrectional war, and that it was during this stage that the regular army units should be formed.[103] Stolzman envisaged the gradual extension and enlargement of the partisan units, thereby transforming the actual character of the war. Bem too advocated a mixture of partisan and regular warfare: first there should be a general insurrection but most of the fighting would be done by big mobile columns attacking the main concentrations of the enemy (Mao's transition to mobile warfare). Chrzanowski was aware of the likelihood that in a people's war a great many civilians would be killed—"*ein National Krieg ist ein Vernichtungskrieg.*" Kamienski's theory was the most explicit: he differentiated between four phases of the war of liberation. It would begin with a chaotic insurrection, there would be a gradual increase of coordination between the local patriotic forces until regular army units were formed to do most of the fighting according to the principles of modern strategy and tactics (meaning the French revolutionary wars) without renouncing the experience gained during the earlier period of partisan warfare. (Again, the similarity with the Maoist concept is quite striking.) Kamienski was aware of the fact that this kind of war would be impossible unless the peasants were liberated. Lastly, there were the skeptics, Nieszokoc and Mieroslawski in particular.[104] Nieszokoc was opposed to partisan adventures in principle ("*ad finitum*") while Mieroslawski attributed them limited importance at best. Following a brief initial phase of guerrilla war there should be a transition as quickly as possible to "real war," namely mass armies attacking the enemy. Mieroslawski wrote about Stolzman that his concept was a mixture of Italian conspirational folly and Polish aristocratic flippancy.[105] In his view Kamienski's theories were an absolute negation of military experience, and would render the conduct of real war impossible.

Given the political and geographical differences between the two countries there is astonishing similarity between the views of the Polish and Italian authors on partisan warfare. To the extent that the Italian military writers of the period were influenced by foreigners, it was Le Mière's book more than any other which became their catechism. The Poles, on the other hand, being cosmopolitans, did follow events in Italy closely. There was something like an organic link between Poland and Italy at the time: the "Legions" had fought in Italy, Chrzanowski was the chief of staff of the Piedmont army at one time and Mieroslawski, too, commanded Italian troops. Stolzman with Carlo Bianco helped to prepare Mazzini's ill-fated invasion of Savoy. F. Raquillier, a Polish general in Italian service, published in Florence in 1847 a "Practical Guide for the Perfect Partisan" in which he sharply criticized the idea put forward by some contemporaries that the partisans should fortify themselves within the main urban centers of the country and defend themselves from there.[106]

Why were the nineteenth-century theorists of guerrilla warfare so completely forgotten, despite the fact that their writings preempted twentieth-century guerrilla doctrine in so many essential points? The short answer is that the theories were discarded because reality refuted them. The Poles were backward-looking with regard to weaponry and tactics. The scythes of the peasants of Raclawice, where Kosciuszko had defeated the Russians fifty years earlier, blocked the progress of military technology. They failed to provide sufficient motivation to the peasants and they ignored the foreign political constellation. The Poles were invariably defeated and if the Italian struggle for unity was eventually crowned with success, it was not as the result of a partisan war as envisaged by Bianco, Mazzini, Pepe and others. It was only after many decades and in distant countries that many of the strategies first voiced in the 1830s and 1840s were to reappear.

BLANQUI

Auguste Blanqui (1805–1881) entered the annals of revolutionary history as the fearless fighter who always failed; Marx did not take him very seriously, Lenin regarded him as an adventurer and Trotsky said about him that he did not know the difference between revolution and insurrection. These judgments hardly do Blanqui full justice, because most of his theoretical writings on insurrection were not published in his lifetime and only became known around 1930.[107] Far from being a blind believer in violent action as the only way to revolution, Blanqui wrote after 1848 that conspiracy, which he had thought of as a civic duty under the monarchy, he regarded as a public offense under the republic, and that only the abolition or the abuse of the franchise would "compel us to convert the ballot into a cartridge."[108] In many respects Blanqui was a modern thinker; for instance, he believed in the need of an avant-garde consisting of déclassé intellectuals who would lead the masses onto the right path to progress. He thought that there was a latent revolutionary situation in France and that only a spark was needed to set the whole edifice on fire. His first major attempt at insurrection (May 1839) ended in total failure. True, there was a crisis, political and economic, the monarchy was discredited and the people, it seemed, were only waiting to join in an insurrection. Detailed preparations were made a long time ahead. The rising was fixed for a Sunday at noon: the army officers would be at the races, the new regiments just arrived in Paris would not yet have familiarized themselves with the geography of the city.[109] Some five hundred to eight hundred insurgents were to attack police headquarters and occupy the Cité; barricades would be erected all over the town. But police headquarters resisted the attack and despite some local successes in other parts of Paris, the soldiers did not go over to the insurgents, nor was there much response from the population. Within twenty-four hours the coup had failed.

In his *Instructions* written thirty years later Blanqui drew the lessons from his failure — and also from the experiences of June 1848. In brief, his conclusions were that though the political constellation had been most auspicious, with the government in a state of disarray and the troops demoralized, the six hundred barricades had been erected without any proper plan of coordination. Some barricades were manned by ten, others by a hundred men who had spontaneously assembled. Some were deserted altogether because those who had built them went to collect weapons, sleep, smoke a pipe, have a drink in a nearby restaurant, or perhaps join some other barricade.[110] Frequently no one was in command, everyone acted as he saw fit. There was constant coming and going, based on the assumption that if everyone did his own job, all would be well. No one knew what was happening anywhere else, no one came to help defend the barricades attacked by troops. In fact only thirty of the six hundred barricades were eventually attacked, but their defeat proved decisive.[111] Two or three regiments would attack one barricade after another and kill the few defenders. There was no coordination and thus (Blanqui concluded) despite their intellectual and moral superiority the defenders were easily defeated. True, the insurgents had been successful in 1830 and again in February 1848. But these had been lucky coincidences: in 1830 the government was totally taken by surprise and panicked, in February 1848 Louis Philippe made no serious effort to defend himself. In June 1848 the rising collapsed despite the fact that the insurgents faced the most miserable of governments which entirely lacked self-confidence. What if the insurgents had faced brutal, militaristic rulers who might have used the most recent technical inventions against them? Blanqui, unlike some other revolutionaries, was not unduly worried by the broad boulevards built by Haussmann: though they facilitated the movements of government troops, they also exposed them to rifle fire. According to Blanqui, the rifle would remain the decisive weapon in street fighting; artillery only made a great deal of noise, and hand grenades were hardly more effective than paving stones. But above all, the revolutionaries needed organization, discipline and a central command. Never again should there be those stormy and totally disorderly risings of ten thousand men.

Two years after these lines were written, shortly after the outbreak of the war with Prussia, Blanqui tried his luck again. At the head of three hundred men he attempted to storm the firemen's barracks in La Villette quarters. But the firemen refused to hand over their arms and, devoid of any other signs of revolutionary enthusiasm elsewhere in Paris, Blanqui's men were sent packing. The next revolutionary rising, the Commune, was only a few months off. In Blanqui's strategy immediate success or failure would depend on the first phase of the fighting. In fact, he only envisaged street battles of short duration; either the masses and the army would join the insurgents, in which case the war was won, or the revolutionaries would remain isolated, in which case they might as well disperse and wait for a more auspicious occasion. As Bakunin put it in conversation with a friend: even if the insurgents were defeated twenty times they might receive popular support on the twenty-first occasion. Each revolt, however unsuccessful, had its uses; hence Bakunin's theory of "propaganda through action" as the only possible way to revolutionize the masses, and his glorification of the *Lumpenproletariat* and bandits as the social elements most likely to overthrow the existing order.[112]* Hence the demand that the professional revolutionary should be ready to engage in violent, even desperate action at any time; Satan in contemporary reincarnation is the spirit of rebellion.

Mazzini, like Blanqui, believed in his more sanguine moments that once the call for a rising had been sounded it would be echoed everywhere: if there was no response, one had to try and try again until one finally succeeded. This belief led him into operations that were exceedingly amateurish. On one occasion he prepared for the conquest of the Kingdom of Naples with the help of twenty-two men, on another he wanted to invade Italy with two hundred patriots, despite the fact that all the police forces of Europe were familiar with the details of his plan. In his writings Mazzini frequently referred to historical examples that were of no relevance to his countrymen, such as the war of the Dutch against the Spanish. He never claimed to be a military expert and when toward the end of his life, he was presented with a sword by his admirers, he said: "I am not a soldier, and I do not like the soldier's trade."

Nevertheless Mazzini was the author of two detailed blueprints concerning the establishment of guerrilla bands. In his view, the guerrillas were the precursors of the nation, which they would rouse to insurrection, but they had no right to substitute themselves for the nation.[113] He assigned the guerrillas a fairly narrowly circumscribed role: they were not entitled, for instance, to punish those among the population who were guilty of collaboration; they had to give account of their operations to a nationwide Center of Action.

* Similar ideas had been advocated even before by the first German theoretician of Communism, Wilhelm Weitling, in his *Garantien der Harmonie und Freiheit* (1842). The existing social disorder had to be deliberately exacerbated through organized theft with the urban *Lumpenproletariat* as the chief revolutionary agent: the poor were to enjoy the growing disorder as the soldier enjoyed war.

They had to avert the enemy wreaking vengeance on small localities, and when they passed through such places they should seek to repress, rather than promote, a revolutionary demonstration on the part of the population.[114]

Such political guidelines apart, Mazzini provided detailed advice which he had, no doubt, gathered from earlier writers on partisan warfare: for example, that a retreat should always be left open ("A band that is surrounded is lost"), that attacks should take place in twilight, that a quarter of the band be kept in reserve at the time of attack, that there be a rifle range of three hundred yards before shooting at an adversary. "Much may be learned by listening with the ear close to the ground, and it does not require much practice."[115] Critics of Mazzini would argue that, figuratively speaking, he did not keep his ear sufficiently close to the ground, and that if Italy was eventually liberated, the guerrilla bands to whom he freely gave advice did not play a prominent part in the process.

MARX, ENGELS AND THE ARMED STRUGGLE

Much as Marx and Engels were preoccupied with the problem of revolutionary violence, they never accepted Blanqui's strategy of street fighting carried out by a few hundred, or at most a few thousand determined followers. Insurrection, as Engels wrote in his comment on the German experience of 1848/49, was as much an art as regular warfare, with its own rules of procedure that, if neglected, would lead to defeat and ruin. Engels's basic advice was never to play at insurrection unless fully prepared to face all the consequences which might ensue. Secondly, once an insurrection had been started, it was absolutely essential to maintain the offensive. Only thus could wavering elements be won over, and the enemy dispersed before it could gather its strength. There should be daily successes however small, since to be on the defensive was the death of every armed rising.[116]

The military experience of the revolutionary party in 1848/49 had been discouraging and Marx and Engels were not to pin their hopes again on another armed insurrection in the style of those that had failed. Although they wrote a great deal on military affairs (Engels's comments on contemporary wars fill several volumes), guerrilla warfare preoccupied them only rarely; they thought it, on the whole, to be of limited applicability. Commenting on the Carlist wars in Spain for his American readers, Marx on one occasion recalled Napoleon's experience. He noted that the guerrilla bands had been most successful while they remained small, and that once they started to "ape" a regular army they frequently suffered defeat; corps of three to six thousand men could no longer hide easily and disappear suddenly without being forced into battle. It was in the first and second stages of the fighting against Napoleon's armies that the guerrillas posed the greatest menace to the French. Marx quoted the description of the Abbé de Pradt of how the French forces were exhausted by the incessant molestations of an invisible enemy who, if pursued, disappeared among the people out of which he would immediately reappear with renewed strength: "the lion in the fable, tormented to death by a gnat gives a true picture of the French army."[117]

According to Marx, Mina, the Empecinado and their followers were among the most revolutionized sections of Spanish society. But in the light of subsequent events in Spain he showed awareness with regard to the dangers of "guerrillaism":

> . . . it is evident that, having for some years figured upon the theatre of sanguinary contests, taken to roving habits, freely indulged all their passions of hatred, revenge and love of plunder, they must, in times of peace, form a most dangerous mob, always ready at a nod in the name of any party or principle, to step forward for him who is able to give them good pay or to afford them a pretext for plundering excursions.[118]

If guerrilla warfare had been effective under specific conditions in the preceding fifty years, Engels still doubted whether it had a future. His skepticism emerges from his comments about the Spanish colonial experience in North Africa: even on broken ground (he wrote), a regular infantry force should easily gain on irregulars. The modern system of skirmishes along an extended line, behind which stood support groups and reserves, the tactics of concentrating troops against a common target, all this entailed, in Engels's view, certain defeat for the irregulars — even if they had a two to one superiority. That the Spanish took so long to advance on Tetuan could be explained by the fact that they had not yet mastered the technique of modern warfare, and that their army had been dispersed over too wide an area.[119]

During the later stages of the Franco-Prussian War of 1870–1871 irregular units played a part of some importance. Popular resistance continued after the regular French armies had been defeated and had virtually disappeared. New troops and *franc tireurs,* fighting behind barricades and *embrasures,* using night attacks and various guerrilla tactics, prolonged the opposition to the German invaders. For a few weeks Engels thought that it might be Spain of 1809 all over again. If real national enthusiasm were revived among the French, everything could yet be won, he wrote in October 1870; in November: "In the course of the last six weeks, the character of the war has markedly changed. . . . The ubiquitous 'four Ulans' are no longer able to ride into a village or town outside their own lines, demanding absolute obedience to their orders without incurring the danger of being taken prisoner or killed. . . . The German positions are surrounded by a belt of no-man's-land and it is precisely there that popular resistance is most palpably felt." And on 26 November: "Once the spirit of popular resistance is awakened, even armies of 200,000 men can no longer make rapid progress, they soon reach the point when their detachments are weaker than the forces opposing them; it depends entirely on the intensity [the *élan*] of popular resistance how soon this stage will be reached."[120] What if every citizen became a soldier, every village and town a fortress? But from the very beginning Engels had doubts whether a people's war was still possible in Europe of the second half of the nineteenth century. Once, many years earlier, he had written that a people who wanted to gain independence could not be restricted to conventional warfare. *Levée en masse,* revolutionary war, guerrillas everywhere — these were the only means by which a small people could defeat a bigger one, an army could resist its stronger and better-organized opponent.[121] But even when these lines were written, at the time of the Austrian-Italian war of 1849, Engels did not really expect that a monarchy could advocate "revolutionary terrorism." Engels knew, of course, about the Vendée, about Spain and Russia, but he preferred to invoke the shining example of 1793.

Engels sadly concluded that there really was not much hope for a people's war in Europe: "such fanaticism and national enthusiasm is not customary among civilized nations. One may find it among Mexicans and Turks but no longer in money-making Western Europe." The same view was expressed on another occasion in December 1870: "It is a fact that men have lost all recollection of a real war . . . the right of real self-defence is granted only to Barbarians."[122] Truly national wars, in Engels's view, had been fought in Algeria and the Caucasus. It was now expected of civilized nations that they would not continue a struggle once the "official nation" had surrendered.

As regards the German armies in France, they considered the *franc tireurs* as assassins and robbers. Accordingly, civilians found carrying arms were shot, and villages were burnt if there was a suspicion that it was from their direction that German units had been fired upon. The path of the German armies in France was marked, as Engels noted, with fire and blood. The Germans had a short memory, for according to Scharnhorst's *Landsturm Ordnung* of 1813 the more effective the means used against the French invaders the better. It was stated *expressis verbis* that the *Landsturm* should not wear any uniform whatsoever, so that they could turn civilian at any moment, thus making it impossible for them to be recognized by the enemy.

Engels's skepticism about the efficacy of guerrilla warfare was based both on his own experiences in the fighting in Baden in 1849, and on an analysis of historical precedents. National insurrection and partisan warfare, he wrote in 1857, were possible only in the mountains. In this context he referred to the Tyrol rising, the Spanish guerrilla war against Napoleon, the insurrection of the Carlist Basques, and to the struggle of the Cherkessians in the Caucasus.[123] But the Tyroleans and the Spanish guerrillas had only been effective because of outside help, the Basques were able to resist for so long because of the almost total disarray of the Spanish army, and the Caucasians, with their greater mobility, were successful only as long as they attacked the Russian rear and ambushed their columns. Whenever the Russians counterattacked, they were victorious.

If the prospects for guerrilla warfare in Europe were not propitious the chances in Asia seemed a little better. Commenting on the Indian mutiny, Engels did not exclude the possibility that guerrilla warfare, involving the dispersion of insurgent units in inaccessible forest and jungle, could cause far more attrition and losses to the British than a battle or siege. But he doubted whether the Indians were able to do this. Their military record during the mutiny had been poor: they had failed to cut the British supply lines and to organize an active small war. He thought more highly of the Chinese capacity to inflict damage on foreign invaders. They might not be able to hold their own against Anglo-Indian forces on the battlefield, but they could poison food (as they had already done in Hong Kong), launch night attacks, kill Europeans on board ships and, generally speaking, engage in the most unconventional warfare. It was pointless to complain about their fanaticism and barbarity; every nation fought in accordance with its level of civilization, and, anyway, the British, too, had behaved in a barbaric manner.[124]

In short, Engels's view on guerrilla war was that it could only succeed in Europe in conjunction with regular army units; on one occasion he argued that the Spanish guerrillas had been able to hold out only because of "a great number of fortresses [*sic*]" to which they had been able to retreat. Small and outmoded as these fortresses had been, they could not be captured short of a regular siege operation.[125] Outside Europe he saw the conditions for guerrilla warfare as more propitious.

Nor were Marx and Engels oversanguine with regard to the prospects of urban insurrection. While they commented in great detail on the political lessons of the Paris Commune they did not provide a similar analysis in depth of the military lessons. But the military experience of the Commune was not lost on them and it contributed to Engels's skepticism which became almost total towards the end of his life. He did not, of course, despair of the victory of socialism nor did he think that future revolutions would necessarily be nonviolent. Precisely because armies had become so powerful did they carry within themselves the seeds of their own destruction: they had become so costly to maintain that they made financial catastrophe virtually inevitable. Furthermore, the "armies of princes" could be rapidly transformed into people's armies, thus bringing about the collapse of militarism from within (*Anti-Dühring*). Since 1848 the techniques of warfare had completely changed and the revolutionary socialists would have to adjust their own tactics accordingly. The main reason for the occasional success of the insurgents prior to 1848 was that a civilian militia had stood between the army and the insurgents and either it had taken the side of the revolution or else was so lukewarm and indecisive as to make the regular army units vacillate likewise. In every case of revolutionary success the fight had been won because the troops failed to obey. The effect of the barricade, in any event, was felt mainly on morale. By 1849 the chances for success in an old-fashioned urban uprising had already become "pretty poor."[126] The spell of the barricade was broken, the soldiers saw behind it not so much "the people" but rebels, plunderers and the scum of the earth. Army officers had learned the art of street fighting, they no longer tried to take barricades by frontal assault — but outflanked and seized them with a little skill in nine cases out of ten.

The changes that had taken place were all to the advantage of the military. Garrisons in the capital cities had become bigger, it was now easier to concentrate troops, and rifles and heavy guns were proving more effective. In 1848 sappers had to use pickaxes to break through walls, fifty years later they had dynamite cartridges at their disposal. On the other hand the insurgents found it more and more difficult to concentrate their forces, and almost impossible to arm them. In 1848 it had still been possible to use homemade ammunition of powder and lead, in the 1890s this was no longer feasible. The workers' enemies wanted the revolutionaries to engage in precipitate military adventures that would lead to their defeat; the ruling classes were far more afraid of the results of elections than of rebellions.

Engels did not altogether rule out street fighting in a revolution but "only if the unfavorable situation is compensated by other factors": the insurgents ought to be numerically stronger and would have to opt for attack rather than passive barricade tactics.

Engels's views were shared by Marx who, though hailing the Paris workers (after the event) for daring to "storm heaven" in 1871, had warned them against a premature revolt and was not in the least surprised by the outcome — the fall of the Commune. The barricade was a mere symbol: only if the enemy forces yielded to moral (i.e., political and psychological) factors would the insurgents win. If, on the other hand, the self-confidence of the ruling class remained unbroken, if it did not panic, the insurgents would easily suffer defeat. This would apply even if the military were in a minority, because their better equipment, training, discipline, unified leadership and organization would outweigh any numerical superiority. The insurgents would win, in other words, only if at least part of the army joined them, and this could happen only in a grave crisis, perhaps after a defeat or a split in the ruling class, a loss of its self-confidence, a failure of the ability and will to exercise the power in its hands. Engels preferred not to speculate about such eventualities, nor did he and Marx reveal much interest in such guerrilla warfare as was likely to occur outside Europe. He was by no means opposed to guerrilla warfare; he simply believed, like almost all military thinkers at the time, that it was not likely to be of great practical importance. The same reasons that made barricade fighting obsolete made guerrilla warfare so much more difficult — except perhaps in the most distant parts of the globe. He did not belittle the importance of the colonial wars, but he would have found it difficult to accept that the fate of the world would be decided in the jungles of Asia or Africa. The revolution would occur in the highly industrialized countries where he foresaw no scope for guerrilla warfare.

MOST

It is one of the ironies of history that Marx and Engels, who showed little enthusiasm about the prospects of guerrilla warfare, nevertheless became the idols of subsequent generations of guerrillas, whereas Johannes Most, the nineteenth-century German socialist who provided an elaborate strategy for conducting "urban guerrilla" warfare, has been virtually forgotten. Born in 1846, a bookbinder by profession, a man of little formal education but of wide reading, he became one of the most successful Social Democratic propagandists of his day. He was one of the first German Social Democrats to be elected to the Reichstag (1874) but had to leave Germany as a result of Bismarck's antisocialist emergency laws. Settled in London, he became editor in chief of *Freiheit*, which gradually departed from Marxism, extolling the "propaganda of the deed," i.e., terrorism. In 1880 he was expelled from the German Social Democratic Party. Following the publication of a paean on regicide, he was sentenced by a British court to sixteen months hard labor.[127]

After his release from prison (1882), Most moved to the United States and in 1884 his *Science of Revolutionary Warfare* was published with the subtitle "A handbook of instruction regarding the use and manufacture of Nitroglycerine, Dynamite, Gun-Cotton, Fulminating Mercury, Bombs, Arsons, Poisons etc."[128] Modern explosives, he wrote in the introduction, were to be the decisive factor in the future social revolution; revolutionaries of all countries should therefore acquire them and learn how to use them. Terrorist acts were to be carried out by individuals, or at most by small groups, so as not to endanger the organization. Bombs should be put into public places such as churches and ballrooms; the whole "reptile brood" should be extirpated, and science was providing the means to accomplish this task. But it was not only the rulers, the nobility, the ministers, the clergy and the capitalists that Most wanted to annihilate, "pigs" too should be liquidated. Murder, as Most noted, was defined as the willful killing of a human being and he had never heard that a policeman was a human being.*

* Most also pioneered the idea of the letterbomb (*Revolutionäre Kriegswissenschaft*, 1873/4). Dynamite had been invented by Alfred Nobel in 1867; it was first used by Russian terrorists in the late 1870s. Previously bombs had consisted of nitroglycerine (invented in 1845 by Salvero) or even earlier of fulminate of mercury (discovered in 1805 by the Reverend Alexander Forsyth but first applied in 1816 by an American sea captain, J. E. Shaw).

For several years Most had a substantial following among American workers, but after the Haymarket affair he gradually lost influence. Liebknecht thought that Most was a madman, Eduard Bernstein on the other hand called him a genius not amenable to discipline. Bebel wrote in his memoirs that if Most would have remained under the influence of men able to guide him and restrain his passionate temper, the party would have found in him a self-sacrificing and indefatigable fighter. But under the antisocialist laws he went astray and although he had once been a model of abstinence ended in the United States as a drunkard.

Most's propaganda for "direct action" was based on the assumption that more and more bombs would have to be thrown and more "reptiles" killed before the enemy would collapse. Unlike Blanqui he was not interested in mass action because he felt that the army and the police would always prevail in a confrontation of this kind. Unlike other socialists of his time he thought that the development of modern science favored the revolutionary terrorist — provided the fruits of science were correctly applied.

The technique of small warfare hardly changed between 1750 and 1900, nor did the practical advice given by the military thinkers of the period. The definition of partisan warfare, in a famous early-nineteenth-century textbook, covers the Pandurs and Croats, the *franc tireurs* of 1870–1871 and the Boer commandos:

> A detachment is partisan, when it operates detached and isolated from the army, and under the genius of its leader, which is not controlled except by orders given in a general manner. . . . The profession of a partisan is a hazardous one. It can only be properly carried out by a skilful, rapid and bold leader, and by a body of men resembling him. . . . The war which he carries on is piratical. The strength of his warfare lies in surprise.[129]

In the period under review a fundamental change took place in the function of partisan warfare. In eighteenth-century military doctrine, even among the advocates of the small war, the partisan always appeared in a supporting role, never at the center of the stage.

The battles of the Spanish War of Succession had been very costly for both sides. At Malplaquet the losses on both sides, killed and wounded, were 36,000, one-fifth of the total strength of the two armies; at Blenheim the percentage of casualties was even higher.

The battles of the Seven Years' War were equally bloody; at Zorndorf one-third of the participants were killed or wounded, and at the end of Kunersdorf, Frederick the Great was left with 3,000 soldiers out of an army of 48,000 with whom he had entered battle. In view of these losses, more and more critics maintained that a small war was less risky, that it could save a great deal of bloodshed — and achieve the same result in the end. It was also, incidentally, a more interesting kind of warfare, allowing greater scope for individual initiative and inventiveness.

But the advice was not heeded. The French revolutionary armies used a new system of operations which had been advocated, however, by certain military thinkers for several decades previously. Its main features were rapidity of movement, flexibility, the use of *tirailleur* tactics, dispersing forces for maneuver and concentrating them for decisive action. Napoleon perfected the "new warfare"; his genius was in the performance, for the basic rules were few and simple: "if I were to give my principles formal expression one day," he told Saint Cyr, "their simplicity would appear surprising." It was all a matter of the economic use of force, a concentration of the greatest number of troops where a strike was intended, a sudden move against the enemy's rear, cutting off his line of retreat, swinging round toward him, encircling and destroying him.[130] Such tactics had been used by military commanders of genius throughout history. Napoleon applied them on a more massive scale than ever before. The armies had substantially increased in number; in the biggest battle of the Thirty Years' War, Breitenfeld (1631), 70,000 soldiers were involved, at Malplaquet (1709) the number of combatants was 183,000, and Napoleon invaded Russia with a *Grande Armée* of some 612,000 men. In Napoleonic strategy, there was no scope for small, let alone guerrilla, warfare; his ideal was Caesar not Vercingetorix. In his comment on Caesar's wars he wrote: "Every nation which loses sight of the importance of a regular army always in a state of readiness, and which opts instead for levies or 'national armies' [i.e., militias] will suffer the fate of the Gauls."[131] In the Napoleonic era, as the Prussian General Berenhorst noted, the small war was swallowed by the big war.

The revolution in warfare lay not so much in the application of new methods and novelty of approach as in the changed nature of war itself. The regular, professional armies of the absolutist age were replaced by citizens' armies and the *levée en masse*. The war between kings became a war between nations, a people's war. The military system of the absolutist age was based on pressure, threats and punishment. Drilled to behave like an automaton on the parade ground, the soldier fought because he was more afraid of his corporal than of the enemy. With the wars of the French Revolution a new soldier appeared, spurred on by patriotic enthusiasm.

Guerrilla warfare in its modern form was first waged not by the armies of the French Revolution but against them in the Vendée and, later on, against Napoleon, in Spain, the Tyrol and Russia. It was a system of warfare chosen instinctively, without the benefit of preconceived doctrine. The guerrilla detachments in the Vendée, in Spain and in the Tyrol were no longer operating within the framework of a regular army and subject to its command but, for most of the time, acted altogether independently. Their war was insurrectionist and chaotic, revolutionary and subversive, not in its aims but in its implications for the future. It was, therefore, highly suspect to the European monarchies. The Russian partisan units were dissolved the moment the French invaders had been defeated; Gneisenau's memoranda of 1808 and 1811 about preparations for a popular rising (in which all means to defeat the hated enemy would be permissible) and the famous *Landsturm* edict of 1813 remained largely a dead letter.

Yet partisan activities in the Vendée, Spain, the Tyrol and in Russia remained isolated episodes and the doctrine of the great war continued to prevail. The emphasis was on large armies and big battles; combat between small detachments was regarded as the exception. The results were sometimes unfortunate, as the Prussian General Willisen, author of a theory of grand strategy, learned to his cost in the war against Denmark (1864). Guerrilla wars continued to be fought throughout the nineteenth century, not in the main theaters of war, such as the Crimea, but in a different context altogether: in the Polish uprisings of 1831 and 1861, the struggle of Caucasian tribes against the Russian invaders, and the campaigns of Abd el-Kader against the French in North Africa. Guerrilla war became an integral part of national wars, fought mainly against colonial powers or other foreign occupiers, by nations (or tribes) without regular armies of their own. Lastly, there was an abiding interest in the theory and practice of urban insurrection among socialist revolutionaries all over Europe. But there were essential differences between the guerrilla wars on the fringes of the European colonial empires, long-drawn-out affairs in which quick successes were neither sought nor achieved, and the revolutionary uprisings in European capitals which could only prove victorious by rapid and decisive action. They were subject to totally different strategies and tactics: the Circassians and the North Africans could win a campaign by a series of escapes, by wearing out their pursuers; the urban revolutionaries would face defeat unless they managed to overthrow the regime within a few days.

4

The Twentieth Century (I): Between Two World Wars

The fortunes of guerrilla warfare had reached a low ebb in the early years of the twentieth century; they did not improve in the period spanning and embracing the two world wars. Victory in these wars went to the stronger battalions, the decision determined in massive battles between vast armies. In World War I guerrilla tactics were hardly applied at all, in the second they played a certain limited role in some countries in the struggle against the foreign occupier. The first third of the century witnessed civil wars and warlordism spreading whenever central state power broke down, as in Mexico and later in China. But these small wars were quite often big wars *manqués,* in that the military chiefs operated as though handling regular armies, imitating with varying success their strategy and tactics. There were national uprisings in Africa and Asia, such as Abd el-Krim's struggle and the Palestine insurrection in 1936–1939, but less frequent and intense than in the nineteenth century.

Guerrilla wars were always patriotic; sometimes they gravitated to the right, sometimes to the left, sometimes toward Fascism and sometimes toward Communism, on occasion betraying traces of both. But guerrilla politics were usually inchoate, unless, as in China, for instance, a political party had sponsored the struggle in the first place. In the Russian Civil War true guerrillaism was to be found neither in the Red Army nor among its White opponents but in the bands of independent freebooters in Siberia and the Ukraine. The international climate was as yet inclement for a general guerrilla upsurge; the European powers, though weakened as a result of World War I, were still strong enough to hold on to their colonial empires. Seen in retrospect, the emergence of guerrilla bases in northern China in the 1930s foreshadowed the guerrilla upsurge after 1945. But not much was known at the time about events in these secluded areas, and the Chinese situation in any case appeared unique, as indeed it was. The country had been in a state of semianarchy for a long time, small wars had been going on incessantly and the Communists were just one of the forces in this imbroglio. Again, the Irish case — "urban guerrilla" — was *sui generis* and seemed to offer few lessons to other countries. Guerrilla movements, almost by definition, could succeed only if the internal or external enemy was weak, or in the larger framework of a prolonged war. But the state was still predominant; even in Mexico, after many years of anarchy the guerrillas were suppressed. Political theory and military doctrine, "bourgeois" and Leninist alike, accorded to guerrilla warfare only a subordinate role. No new theories of guerrilla warfare emerged during this period; in fact, the neglect of the subject was almost total. Military thinkers were almost exclusively concerned with the *Vernichtungsschlacht* (Cannae), *Blitzkrieg,* with tank warfare and the impact of air power. The new weapons which had first been used in World War I seemed to tip the balance even farther against guerrilla warfare, leaving little, if any, scope for it in the future. Armored cars had been used by the British raiders against the Turks and the Germans in East Africa, though as yet not widely, and without much effect because of the unsuitability of the terrain.[1] Aircraft were also used for the first time in support of Lawrence's raiders, the Serbian insurgents, and against Lettow-Vorbeck for reconnaissance, dropping agents behind enemy lines, and for providing urgently needed supplies. All this was still, however, on a small scale and the evidence about the effects was contradictory. Lettow-Vorbeck wrote that the British planes never really bothered him, and once his forces had entered the jungles of Mozambique, the reconnaissance flights served absolutely no practical purpose.[2] He himself had no air support; the Germans had tried to supply their forces in East Africa by air but failed. A Zeppelin with badly needed supplies was once dispatched from Bulgaria in November 1917, but was called back after reaching Khartoum. Nonetheless, there was little doubt that with the improvement in aircraft technology and the perfection of tanks and other armored vehicles, these new weapons would play a very important part in future conflicts — and it looked as if the guerrillas would have no answer against these threats. Airplanes were extensively employed in the North West Frontier fighting. "It is impossible to overestimate the value of aircraft in tactical cooperation with other arms. Their presence alone greatly raised the morale of our troops."[3] Information gained through aerial photography was of great value, even though there were as yet few trained observers. It could surely be only a question of time before the guerrillas lost their last secrets — their hideouts would become known, their movements be detected and the element of surprise, the main source of their success, would evaporate. In 1918, in short, the prospects of guerrilla warfare loomed less than brilliant.

WORLD WAR I

Apart from some minor actions of short duration, such as the activities of small bands of Serbian irregulars, guerrilla operations during World War I were restricted to two theaters of war, the Arabian peninsula and East Africa, where Lettow-Vorbeck with a minuscule force managed to contain for over four years a total force "considerably larger than Lord Roberts' whole army in the South African war."[4] The operations in Arabia were in Lawrence's own words a sideshow within a sideshow. Nevertheless, in later years his raids were to attract infinitely more attention than the war in East Africa. The Oxford don, Arab headgear and all, was a flamboyant personality in the great tradition of British adventurers and explorers in the Orient. He quarreled with orthodox military authority and underwent incredible hardships; his complicated and tortured mentality fascinated the intellectuals and his books were received with enthusiasm by the avant-garde critics of the 1920s. Hailed as a genius by some, derided as a charlatan by others, this elusive figure was to intrigue, more perhaps than any other hero of the Great War, both his own generation and the ones to follow, still providing until today inspiration for biographers, moviemakers and amateur psychoanalysts. Appearing to shun publicity, he attracted it beyond a single other contemporary — among his friends were the leading writers of the period, unorthodox strategists, such as Liddell Hart who compared him to Napoleon, and leading American practitioners of the new art of public relations. Even those who bitterly and sometimes unfairly attacked him, Richard Aldington for one, added to the Lawrence myth.

Lawrence's courage and qualities of leadership are beyond all doubt, but his originality and the importance of his exploits have certainly been magnified; seldom in the history of modern war has so much been written about so little. It was neither the first nor the last time in the history of guerrilla warfare that the measure of attention paid to a particular campaign depended less on its military importance than on the accident that a gifted writer wrote about it. But for Euclido da Cunha, Canudos would rank at most as a footnote in Brazilian history; but for Balzac (*Les Chouans*) and Tolstoy (*Hadji Murat*), the Vendée and Shamil's wars would be less well remembered. Ernst von Salomon's books helped to popularize the German *Freikorps*, and *For Whom the Bell Tolls* made many readers believe that guerrilla warfare was a major element in the Spanish Civil War, whereas in reality there was little of it. As a partisan commander, Lettow-Vorbeck stood head and shoulders above Lawrence, but his personality was neither particularly interesting nor attractive. A Pomeranian Junker by birth and a diehard reactionary, he was a typical product of the German officer caste; and not one of the brightest at that. Aged forty-four when war broke out, he had not advanced beyond the rank of lieutenant-colonel. His transfer to East Africa was professionally a dead end; there was no plan to defend the German colonies in the event of war. He was, to put it mildly, an indifferent writer; his books, published after the war, are mere variations on the same theme.[5] In short, it is hard to envisage a greater contrast than that between the brilliant, unorthodox British amateur soldier and the dull, conventional, unimaginative and unattractive German professional. But Lettow succeeded brilliantly in adverse conditions against a vastly superior force; Lawrence's few raids over a much shorter period against small enemy units were not in the same class. Military experts have acknowledged this — the official British history of World War I devoted two volumes to Lettow's operations in East Africa; one would look in vain for the name "Lawrence" in the memoirs of German and Turkish commanders in the Near East.[6] It is not certain that they were even aware of his existence. There has been in recent years a modest Lettow-Vorbeck revival, but as far as the general public is concerned, Lawrence figures as one of the central figures in guerrilla warfare while Lettow has passed into oblivion.[7]

Lawrence's guerrilla operations, which began in late 1916, were part of a general blueprint for an Arab rising against the Turks. The insurrection had in fact started even earlier, in June 1916, but the attempts to seize Medina and other places occupied by the Turks were abortive. Despite a numerical superiority of more than three to one, the Bedouins, untrained and ill equipped, could not defend a line or a point, let alone launch a massive attack against regular troops. Lawrence shrewdly realized that he should concentrate his attacks against the Hedjaz railway in such a fashion that the Turks could just about keep it working with the maximum of loss and discomfort, compelling them to strengthen their posts beyond the defensive minimum of twenty men.[8] To this end he established a small, highly mobile and highly equipped striking force. This approach worked well. Wejh was taken by the Arabs in January 1917, they were successful at Abu al Lissal and Auda, and in July 1917 they entered Aqaba after the Turkish garrison of three hundred had surrendered.[9] In these encounters the Bedouins were usually stronger in numbers; a party of ten thousand was dispatched to Wejh, halfway between Medina and Aqaba, which was defended by only about two hundred Turks. Subsequently it appeared that an assault by five hundred Arabs (landed by British ships) was sufficient to seize the place. Military actions after the capture of Aqaba were no longer along unorthodox lines; the Arab forces continued to march on Damascus, but this was coordinated with the general advance of the Allied armies.

Lawrence found coping with his desert warriors anything but easy; the Arabs had no artillery and they were frightened by the sound of the Turkish guns. "They thought weapons destructive in proportion to their noise."[10] Discipline was nonexistent, and when the time came for looting, "they lost their wits, were as ready to assault friend as foe." Lawrence's superiors called them cutthroats; he commented with pride that "they would cut throats only to my order," a somewhat rash boast in the light of his own recounting that at Mudawwara he had to defend himself three times against his own men who pretended not to know him, that at Al Shalm, in the general looting, the Bedouins attacked their allies, the Egyptians. Lawrence's chief aide during the campaign was one Abdulla al Nahabi — Abdulla the Robber.[11] Eventually the Turks had to evacuate the Arabian peninsula, but since the British army operating from Egypt had meanwhile occupied Sinai and had reached Khan Yunis in February 1917 and Gaza in March, they would have been withdrawn anyway as their presence there no longer served any useful purpose and they were in danger of being cut off. Thus, Lawrence's guerrilla doctrine about winning campaigns without giving battle — to be amplified later — was not really tested.

When Lettow-Vorbeck took over the command of the *Schutztruppe* in East Africa in January 1914, the outlook was bleak. There were altogether approximately six thousand Germans living among eight million Africans. The German colonial record, though not worse on the whole than that of other European powers, was certainly no better; in the suppression of the big Maji Maji revolt in 1905–1906 tens of thousands of Africans had been killed. Lettow's force consisted of two hundred and sixty German officers and NCOs and two thousand native soldiers (*askaris*). Unlike in the fatherland, there was no enthusiasm among the local Germans when the war broke out; they had believed that a European war would not affect Africa.[12] Lettow's position was impossible; he could not expect any help or supplies from outside. He had to fight not only the British and the Belgians, but had to carry on a running battle with Schnee, the civilian governor who was nominally commander in chief. Lettow ignored his orders and Schnee threatened to have him court-martialed after the war. Lettow may have hoped that the war in Europe would be over within a few months and that, with a little luck, he could hold out that long, inflicting maximum damage on the enemy. He had some previous guerrilla experience, gained in the Hottentot war in German Southwest Africa; the black people in this war were led by Jakob Morenga, a Herero, an exceedingly able commander, who had learned the art of commando warfare from studying de Wet.[13]

There were four phases in the East African war; in November 1914 a first British attempt to land Indian troops at Tonga ended in disaster. Throughout 1915 there was stalemate which the Germans used to attack the seven-hundred-kilometer-long, vitally important Uganda railway. The next phase opened in March 1916 when the British, having built up their forces with South African assistance, penetrated deep into German East Africa. By the end of 1916 two-thirds of the German territory had been occupied by the Allied forces led by General Smuts. But the Germans had not been eliminated, and when Lettow's unit crossed in 1917, first into Mozambique and subsequently into Northern Rhodesia, the British were unable to pursue him in strength for almost a year; the logistic difficulties seemed insurmountable. Lettow surrendered at Abercorn in Northern Rhodesia after he had been informed of the armistice in Europe. His force consisted at that time of 156 Europeans, 1,168 *askaris* and some 3,000 native carriers. Altogether, three thousand Europeans and eleven thousand *askaris* had at one time or another served in its ranks, the basic unit being the company of which there had been sixty. But the Germans had suffered losses during their retreats and, when Lettow crossed into Portuguese Africa, he decided to leave part of his force behind. Like all generals, he was prone to exaggerate the numbers he faced; he asserted that on one occasion twelve thousand armored cars had been brought into action against him.[14] And another thing that should be borne in mind is that the enemy troops he was countering were not exactly the flower of the British Empire. "It was too piteous to see the state of the men," Meinertzhagen wrote after the battle of Tonga.

> Many were jibbering idiots, muttering prayers to their heathen gods, hiding behind bushes and palm trees with their rifles lying useless beside them. I would never have believed that grown-up men of any race could have been reduced to such shamelessness. I do not blame the men, still less their officers. I blame the Indian government for enlisting such scum.[15]

Lest Meinertzhagen be accused of racism, it may be useful to quote his comment on the quality of British leadership:

> I think the worst and most expensive error of the campaign was the employment of generals who were not first class; their quality was lamentable. I have no hesitation in saying that if we had had a general of the calibre of General von Lettow-Vorbeck, and if the Germans had had an Aitken, Wapshare, Stewart or Malleson — or even Smuts — the East African campaign would have been over by the end of 1914 and hundreds of valuable lives and millions of pounds would have been saved.[16]

While Lettow succeeded in containing in East Africa a force so numerically superior to his own, one can but stress again its, on the whole, inferior fiber; the great majority of soldiers (South Africans, Indians, Africans) and superannuated generals would not have been employed in Europe anyway. The British war office was giving this campaign low priority. During 1915 it opposed any major attack against the Germans. All this does not, however, detract from Lettow's achievements. The difficulties he encountered at every level were formidable. Cut off from outside supplies, everything needed by his troops had to be locally produced in the most primitive conditions; shoes, shirts, quinine (of which great quantities were a requisite) and even some *ersatz* gasoline. The Germans revealed great ingenuity in this respect, and it was perhaps no idle boast when Lettow wrote after the war that "we could have continued for years."[17] The local *askaris* fighting with the Germans were more promising human material than the Indians — a fact gradually realized by the British that was to induce them to expand their own King's African Rifles quite considerably — but Lettow still had to train his recruits and mold them into an efficient and disciplined little partisan army. Only soldiers of stout caliber would be able to survive four years of long marches through deserts and bush, lack of food and water, wild beasts, disease, and of course enemy attacks.

The East African experience tends to disprove much that has been written about the preconditions for guerrilla warfare. Lettow-Vorbeck was no fanatic, just a tough regular soldier who thought his duty was to fight as long as he possibly could. He was in no way a charismatic leader, able to generate enthusiasm among his subordinates. True, he had acquired a smattering of local tongues and studied the native customs, but his attitude toward his men was old-fashioned, paternalistic, if not downright authoritarian. He could not promise them anything, nor influence them other than through the example of his own behavior. The *askaris* had no particular reason to love the Germans or to support them in the war. Furthermore, by 1916 at the very latest it must have appeared doubtful whether Germany could be the victor. The *askaris* did not get their pay for more than four years, and there was hardly any booty. Lettow could neither cajole nor threaten them; they were free to desert at any time. Yet against all odds, he inculcated in them a pride in their uniform and their units, discipline and a fighting spirit, so that they fought exceedingly well for a cause which was not their own.

Lettow developed his guerrilla tactics only by trial and error. His first skirmishes with the British, while successful, were too costly for the Germans in men and supplies and it was this that decided him to subdivide his little legion into smaller units, sometimes of no more than ten men, who were sent on special missions. He had a few guns, dismantled from a ship which had been destroyed by the British; they were carried through bush and jungle — whether to any great effect is not certain. The machine gun was the most important weapon in the bush; Lettow's great problem all along was lack of ammunition. His soldiers had standing orders to regard the acquisition of ammunition from the enemy as their foremost function; they had to have more bullets at the end of a battle than at the start.[18] When they surrendered their weapons in November 1918, it appeared that their rifles, almost without exception, were of either British or Portuguese origin.

The British adapted far less well than the Germans to local conditions; it is difficult to visualize a British general cutting his own shoes out of deerskin, as Lettow did. They should have used native soldiers rather than Europeans and Indians from the beginning. In 1916 their numbers began to tell, but the very size of the British army hampered its mobility, the organizing of supplies proving a mounting problem the farther they went. Lettow figured rightly that he would still have local superiority in Mozambique, and that while the Germans would be independent of supply dumps, the British would be increasingly reliant on them to keep their immense quantity of men and materiel sustained and maintained over the ever-growing distances. All this made it possible for Lettow to retain the tactical offensive throughout the war, despite the fact that he was strategically on the defensive.

Four decades later, Kenya, Uganda and Tanzania had become independent states. Von Lettow-Vorbeck survived not only Wilhelm II, but even Hitler, and almost outlived Adenauer. He had never been a member of the Nazi party, perhaps because they were not monarchists, perhaps because he disliked their socialist slogans. In the early 1960s he still pursued his favorite sport of hunting. He died in Hamburg in 1964. He was ninety-four years old.

THE POSTWAR CRISIS

Following the breakdown of the old order in Central and Eastern Europe and the disintegration of the Russian, German and Austrian armies, irregular units began to emerge, to become for several years a factor of some military importance. There was no lack of recruits; the young officers and soldiers had been fighting for four years without interruption, it was the only kind of job at which they were proficient. The survivors of the greatest concentrated mass slaughter in modern military history had few scruples about shedding more blood; human lives counted for far less after 1918 than before World War I. There was plenty of arms and ammunition; all the ingredients existed for bloody and prolonged civil wars.

Russia was in many respects ideally suited for partisan warfare; guerrillas were more numerous on the right than on the left, but most frequent were the nonpolitical, "a-plague-on-both-your-houses" bands. The Bolsheviks would use partisan tactics from time to time, but neither Lenin nor Trotsky was a great admirer of this particular form of warfare; they rejected the idea of a militia, even though this concept had figured prominently in their political program before the revolution. Having seized power, they realized that their need was for a well-trained, well-organized regular army, not enthusiastic amateurs. Partisan bands had been of some use in the pre-revolutionary period, but once the Bolsheviks ruled large parts of Russia, the main task was to maintain their hold, and this was not the work for partisans. Besides, Bolshevik influence was strongest in the towns, weakest in the countryside — the party thus lacking steadfast bases most fit for the launching of guerrilla warfare.

The main areas of partisan warfare were Siberia and the south of Russia. According to Soviet sources, some hundred to a hundred and forty thousand guerrillas operated in Admiral Kolchak's rear.[19] The number is almost unquestionably too high; figures in guerrilla wars are always inflated. Tens of thousands of partisans did fight the White armies — but many of them fought the Red Army as well. Local conditions favored partisan warfare: the vast stretches of forest provided excellent cover. There was no communications network save for the Trans-Siberian railway, which constituted an easy target for partisans. Kolchak and his Czech allies were holding the railway, but the rest of Siberia was in a state of anarchy. The Siberian partisans, mostly smallholders, whose individualism was proverbial throughout Russia, held no brief for the Bolsheviks, but the outrages and the systematic looting perpetrated by some of the White partisan units, led by bandits such as Ungern-Sternberg, Kalmykov (*ataman* of the Ussuri Cossacks) and Semyonov, a Cossack of part Mongol extraction and a Japanese agent, eventually drove them into the Soviet camp.

The war between the Red and White armies, like all religious and political wars, was callous, sanguinary and claimed a host of civilian victims. It is difficult to establish which side was answerable for more bestialities, particularly since the situation varied from front to front. Insofar as the irregulars were concerned, the White partisans undoubtedly had the edge. The American General Graves, an observer in Siberia and a professional soldier not given to squeamish overstatement, wrote of Kalmykov that he was the "worst scoundrel I ever saw or heard of and I seriously doubt, if one should go entirely through the Standard Dictionary, looking for words descriptive of crime, if a crime could be found that Kalmykov had not committed."[20] Semyonov was a bird of similar feather, while Ungern-Sternberg had received a blow on his head during the war which seems to have unhinged the mind of an officer who had not been too stable in the first place. All three were cavalry officers who had fought in the Carpathian campaign.[21] If Kolchak was defeated, it was partly owing to the activities of these savage bands only nominally under his command.

There were not many Communist partisans outside cities such as Tomsk, Omsk, Irkutsk and the towns of the Far East.[22] The anti-White partisan units behind Denikin's lines in southern Russia consisted mainly of deserters (the so called red-green units), and as in Siberia, the Bolsheviks were not too effective outside the cities. Politically the south was, on the whole, hostile territory from the Communist point of view: the Ukrainian peasants were strongly nationalistic; most of the Cossacks in the southeast and the Caucasus, who constituted almost half of the rural population, were supporters of the old regime. Among the anti-Bolshevik Cossack irregular units, Shkuro's from the Kuban (the "wolves") and Grigoriev's band were two to acquire great notoriety. Shkuro had little to learn from the Kalmykovs and Ungern-Sternbergs who devastated Siberia; Grigoriev had originally cooperated with the Bolsheviks but turned against them, heading a mutiny in May 1918 which almost overthrew Soviet power in the Ukraine.[23] Grigoriev was shot while negotiating with another major partisan leader, Nestor Makhno, a Ukrainian anarchist, the most colorful of them all. Makhno came from a poor peasant family and had spent years in a Tsarist prison; he first made his name as a resistance leader against the German occupiers of the Ukraine and then, for about eighteen months, collaborated closely with the Bolsheviks. At its zenith, in the late autumn of 1919, Makhno's movement numbered between twenty-five and fifty-five thousand adherents.[24] His chief base was his native village of Giulai Pole in the Ekaterinoslav district. A man of small stature, he made his motley force of deserters — bona fide anarchists, landless peasants, adventurers and bandits — into a formidable fighting force. He was perhaps the greatest guerrilla fighter of the lot, developing techniques of fighting dependent on swift dispersal and assembly, together with rapid movement by carts and captured gun carriages which were, when necessary, lifted onto flatcars and moved by rail.[25] He was a leader of great cunning and many ruses; on one occasion he dressed his units up as Ukrainian police units, on another as Red Army battalions. His command was the only truly radical one in the civil war; the Red Army employed former Tsarist officers, but no one of middle-class or aristocratic origin could serve as an officer in Makhno's armies. His soldiers killed the habitual numbers of Jews in their pogroms, but there were also Jews among his closest collaborators; Grigoriev, when visiting him, complained at the presence of Jews in Makhno's camp. Like many partisan commanders, he was a heavy drinker; at one point his partisans passed a resolution that orders of the commanders had to be obeyed only if they were sober when issuing them.[26]

Makhno was a genuine anarchist, who believed in the abolition of the state; wherever he went, prisons were destroyed, and the banknotes printed on his behalf advised that no one would be prosecuted for forging them. His movement was bound to fail because it was merely regional in character and could not link up with similar groups in other parts of Russia. In August 1921 Makhno gave up what had become an unequal struggle and with his two hundred and fifty remaining followers crossed into Roumania. If Grigoriev was a mere brigand who could switch sides in the civil war without compunction, Makhno was a political leader, albeit a very confused one. For some anarchists he was to become a patron saint, whereas the Communists dismissed him as nothing but a bandit. The truth, as so often, lies somewhere down the middle, simply demonstrating yet once again that in guerrilla warfare the distinction between patriotic and revolutionary leader and marauder is easier to draw in theory than in practice.

By 1920, with the execution of Kolchak and the flight of Wrangel, the Red Army had at long last defeated the White armies. Partisan warfare was to continue, however, though on a smaller scale, in various parts of the Soviet Union. In the eyes of the Soviet leaders these partisans were of course no more than another brand of plain brigands, to be handled as such — just as the White generals had treated the Bolsheviks as criminals, just as all governments through the ages have denied their irregular opponents political motivation and status. That the anti-Soviet partisans were marauders no one would deny, but it must equally be allowed that their inspiration was largely political and social — there could scarcely otherwise have been any accounting for their mass support. Antonov for one, the leader of the Tambov guerrillas, had been jailed for years by the Tsarists for acts of violence during the 1905 revolution. He called himself a social revolutionary and with the backing of angry peasants first set up a partisan band in 1919. By early 1921 he had as many as twenty thousand volunteers, almost exclusively peasants; the land of some had been taken away to establish state farms, others had been hard hit by the requisitions carried out by order of Communist officials.[27] The political demands of the insurgent peasants were radical ("The land to the toiling peasants!"). In suppressing this counterrevolutionary insurrection the Bolsheviks behaved as the White armies had done; houses and farms were burned, hostages were taken and sometimes executed. Antonov achieved great popularity because his force, as a matter of principle, plundered only state farms while the Red Army lived off requisitions from the peasants, but his movement was defeated in late 1921; it was a purely regional uprising that could not hold out against vastly superior regular forces. The Soviet authorities did not, however, rely entirely on military repression and for a time discontinued the nationalization of land, reduced requisitioning to a minimum and introduced the more liberal New Economic Policy (NEP). Thus peasant riots gradually abated; fighting of one sort or another had continued for seven years, and the peasants were only too happy to work the land again.

The one exception was Central Asia, where guerrilla warfare continued up to the early 1930s. The Basmatchi, the Soviets' main opponents, were made up of partisan detachments, almost always on horseback. They were elusive and, in the words of a Soviet eyewitness, often dissolved in the neighboring villages "literally before the eyes of our troops, who would immediately undertake a general search of the villages but without any result."[28] According to Soviet sources, the Basmatchi, who first appeared in the Fergana valley, the rich center of cotton plantations, and subsequently spread to other parts of Central Asia, were professional bandits who had made common cause with the local reactionaries, the Beys and the Mullahs. (The origin of the term *Basmatchi* is not clear; it has been variously translated as "raider," "robber," and "downtrodden.")[29] Again, there is no denying that there were robbers among the Basmatchi, if perhaps more at the start than in the later years. But, again, banditry would hardly explain the widespread support they enjoyed among the local population, making it that much more difficult for the Soviet authorities to destroy them. The marauders' popularity and strength lay in their constituting simultaneously a movement of national resistance against the Russians who had, to put it mildly, shown little tact in their dealings with the natives. By the same token, *Basmatchestvo* was also a social movement, reflecting peasant protest against requisitioning and collectivization. The Basmatchi were weakened by internal divisions; the Uzbeks did not cooperate with the Kirghiz, and the Turkmens would not collaborate with either. For a short while in 1921–1922 it appeared that all the bands might unite under the leadership of the

Turkish leader, Enver Pasha, who had cooperated with the Soviet government but then switched his political allegiances to the Panturks of Central Asia. He failed, however, in his attempt to consolidate all these peoples and tribes and was killed in a skirmish with Soviet forces in August 1922. Enver was no outstanding partisan leader; he had been accustomed to giving orders to armies and found it hard to adjust himself to commanding bodies of only three thousand men.

The influence of the Basmatchi dwindled as the Soviet authorities rescinded some of the harshest abuses of power and as they made religious and economic concessions to the local populations. Nevertheless raids continued in the Samarkand region up to 1924 and aircraft and tanks had to be used against the insurgents. There was resistance in the Fergana valley as late as 1926, and small raids from across the Iranian and Afghan borders over which the Basmatchi had escaped were reported even in the 1930s. The Soviet border was long, complicated to control, and whenever the Basmatchi crossed into Soviet territory they apparently had no problem hiding among sympathizers. They had ceased to be a real military and political menace much earlier, but it is interesting that even a totalitarian state with its unlimited means of repression needed almost a decade to stamp out the last remnants of armed resistance.

Soviet military thinkers were very much preoccupied during the 1920s and 1930s with topics such as the future of tanks, artillery, aircraft, and the character of a future war in general. They were not concerned with the prospects of partisan warfare. At best, insofar as they thought in terms of it at all, their attitude was ambivalent. Nevertheless, a strong claim could be made for regarding Marshal Tukhachevski as one of the originators of the theory of modern counterinsurgency. In a series of articles published in 1926,[30] he reveals with great candor the difficulties Soviet power was facing in its struggle with counterrevolutionary bands in European as well as in Asian Russia; a rebellion, he points out, was not necessarily crushed when the band had been destroyed, military measures had to be closely linked with political and economic steps, and even then success would not necessarily be immediate. Tukhachevski argued that surrounding guerrillas was time- and manpower-consuming and very often ineffective. His definition of "national banditry" is of interest — a "peasant rebellion . . . organized by the kulaks which attracts the poor elements in the villages."[31]

THE FREIKORPS

Following the dissolution of the German army after the armistice in November 1918, some hundred and twenty *Freikorps* (free corps) came into being, numbering altogether about two hundred and fifty thousand men. They varied in their status, size, function and political orientation. Some were more or less legal, that is, recognized both by the Allies and the German government of the day, others were semilegal, being recognized only by the German government, while yet others were altogether illegal. Some went on fighting, with short interruptions, for several years, others existed for a few days only. Some had the strength of several divisions, whereas the Freikorps Gross Thüringen consisted of one lieutenant and thirty-two soldiers.[32] The strength of the average free corps was that of a battalion or a brigade, and they were frequently called after their commander (Ehrhardt, Rossbach, von Loewenfeld). A very few were republican in orientation, but the great majority were right wing, or even semi-Fascist; the Baltikum Freikorps was the first to display the swastika on its helmets. The Social Democratic government tolerated some of the Freikorps because it needed military units both against external enemies who had penetrated German territory — the Poles in the east — and against the Spartacists who tried to overthrow the Social Democratic government. The government would have preferred a fighting force of reliable republicans, but there had been few, if any, republican officers in the imperial army, and if the Bolsheviks had a few months to forge a new one, the German Social Democrats had only a few days.

Some Freikorps joined forces with the White armies against the Bolsheviks, others provided cover for the retreating German armies from the east, others again served as border police, or fought against Communist paramilitary units inside Germany. Many free corps had official recruitment offices in the major towns, this leading to frequent abuses, such as new recruits enlisting in several units at one and the same time. The general atmosphere reminded observers of Wallenstein and the age of the Thirty Years' War.[33] The activities described so far would have been those normal to regular army units, the police or border guards. But in addition, there were operations of traditional guerrilla character — in Upper Silesia against Polish units, in Carinthia against the Yugoslavs, in the Ruhr in 1923 against the French occupiers, and in the Rhineland against the local separatists.

The fighting in Upper Silesia was the heaviest and in many ways the most confused because it was carried out by partisan units on both sides; on the German side the Bavarian "Oberland" Freikorps was prominently involved, while the Poles were led by Adalbert Wojciech Korfanty, a former member of the German Reichstag, a gifted and very ambitious politician and propagandist, who later became deputy prime minister in Poland.[34] The Allied statesmen had left the fate of Upper Silesia wide open and Korfanty, with the discreet help of the Polish government, tried to maneuver as many *faits accomplis* as possible before a plebiscite took place. He had earlier successfully engineered an insurrection in Poznan, but he found the going in Silesia much rougher. The Poles were a minority except in some mining and rural districts; besides, not all Polish-speaking Silesians supported the Polish cause. The German irregulars, while badly equipped, were more numerous, and to make matters still worse for him, coordination between Korfanty and his officers was deficient. Both sides committed acts of senseless terror. The Germans assassinated a senior French officer, the Poles killed some forty Italian soldiers who were to supervise the plebiscite. But whereas the French supported the Poles anyway, the Italians and the British, who had been neutral in the dispute, were incensed by the Polish attacks. Since the Polish government very much depended on Allied goodwill, it had to dissociate itself eventually from Korfanty. Meanwhile, in May 1921, a major battle took place at Annaberg in which the Poles were routed. Some Polish officers wanted to fight on, but Korfanty accepted an armistice and later a political decision which gave Poland the more important part of the Upper Silesian coal mines. Altogether, some sixty thousand Poles and thirty thousand Germans were involved in the fighting in Upper Silesia.[35] It was to a large extent war by proxy; Germany still had a regular army but it could not be used for fear of French intervention. For different reasons, Poland could not employ its new armed forces. Thus, military operations in Upper Silesia on both sides turned into partisan warfare, with the local population the principal victim.

In Carinthia, operations were on a more restricted scale. German-speaking peasants organized themselves into small units but the conflict was no less bitter, because it was waged between neighbors, dividing little villages and even hamlets into two armed camps. The struggle against the French occupation of the Ruhr had the support of all German political parties. It took for the most part the form of passive resistance, which still did not inhibit the occasional terrorist act, such as the mining of the railway line between Duisburg and Düsseldorf. This sabotage was organized by Albert Leo Schlageter, an early member of the Nazi party who had fought with the Freikorps in Upper Silesia. Apprehended by the French, he was executed in May 1923, thus becoming the earliest martyr of Hitler's Third Reich, a "fighter for national liberation who had paid the supreme penalty for his patriotic idealism."

The free corps consisted chiefly of former officers and soldiers of the Imperial army (some Freikorps consisted entirely of young officers), but students who had been too young to fight in the Great War also volunteered. The veterans were quite familiar with the tactics of fighting in the open country, but they were not accustomed to street battles and they learned only by trial and error the technique of crowd control.[36] The great majority of the soldiers of the Freikorps were right-wing activists, many of them becoming even more radical in their opposition to the Weimar Republic as the fighting continued. But traditional labels are of only limited help in explaining the Freikorps phenomenon. Bitterly opposed though they were to Communism, they hated the Poles and the French even more; not a few of them were enthusiastic advocates of a German-Soviet military alliance against Poland and the West. The spirit of Tauroggen, the anti-Napoleonic convention of 1813, was again conjured up. There are many illustrations of the anti-bourgeois and anticapitalist spirit prevailing in these units. They despised the "fat, cowardly bourgeois" and all he stood for; and they made it known, time and time again, that they had not the slightest wish to fight for the preservation of this social order. They had far more respect for their enemies, the Communists, and the Communists tried hard to attract members of the free corps to their ranks. Karl Radek devoted a friendly essay to the memory of Schlageter. Schlageter and his comrades were, so he wrote, men of goodwill, confused or misguided nationalists, who could be swayed either way. They were uprooted men, radicals who shared with the Communists the militancy, the desire to overthrow the political system. "I cannot go home and start the old life," one of them wrote later, "my Germany is where the Verey lights illuminate the sky, where the time of day is estimated according to the strength of the artillery barrage. It ends where the train for Cologne departs."[37]

The radicalism of the Freikorps also found expression in their way of life. Former colonels served under the command of lieutenants, and there was equal pay for all, from general to the youngest recruit.[38] There was little marauding in these campaigns in comparison with other guerrilla wars. Individual banditry was not in the Prussian tradition, it was detrimental to discipline; the state, the collective, was entitled to maraud on a grand scale, but not the individual. Many members of the free corps joined the Reichswehr in later years. Many became supporters of the Nazi party, but only a few rose to its top leadership. There was a tendency in the Third Reich to play down, not so much the historical role of the Freikorps in general, but that of those who had taken a leading part in them. Some former Freikorps men were killed in the Nazi purge of June 1934, others deviated from the Nazi cause in time and were imprisoned or executed. The volunteers of the Freikorps fared a little better under Hitler than the Old Bolsheviks under Stalin, but not by very much.

THE EXPERIENCE OF WAR AND REVOLUTION: REVOLT IN THE DESERT

The war years were arid insofar as the development of guerrilla doctrine is concerned. The only original contribution was made by T. E. Lawrence in an essay published in the *Encyclopaedia Britannica*, a curious mixture of brilliant insights, of stating the obvious, and of arrant nonsense. It is marred by pretentious neologisms ("bionomics," "diathetic"), and deliberately paradoxical formulations which do not survive critical analysis. ("The Turkish army was an accident, not a target." "In irregular war, if two men are together one is being wasted.") Lawrence believed guerrilla warfare could be proved an exact science, granted certain factors and if pursued along certain lines. These factors are an unassailable base, a regular army of limited strength that has to control a wide territory, and a sympathetic population. The guerrillas, Lawrence argued, must have speed and endurance and be independent of lines of supply. They also need the technical equipment to destroy or paralyze the enemy's supply lines and communications:

> In fifty words: granted mobility, security (in the form of denying targets to the enemy), time, and doctrine (the idea to convert every subject to friendliness), victory will rest with the insurgents, for the algebraical factors are in the end decisive, and against them perfections of means and spirit struggle quite in vain.[39]

Lawrence maintained that the Turks would have needed six hundred thousand men to control the Arabian peninsula, but as they had only a hundred thousand they were bound to fail. (They had, in fact, far fewer.) They were low, besides, on war materiel, consequently "the death of a Turkish bridge or rail, machine or gun, or high explosive was more profitable than the death of a Turk." The enemy soldier should never be given a target; many Turks on the Arab front "had no chance all the war to fire a shot." To achieve this, the guerrilla needed infallible intelligence.

The enemy, again in terms of Lawrence's thesis, should be encouraged to stay in harmless places in the largest numbers. Propaganda is important: "the printing press in the greatest weapon in the armory of the modern commander." Range is more to strategy than force, "the invention of bully beef has modified land war more profoundly than the invention of gunpowder." Guerrilla tactics should be what they had been in the Arab peninsula, "tip and run, not pushes but strokes." The smallest force is used to reach the farthest place in the quickest time. Lawrence's concept of guerrilla warfare was based entirely on his own experience in Arabia; he seems not to have been aware of the lessons of guerrilla warfare elsewhere, let alone of the existing literature on the subject. His generalizations are of limited worth only, valid in certain circumstances, inapplicable in others. Nor should one look for consistency in his writings. Thus, in a letter to Wavell he noted that if the Turks had mounted machine guns on their touring cars and patrolled the desert, they would have put a stop to the Arab camel parties and so to the whole rebellion. "It wouldn't have cost them twenty men or £20,000. . . . They didn't think hard enough." This observation is certainly at variance with his statement that the Turks would have needed six hundred thousand men to control the peninsula. The Turks, to put it somewhat crudely, lost out not because of any "algebraical factors," but because they did not have armored cars. Another time he argued that bombing tribes was ineffective, that guerrilla tactics were a complete muffing of air force, a statement of doubtful validity with regard to desert warfare. He wrote that guerrilla war was essentially a moral contest and that counterpropaganda was never effective when conducted on the conservative side. But in the same breath he declared that Turkish intelligence was miserable, that one well-informed traitor could spoil a national rising, and that the Turks had failed because they did not go to the effort of buying a few venal men.

Lawrence succeeded on a modest scale because, like Lettow-

Vorbeck, he understood that he had to go for the main weaknesses of the enemy, and that warfare had to be adapted to local conditions, human as well as geographical. Compared with Lettow's precarious position in East Africa, his situation was much more advantageous; he had money, almost unlimited supplies, and there was a separatist movement that could be mobilized against the enemy. The Turkish army in Arabia was overextended and for that reason Lawrence's "rapier play" could have pointed thrust. But to argue, as he later did, that this approach was generally applicable and preferable to Allenby's "wood-chopping tactics" was simply not true; rapier play and evading battle would not have worked in Palestine, let alone in the European theaters of war against heavy army concentrations. Lawrence's ideas were rejected at the time for the wrong reasons by orthodox military thinkers. Liddell Hart, on the other hand, popularized his views because they fitted in so well with his own concept of the strategy of the indirect approach. He lived to regret his enthusiasm. In essence, however, Lawrence's theories gained their generally wide currency and appeal because he was a romantic figure and had little, if any, competition. The few other contemporary practitioners of the art of guerrilla warfare were not literary men or, as already remarked, simply did not bother to put down their experiences in writing. If they did, they refrained from engaging in generalizations.

THE EXPERIENCE OF WAR AND REVOLUTION: LENIN, TROTSKY AND THE PARTISANS

While Lawrence in later years came to be regarded as the great guerrilla theoretician in the West, Lenin was largely, and by most as ignorantly, held as its chief proponent in the East. More than any other major revolutionaries of his generation, Lenin studied military strategy and organization and, of course, the art of revolution. But although he had much of account to say about revolutionary situations and the proper tactics to be employed in each, he certainly offered no new and startling advice on guerrilla warfare. He was more favorably inclined toward it than other radical socialists, but that is saying very little indeed. In the many volumes of his published works there is just one short article on the subject and some occasional references in 1905–1906 and in 1918–1919. Whatever interest he had in it at all was first aroused by the Moscow insurrection in December 1905, and the armed rebellions in Latvia, Poland and in other parts of the Russian Empire during the revolution of the same year. He noted that the Moscow experience had shown that, *pace* Engels, urban insurgency was not altogether obsolete; Moscow had inaugurated new barricade tactics, a fact which had been observed, incidentally, by Kautsky even before Lenin.[40] Unfortunately there had not been enough volunteer fighters, and their arms had been inadequate. Lenin thought that the experience had nevertheless been positive and that its lessons should be spread among the masses. One month later, in September 1906, Lenin directly addressed himself to a consideration of what guerrilla warfare was, in what conditions it could be effective, and what the correct attitude of a revolutionary should be toward a question which "greatly interests our party and the mass of the workers."[41] Lenin defended the guerrillas against their Social Democratic critics who invoked — vainly in his view — the authority of Marx and Engels. Blanquism and old-style Russian terrorism, which had been denounced by Marx and Engels, had been futile because it was an affair of a few intellectual conspirators. Since then the situation had changed. "Today, as a general rule, guerrilla warfare is waged by the worker combatant, or simply by the unemployed worker." (This statement was not altogether accurate; in Latvia the peasants, not the workers, had been in the forefront of guerrilla warfare, and Moscow apart, Latvia had been the main focus of the insurrection.) Thus, reasoned Lenin, guerrilla warfare had to receive the Bolsheviks' blessing; it was an "inevitable form of struggle at a time when the mass movement had actually reached the point of an uprising and when fairly large intervals occur between the big 'engagements' in the civil war."[42] It was not true, as Plekhanov and others had argued, that guerrilla warfare demoralized the revolutionary avant-garde, only the senseless methods of unorganized, irregular bands had that effect. The avant-garde party had to direct the masses not alone in the major battles of the revolution but also in the lesser encounters. There was no gainsaying that guerrilla warfare brought the class-conscious proletarians into close contact with "degraded, drunken riff-raff." But this meant only that the Bolsheviks should not regard it as the sole, or even as the chief instrument of struggle, or ever as anything but subordinate to other methods. It did not mean that guerrilla warfare should be left to the riffraff. Lenin refrained, *expressis verbis*, from prescribing "from our armchair" what precise part guerrilla warfare should play in the general course of the civil war in Russia.[43] He did once, but once only (in 1906) claim that partisan warfare in combination with uninterrupted strikes, attacks and street fighting throughout the country would effectively exhaust the enemy. No government could withstand such a struggle in the long run, it was bound to paralyze industry, demoralize the bureaucracy and the army and create discontent among the people.[44]

This was the sum total of Lenin's prerevolutionary dicta on guerrilla warfare. The insurrection of December 1905, he thought, had demonstrated that armed uprisings could be victorious even when pitted against modern military techniques and organization. But guerrilla warfare was only one of the tools for the revolutionary, and not the most important one. Between 1906 and the revolution of 1917 Lenin did not refer to the subject again, apart from welcoming the Irish rebellion of 1916, and in his polemics against Rosa Luxemburg's thesis that national wars were no longer possible. Lenin felt that such wars were still possible in Europe and inevitable in colonies and semicolonies.[45] On a very few occasions he pointed to the revolutionary potential of the peoples of the East without, however, elaborating. But whereas Lenin sedulously studied Clausewitz and made copious notes which were published posthumously, even the most diligent guerrilla enthusiasts have been unable to discover any further references to guerrilla warfare up to and including the revolution.[46] In 1917 the main task facing the Bolsheviks was to win over as many units of the Tsarist army as possible and to transform, with the help of ex-Tsarist officers, the old army into an instrument of Soviet power. Thus, in the words of a leading historian:

> Guerrilla war and military freebooting held little appeal for Lenin, squeezed dry of any drop of romanticism. . . . The republic could not defend itself with untrained mobs or be held together by wild-eyed guerrillas.[47]

Lenin frequently referred to *partisanshchina* (guerrillaism) after 1917, but always in a derogatory vein. "One should shun *partisanshchina* like fire," he wrote, "the arbitrary operations of individual detachments, the disobedience vis-à-vis the central power. It leads to ruin."[48] Or again in July 1919 in an appeal to intensify the struggle against Denikin:

> The partisan spirit, its traces and remnants, have caused our army more suffering, defeats and catastrophes, more losses in life and material than all the betrayals by [former Tsarist] military experts.[49]

What the Red Army needed above all was iron discipline and central control; guerrilla warfare was at best ineffective in this struggle. Perhaps there had been a justification for it in the first difficult weeks and months while the Red Army was being born, but that once achieved, guerrillaism had to be stamped out "with an iron fist." Trotsky, the architect and first commander of the Red Army, entirely agreed with Lenin; those who obstructed the Leninist approach were Voroshilov, and, to a certain degree, Stalin. But this dispute mainly reflected some of the Old Bolsheviks' resentment of military specialists who had been taken over from the Tsarist army. The Old Bolsheviks wanted a party committee to run military operations, whereas Lenin opposed misplaced "collegiality"; decisions had to be taken by one man, the commander.[50]

The debate about a specific Soviet military doctrine much exercised the Bolshevik experts in the early post-revolutionary period. Some of them contended that the Red Army should launch partisan actions and deep-penetration raids in the enemy rear. Trotsky, in reply, noted that these tactics had in fact been used first by the White armies:

> The first big raid was made by Mamontov [a general in Denikin's army]. Petliura [the Ukrainian nationalist] was the leader of partisan formations . . . the operations of Ungern's and Makhno's detachments — these degenerate, bandit outgrowths of the civil war — were distinguished by great maneuverability. What conclusion follows from this? It follows that maneuverability is not peculiar to a revolutionary army but to civil war as such.[51]

Guerrilla warfare, as Trotsky said on another occasion, was the "truly peasant form of war," but this was not meant as a compliment, for he added, "Similarly in religion the peasantry is unable to go beyond the sect" — a generalization which will hardly be underwritten by historians.[52] Guerrilla war, as he saw it, was a primitive form of warfare, inevitable perhaps in some cases, but devoid of any specific revolutionary character. His fairest assessment of the value of guerrilla warfare was made in a speech in 1923:

> The guerrilla movement had been a necessary and adequate weapon in the early phase of the civil war. The revolution could not as yet put compact armed masses into the field, it had to depend on small independent bodies of troops. This kind of warfare demanded self-sacri-

> fice, initiative and independence. But as the war grew in scope it needed proper organization and discipline and the guerrilla movement then began to turn its negative pole to the revolution.[53]

Stalin had been one of the early "guerrillaists" in the Red Army and when the civil war was over he stressed in a speech the importance of the rear in the fight against the White generals. But what has been interpreted by some writers as a manifesto of guerrillaism turns out to be, on closer scanning, nothing more than a reference to the "tacit sympathy, which nobody hears or sees" — scarcely a characteristic of the armed struggle.[54] Stalin, the archdisciplinarian, forever suspicious of independent initiative, the leader who wanted to concentrate all decisions in his own person, was bound to regard partisan warfare with disfavor — except *in extremis*, and under the close supervision of the party and the secret police. When war broke out in 1941, great efforts had been made to strengthen Soviet artillery, tank units, the air force and other parts of the regular army. No such preparations had been made for partisan warfare.

In the light of these facts, the emergence of a cult depicting Lenin and Trotsky as great guerrilla strategists is difficult to understand and impossible to justify. Mao Tse-tung did not join this chorus; in his speeches and writings he did not attribute any special significance to the lessons of the Russian Civil War. Some of the blame for the cult rests with Western military historians, and theorists who "discovered" Lenin in this context in the 1950s and 1960s. In their search for the key to the mysteries of revolutionary warfare, they failed to discriminate between the various modes of revolutionary struggle in different ages, countries and societies. For both ideological and practical considerations, the Soviet approach to guerrilla war was as ambivalent as the Tsarist attitude. While not entirely ruling out its applicability in certain extreme situations, a political regime such as the Soviet Union, based on centralized control, order and discipline, could not tolerate an inherently disorderly system of warfare based on lack of central control and on individual enterprise. Furthermore, Soviet military thinking has always been oriented towards the concept of masses and large numbers, not the feats of small groups of intrepid men. Bolshevism derived some of its inspiration from the Jacobins, and the reliance on mass armies was part of this inheritance. In the 1920s and 1930s books on the civil war partisans were not uncommon, and there were films dedicated to the exploits of Shchors and Chapayev. But the patron of guerrilla warfare in the Soviet Union between the two world wars was not the army, but the secret services.[55] Guerrilla warfare was interpreted as one specific aspect of intelligence and sabotage work behind enemy lines, to be carried out by highly trained individuals, or very small teams. The idea of a people's war on guerrilla lines was rejected as unfeasible not for the Soviet Union alone, but for the advanced capitalist countries as well; it was sporadically entertained as one of the forms the revolutionary struggle might nonetheless take in Asia, Africa and Latin America.

Thus, as an instance of this occasionally qualifying attitude, the sixth Congress of the Comintern (1928) recommended that, "the situation permitting," Communists should proclaim slogans calling for national-revolutionary uprisings and the immediate formation of national-revolutionary guerrilla units. But the situation did not permit, the third (radical) period of the Communist International was followed by the conciliatory popular front era, and save for the Chinese Communists, acting quite independently of Moscow, no one heeded the Comintern resolution of 1928. Manuals were published and courses instituted in the twenties and thirties for tutoring and training in conspiratorial work and insurgency technique in major cities, but not one on guerrilla warfare.

Of the twelve chapters in a Soviet guide to insurrection published between the two world wars only one, the very last, deals with "revolutionary guerrilla methods."[56] It was written by a young, "friendly, unassuming Indochinese revolutionary" named Ho Chi Minh. Ho argued that in the overall pattern of the class struggle guerrilla movements play the role of an auxiliary factor; they cannot of themselves achieve historic objectives, but can only contribute to the solution provided by another force — the proletariat. The peasant movement, however large, could not count on any conclusive success if the working class did not move. Ho predicted that Soviet power would initially establish itself in China in some province or group of provinces possessing a great industrial or commercial center such as Kwangtung, Hupeh or Hunan — not in Kansu or Kweichow.[57] Subsequent events in China and Indochina did not bear out these predictions; the guerrilla movement was not just an auxiliary factor, the working class did not move and Soviet power established itself far from the industrial centers. It has been suggested that Ho at the time may have known better but that he had to pay lip service to the collective wisdom of the Comintern. It is far more likely, however, that he changed his views only later, when he realized that Asian revolutionaries could not possibly wait for a working-class initiative and that the peasants would be the main force in the Asian revolution. Ho's essay included some valid observations; conditions for guerrilla warfare varied from country to country; the strength of the guerrilla was not in defense (because they were not strong enough for defensive action); guerrillas had to avoid decisive encounters if the circumstances and the balance of forces were not in their favor. But these insights had been common knowledge for centuries and there was nothing specifically Marxist-Leninist about them. Ho's specific predictions and guidelines were quite wrong; but like Lenin and Mao, he was an opportunist of genius. He was quick to recognize that if the workers were too weak or would not fight, peasant movements led by intellectuals constituted a promising alternative. In 1928 there was only one young Chinese leader who dissented from the collective Leninist wisdom that "the city inevitably leads the village"; but even Mao did not at the time advocate guerrilla warfare.

There were a few articles in 1937–1940 in Soviet periodicals about Mao's experience in northern China, but there is no evidence that Soviet military leaders took notice.[58] Mao's famous treatise on guerrilla warfare was published in Russia only in 1952 as far as can be ascertained, years after it had first been translated into English. Guerrilla warfare was peasant warfare, and the Bolsheviks were, after all, primarily a working-class party.

CONNOLLY AND THE PROBLEM OF STREET FIGHTING

Among the few socialist thinkers of the early twentieth century who gave more than passing thought to military affairs was that highly unorthodox Irish Catholic revolutionary, James Connolly. Analyzing the lessons of the Russian revolution of 1905 as they might apply to his native country, he wrote that the tactics of the Moscow insurgents had been basically right; but for their miserable equipment, they would have seized the army's field guns. The rising was doomed because there were no simultaneous uprisings in other Russian cities and the peasantry had been hostile. But Moscow had shown that a well-defended line of houses was a position of strength. This surely had some important implications in relation to the Irish struggle. Ireland, Connolly acknowledged, was no ideal guerrilla country in the traditional sense, it had no mountainous passes or glens. But

> a city is a huge mass of passes or glens formed by streets and lanes. Every difficulty that exists for the operations of regular troops in mountains is multiplied a hundred fold in a city. And the difficulty of the commissariat which is likely to be insuperable to an irregular or popular force taking to the mountains, is solved for them by the sympathies of the populace when they take to the streets.[59]

Connolly's observations, while superficially plausible, were grounded, in fact, on the old fallacies of nineteenth-century barricade fighting. Street fighting was admittedly difficult for regular troops, especially if they were untrained for this purpose and if the population were hostile. But a frontal collision constituted the very antithesis of guerrilla warfare. An insurgency such as Connolly envisaged would either lead within a few hours to victory or, more likely, to defeat, as it did in Dublin in 1916. A later generation of Irish insurgents, having digested the lesson, did not opt for street fighting. Looked at in retrospect, the Moscow rising of 1905 provided fresh hope for some revolutionaries, but far from being a panacea, it could well result in failure and ruin.

IRA

The history of the Irish struggle for independence from the time of Wolfe Tone's "United Irishmen" until the Easter Rising of 1916 is a chain of abortive conspiracies and defeats. Ireland was indeed, as Connolly had recognized, bad guerrilla country and the Irish insurgents were ill prepared to conduct that kind of fighting. Britannia still ruled the waves and could easily prevent arms reaching Ireland from the United States and the Continent. The Fenians were internally split and they could not keep a secret; forthcoming operations were widely discussed. The British secret service had effectively penetrated the ranks of the Irish nationalists; one of its agents operated for almost a quarter of a century in the leading councils of the American *Clan-na-Gael*.[60] Last but not least, the clergy, however patriotically inclined, did not support military action; as Bishop Moriarity of Kerry said of the Fenians: "Hell is not hot enough nor Eternity long enough to punish such miscreants."

In the 1860s the Fenians enlisted several American officers of Irish extraction who had gathered guerrilla experience in the American Civil War. Conspicuous among them was a Captain

McCafferty who had been one of Mosby's rangers. According to a colleague, he was essentially a man of action, who thought, quite mistakenly, that insurgent cavalry on the American pattern could achieve a great deal in Irish conditions:

> He began by saying, "I believe in partisan warfare". Probably only O'Reilly and one or two more knew what the word "partisan" meant, but if he had said "guerrilla" warfare they would have understood him.[61]

McCafferty was to be in charge of the attack on Chester Castle in the uprising of 1867, but as so often in the history of Irish rebellions, either the British were forewarned or the Irish failed to gather in time for the operation. Later on McCafferty made an even more daring suggestion — the kidnapping of the Prince of Wales — but this scheme the Fenian leaders rejected.

It was only during World War I and its aftermath that large-scale armed struggle in Ireland became a reality. As Wolfe Tone had noted one hundred years earlier: "England's difficulty was Ireland's opportunity."

The Easter Rising of the Irish Revolutionary Brotherhood was quickly and ruthlessly suppressed by the British armed forces, and fifteen of its ringleaders were executed. The Irish leaders had given up the old idea of guerrilla war in favor of an urban insurrection. But an urban insurrection could only succeed if it was not confined to one town, although in this case its indubitably poor preparation was not the determining factor; it is virtually certain that the rising would have failed even if it had received more widespread support. It should have been clear that for political reasons Ireland's opportunity would come only after the war; revolt at that particular time could have had a chance only if the British army had been in an advanced state of disintegration — which it was not.[62]

But if the Easter Rising of 1916 was a crushing defeat, it still gave fresh impetus to the nationalist movement; it helped to mobilize a whole new generation of activists to the Irish cause. The blood of the martyrs had not been shed in vain.[63] By war's end it was obvious that Ireland would attain self-government, the only question now being how soon and on what conditions. In the December 1918 elections the Sinn Fein Republicans emerged as the strongest party by far; in January 1919 they convened the first Dail (parliament). Independence, however, was still not at hand and was to come only after three more years of bitter fighting.

Outstanding among the leaders of the Irish Volunteers (later the IRA) was Michael Collins, then in his late twenties. Intellectually a self-made man and a military amateur, Collins provided the very qualities which the Irish rebels had palpably lacked in the past, including an appreciation of the paramount need for strict organizational control and secrecy. As head of intelligence and later as chief of staff, he masterminded a strategy of revolutionary terror directed above all against the "G" (intelligence) branch of the British police in Ireland and their informers. Some hundred and twenty policemen were killed and almost two hundred injured in these attacks in 1919–1920; other victims included "collaborators" with the British such as judges and civil servants. In a venturesome raid in April 1919 Collins's men seized all the secret files of the Special Branch in Dublin. British intelligence, which Collins regarded as the single most dangerous enemy, was effectively paralyzed.[64]

Throughout 1920 the IRA continued its campaign of terror, its ambushes and raids against police barracks, the assassination of political enemies. It found it easier to cope with the fifty-thousand-odd British soldiers who had been concentrated in Ireland than with the Black and Tan volunteers and the Auxiliaries, many of them former British officers and NCOs. These anti-IRA irregulars, unlike the army, responded with a campaign of counterterror, militarily quite effective but politically counterproductive. Indiscriminate retaliation drove many waverers into the ranks of the IRA.

Outside the urban centers the IRA set up flying columns, but its members knew little about explosives and their use; they had, as a rule, only one week of collective training. Their task was to "inflict more casualties on an enemy force than those it would suffer."[65] But in the event, the IRA sustained more losses than the British in men killed (about six hundred) between January 1919 and July 1921, when a truce came into force. If, according to Collins, the effective strength of the IRA was three thousand, it had lost almost a quarter of its men, not counting those injured.

In Ulster, the IRA attacks provoked fighting along sectarian lines and a Protestant backlash. Thus, when the war ended, the IRA guerrillas, while appearing all-powerful to the general public, "were in fact almost at the end of their tether. Losses had been heavy and arms were running dangerously short."[66]

The truce did not satisfy the extremists, some of them regarding the acceptance of the Free State perpetuating the division of Ireland as an act of betrayal. In the civil war that followed, Michael Collins was to be assassinated as were many other leading figures, the wounds still remaining open to this day. The IRA was declared illegal in the Irish Free State and, although it continued to exist and to engage in small-scale operations, it was reduced to insignificance for many years to come. It made the headlines in 1939 when bombs were placed in London and Coventry. But since this new campaign coincided with the outbreak of World War II, even its publicity value was short-lived and it petered out in the spring of 1940.

During the war the IRA established contact with Nazi Germany through Sean Russell, its chief of staff, and Frank Ryan who had fought with the Republicans in the Spanish Civil War. A radio link between the German intelligence and the IRA was established; it did not, however, contribute much to the German war effort.[67] The circumstance is of interest only in view of the subsequent ideological development of the IRA, its official leadership in the 1960s veering sharply to the left.

During its first postwar heyday, from 1919 to 1922, the IRA was the military arm of a national movement. It was a genuinely popular little army — workmen of this and that kind, small farmers, shop assistants, employees. Their inspiration was fiercely nationalist and sectarian. It received financial help, as before and after, chiefly from the United States. And if it eventually achieved some of its objectives, it was not alone because the political constellation was auspicious, but also because it had had paralyzed enemy intelligence and created a general climate of lawlessness and fear. Its main weapon was not partisan warfare but individual assassination. It had few full-time soldiers; most of its members continued to pursue their regular civilian work, being mobilized on short notice for a few hours or a few days for special operations. Specific conditions in Ireland dictated the employment of terrorist methods rather than guerrilla warfare. It is doubtful whether IRA partisan bands roaming the countryside would have been able to hold out for very long; the cities, on the other hand, provided conveniently anonymous cover.

To combat a terrorist organization effectively, the British would have needed several more divisions. But after a long and costly war public opinion in Britain would not stand for this. Most people in Britain were sick and tired of the Irish troubles; some Englishmen would admit that the Irish had been wronged and those who had no guilt feelings thought that Britain would be better off without Ireland anyway; they had reaped nothing but ingratitude, insults, and endless murderous attacks. If the Irish preferred secession to belonging to a great Commonwealth, they should be given the opportunity to go their own way. This, in briefest outline, was the psychological and political background to the decision granting Ireland independence. The establishment of the Free State brought the immediate terror to a halt, but, as the years were to demonstrate, by no means eliminated its sources and failed to prevent a major revival in Ulster five decades later.

IMRO

Twentieth-century European guerrilla movements were usually separatist in character and, in view of the geographical dispersal of minorities, this frequently involved them in a three- or four-cornered conflict. It was one thing to appeal for a holy struggle against foreign rulers or invaders; it was far harder to come to terms with neighboring nationalities or minorities who did not share the same aspirations. Thus, the IRA in its clash with the British after World War I failed to establish its predominance in the north. And thus partisan activity in wartime Yugoslavia was hampered by the ethnic antagonism inside the country, and the Macedonian IMRO, which came into being in opposition to the Turks, was fighting Greeks and Serbs as well at one and the same time.

IMRO (the Internal Macedonian Revolutionary Organization) was founded in the 1890s when Christian villagers in Turkey, inspired by the example of the Serbs, Bulgars and Greeks, were roused into a striving for their own national independence. Because they were a small people and since their territorial claims conflicted not with one country alone, they early on elected to integrate into Bulgaria. Their struggle against the Turks began with isolated raids from across the border engineered by Macedonians who had already emigrated to Bulgaria. But the local population was something less than enthusiastic and it was only in later years, with the growth of the movement inside Macedonia itself, that guerrilla warfare became possible. IMRO's motto was "Freedom or Death," its banner a black flag bearing a crimson skull and crossbones. It aimed at a concerted national uprising, which took place on *Ilin Den* (St. Elias Day), 2 August 1903. About fifteen thousand Macedonian irregulars fought a total of forty thousand Turkish soldiers for seven weeks. More than a hundred Macedonian villages were completely destroyed in the course of this insurrection and five thousand Macedonians and Turks found their death. A frontal assault against the Turks was bound to fail, and if IMRO had hoped that the Bulgarians or the Russians would come to its assistance, it soon realized that it had been sadly mistaken. Its guerrilla tactics between 1896 and 1903 had been more effective; these had been small-scale operations, usually carried out by two or three volunteers so as to prevent Turkish punitive raids. The Macedonians had virtually established a state within a state, collecting taxes, even running their own "revolutionary postal service"; Turkish rule was confined to big towns such as Salonika — and only in daytime at that.[68] According to Macedonian sources, 4,375 Turkish soldiers had been killed in 132 skirmishes during the guerrilla war prior to *Ilin Den* — no doubt a grossly exaggerated estimate.[69] IMRO never quite recovered its influence in Thessaly and Turkish Macedonia after the defeat, which resulted in the migration of thousands of Macedonians to Bulgaria. Most of IMRO's subsequent operations took place in Bulgaria and Yugoslavia. The IMRO insurgents fought bravely, but their strategy had been at fault, along with its being based on a mistaken assessment of the international situation.

Initially a genuine movement of national liberation, IMRO degenerated within the next decade into a gang of hired assassins and a tool of foreign powers. With the transfer of its activities to Bulgaria it specialized in bank robberies, drug traffic and extortion, not necessarily for political purposes. It still, in its declarations, invoked the liberation of Macedonia as the ultimate aim, but as an outside observer noted in the late 1920s, a full account of its activities would be to compile a dossier "which would make American gangsterdom look insignificant."[70] Some of the elements of corruption had been present from the very beginning; patriotic robbery, smuggling and extreme cruelty had long been part of the tradition — to bury an opponent alive was by no means considered a particularly vicious way of expressing one's displeasure. IMRO had engaged from its earliest days in indiscriminate bomb throwing in Muslim bazaars and mosques. It was commonplace to kill rivals and enemies within its own ranks. Having no substantial funds and dependent on the outside for supplies, IMRO solicited, and received, both money and arms from Bulgaria, Austria (during World War I), and later from Fascist Italy; at one time there was also some cooperation with the Soviet Union. But there were usually strings attached; to get the Austrian subsidy, IMRO undertook military operations against its fellow Slavs, the Serbs, and this at the time when Serbia was fighting for its very survival. Bulgaria was to become IMRO's chief protector and paymaster; as a *quid pro quo* IMRO, in close cooperation with the Bulgarian police, set about the systematic liquidation of oppositionist politicians in Sofia and other Bulgarian cities.

IMRO was beset by deep splits; the Mikhailovist faction, in the pay of Italy, spent far more time and effort in killing the Protogerovists than in fighting for an independent Macedonia. Occasionally IMRO would stage raids into Yugoslavia from its bases in Bulgaria; according to the official IMRO version, its headquarters were in Yugoslavia, but its leaders resided in Sofia and it is not certain whether they ever set foot on Yugoslav soil. Whenever the Yugoslavs lodged a diplomatic protest following an IMRO raid from across the border, the Bulgarians would indignantly deny any such imputation; as far as they were concerned, IMRO was a partisan army based in Yugoslavia, just as in later years Fatah was officially located in Israel, not in Lebanon. Such total dependence on Bulgarian goodwill had its drawbacks, as IMRO discovered to its detriment when relations between Sofia and Belgrade improved in 1933. IMRO was no longer needed by the Bulgarians and in July 1934 Mikhailov, again a fugitive, crossed the border into Turkey looking for political asylum with the archenemy of his people. Since then the Macedonian issue has cropped up whenever Bulgaria's relations with Yugoslavia have been at a low ebb as, for instance, after Tito's defection from the Cominform. But there has been no revival of IMRO. The tragic history of the Macedonian Revolutionary Organization is that of a small people which, given the geographical facts of life and the balance of political power in the region, had no chance of attaining full national independence. In its attempts to gain support, IMRO became subservient to alien interests and ultimately it lost its political identity altogether.

ABD EL-KRIM

The biggest colonial war in the twentieth-century interwar period was fought in Spanish Morocco, an area which had seen much guerrilla fighting ever since Roman days. It was led by Abd el-Krim, the chief Qadi of the Melilla region, a Kabyle who despised the Spanish and hated the French. His slogan was, "The Rif is poor, we fight to make it rich." Abd el-Krim was an educated man; he had worked in the civil service, had acted as editor of the Arab supplement of the leading local Spanish newspaper and at one stage had served as the first professor of the Berber dialect.

In 1919 Abd el-Krim left Melilla and joined his father in his native mountain village in order to prepare the rebellion. By the spring of 1921, following a cold winter and poor harvest, he had concentrated a little army (*harka*) of about three thousand men.[71] Spanish forces under General Silvestre were ambushed and annihilated in July 1921, the native police and army auxiliaries mutinied and a general insurrection ensued. Within a few weeks the whole of eastern Morocco was in Abd el-Krim's hands. He could have taken Melilla, the capital, but his men were preoccupied with looting and he busied himself instead with the establishing of a Berber state, the Rif Emirate.

All in all the Spanish lost more than ten thousand men in the disasters of 1921, and it was to take them five years, countless military setbacks and domestic crises before, with the help of a hundred and sixty thousand French soldiers and an even larger army of their own, they were able to subdue the Berbers. Abd el-Krim soon acquired the reputation of an inspired guerrilla strategist. But in actual fact, "the disaster was due more to Spanish demoralization than to Berber prowess."[72] The Spanish officers facing Abd el-Krim (Franco, Sanjorjo, Mola, Queipo de Llano — all of civil war fame) had little guerrilla experience, and their army was in a state of advanced decay. General Weyler, the tough old soldier who had seen guerrilla action in Cuba many years before, bitterly attacked the inefficiency, cowardice and corruption which had come to light in Morocco. Lyautey and Pétain watched events from the sidelines in French Morocco with a mixture of concern and *Schadenfreude;* it was only toward the end of the Rif war that France and Spain decided to act in unison, the last thing Abd el-Krim would ever have credited.

Abd el-Krim showed a good grasp of the essentials of guerrilla warfare, such as mobility, but his tactics were not particularly sophisticated; his soldiers would launch sneak attacks against enemy outposts and, if these failed, they would wait until the enemy ran out of food and water.[73] Throughout most of the war the Riffi supply line to Tangiers was kept open. They received money from German, British and Dutch firms who were interested in mining concessions, they were assisted by some European military advisers, and their artillery (including some French 75 mm guns) was equal, if not superior, to the Spanish. Abd el-Krim was the first guerrilla leader in history with some aircraft of his own — though there is some doubt whether any of his planes actually ever took off. The whole conflict was deeply and increasingly unpopular in Spain where the *abandonistas* opposed any further major military effort. As it was, Abd el-Krim's assumption that time was working for him still almost came true. The Spanish government tried hard to find a face-saving formula in its negotiations with him. Had he not insisted on total independence and had he not attacked the French zone in 1925, it is doubtful whether the two European governments would have made common cause to dislodge him. In September 1925, in one last determined offensive, a Spanish force led by Lieutenant-Colonel Francisco Franco landed at Albucamas, in Abd el-Krim's rear, and by early 1926 it was obvious that his defeat could no longer be far distant. He surrendered to the French in May 1926 and was exiled to Réunion. Released in 1947, he died in 1963.

Looking back in later years, Abd el-Krim blamed his defeat on the *marabouts,* the Muslim preachers who, he claimed, had thwarted his plans for national unity. But Abd el-Krim had done little to counteract them while he was in a position to do so. He established a theocracy in which everyone was obliged to pray five times a day. His regime was tyrannical; his men committed atrocities not only against their European enemies but also against other tribes. With all his talents as a guerrilla leader, his great enterprise and energy, Abd el-Krim's ambitions grew beyond any reasonable hope of fulfillment; in effect, his downfall was caused by his own *hubris.* When Spanish Morocco became part of Morocco in 1957, Abd el-Krim was still alive in his Cairo exile. The year after, his own tribe, the Beni Urriaguel, revolted against the Moroccan government, but their rebellion was put down by Rabat with short shrift.

THE PALESTINE REBELLION

Between the two world wars British forces faced armed resistance in various parts of the empire. But these outbreaks of violence were either short-lived (Amritsar in 1919, the Moplar [Malabar] rebellion in 1921, Cyprus in 1931), militarily insignificant (the Burmese rebellion in 1930–1932), or nonviolent in character, such as Gandhi's Swaraj movement. A British authority noted that even in the Burmese insurrection, which lasted for eighteen months, a military police force should have been able to cope with the situation and would in all probability have nipped it in the bud.[74] The army had to be called in only because the police were not strong enough and lacked experience.

Palestine was the one major blot on an otherwise almost idyllic landscape. There had been two previous Arab outbreaks in Palestine, albeit on a minor scale, in 1919 and 1929, directed almost exclusively against the Jews. The third, far more extensive revolt was directed both against British Mandatory rule and the Jewish community. Following the rise of Hitler, Jewish immigration to Palestine had risen by 1935 to over sixty thousand. The number of Jews was still less than half that of the Arabs, nor was it true, as some Arab leaders asserted, that the Arabs had suffered economically from Jewish immigration. Arab resistance was political, or more precisely, national and religious in character; the fact that it was led by the Mufti (chief religious dignitary), Haj Amin el-Husseini, was perhaps not altogether accidental. The Arabs resented the steady influx of foreigners, who, they feared, would one day make them a minority in their own land and whom they in any case considered an undesirable element. Arab spokesmen accused the Zionists of Bolshevism in the 1920s; four decades later they were to be charged with Fascism. The Zionists had not changed, but intellectual fashions certainly had. That Palestine's neighbors had attained independence, or were about to gain it, acted as a spur to the Palestinian Arabs.[75]

The insurrection began with a general strike and some sporadic acts of violence. It had been preceded by increased brigandage, some of it political in nature. The band of Sheikh Izzed Din Kassem, pursued by the French, had infiltrated northern Palestine; Kassem was a religious leader who had apparently taken to brigandage for patriotic reasons, and became a national hero. Shot in a clash with the Palestine police, his funeral in Haifa turned into a great national occasion. Something of glamour had for long attached to those indulging in brigandage; Abu Gilda's exploits in the 1920s have remained proverbial to this day. Not that the heroically selfless reputation of these brigands was always warranted. What is worthy of remark is that some of them — Abu Durra, Aref Abdul Razek, for instance — took a leading part in the rebellion.[76]

Its first phase witnessed small-scale attacks directed chiefly against the country's rail and road network. The British Mandatory administration lost control over Palestine's hilly regions (Galilee, Samaria, and part of Judaea) and this although the guerrillas numbered no more than five thousand at the time. But the police, largely composed of Arabs, could not be trusted, and anyway had no orders to intervene. There were only a few British army units in the country; military command was, in fact, in the hands of the Royal Air Force, and there was no officer of general rank. The civil administration dragged its feet for about a year without taking any drastic action. The Jews, with a few exceptions, did not engage in counterterror but limited themselves to purely passive resistance.

As the months went by with no sign of the rebellion abating, the British government looked for a political solution to the crisis; Jewish immigration was to be drastically restricted and other measures introduced to allay the fears of the Arabs. But these conciliatory steps did not go far enough to placate the Arabs and in November 1937 the rebellion entered a new and more dangerous phase. By that time the rebel bands numbered some fifteen thousand members, supplemented by a still larger host of villagers mobilized as required for special undertakings. The rebel units were mainly concentrated in the north but some operated in the Mount Carmel region, in Samaria and Judaea. The largest unit was commanded by Fawzi Kaukji, also a fugitive from Syria. The rebel high command was in Damascus, but there was in actual fact little, if any, coordination between the bands. The Damascus leaders helped with money and arms, some of which came from Fascist Italy. On the whole the rebels were skimpily equipped; they had no artillery, no heavy machine guns, no motorized transport. Later it appeared that British and Jewish accounts about the quantity and quality of Arab equipment had been considerably exaggerated; the standard weapon of the rebels was the old (World War I) Turkish rifle and they used bombs of a primitive kind.[77] The insurgents would focus their attacks with some effect against highroads, and railway and small Jewish settlements; they would refrain from clashes with the army or police strongpoints ("Tegart fortresses") in Arab territory. Part of the supplies and the money needed by the rebels was collected through taxes imposed on the not always willing villagers. The gang leaders and their followers engaged frequently in settling old personal accounts and tribal feuds. More Arabs were killed at the hands of Arabs than British and Jews put together.

During the summer of 1938 the rebellion spread from the hilly regions throughout the country. "By October 1938 a large part of Palestine was physically under the control of the rebels, and almost the entire Arab population was either giving active support to, or was dominated by fear of, the rebels."[78] As a result, substantial British army units were dispatched to the country and in October of that year the army was officially made responsible for the maintenance of public order. It lacked counterguerrilla experience altogether, but a number of elementary measures were sufficient to break the back of the revolt within three months. These included the imposition of curfews, traffic restrictions, occasional *razzias* and the building of roads into rebel territory. In 1938, the worst year of the rebellion, approximately two thousand Arabs were killed, as well as three hundred Jews and seventy British. (The official figure for Arab casualties, sixteen hundred, was for once almost certainly too low.) Isolated attacks continued throughout the spring of 1939, but when World War II broke out, the rebellion had already petered out — partly as the result of substantial political concessions made by the British government, but mainly in view of the military defeats and dispersing of the bands which were not strong enough to fight regular army units and not agile enough to evade them. Futhermore, the Jews, too, had gone over to active defense in the later stages of the rebellion, and as the tide turned against the bands they found far less support in the Arab villages. They no longer readily obtained supplies and they could not take it for granted that their whereabouts would not be betrayed to the authorities.

The Palestine rebellion was not, as is sometimes claimed, a peasant uprising, even though most of the guerrillas themselves were villagers. The political struggle preceding the revolt had been the work of the urban upper and middle classes and the intelligentsia, but these disappeared from view once armed struggle broke out. With one exception (Abd el Kader Husseini), the leading Palestinian families were not actively represented in the guerrilla movement; many of them moved to Egypt or Lebanon during the "riots" — as these were called locally throughout their three years' duration. The military chiefs were all "lower class"; one of the most respected among them, "Abu Khaled," had been a stevedore at Haifa harbor.[79] But even though it was a popular movement, it was by no means radical by modern standards; it lacked a social program, there was no demand for the redistribution of land, and the general inspiration was nationalist-religious-fundamentalist in the narrow sense. In other words, the aim was to fight foreigners and infidels. Militarily, the guerrillas chose by instinct the correct tactics. They did not try to establish liberated zones, which they would not have been able to hold, but engaged instead in hit-and-run attacks. But they had little military training, there was no overall strategy, no coordination, no outstanding leadership. The country was too small and the bands too exposed for successful partisan warfare. When the revolt was finally put down, it transpired that the guerrillas had been unable to overrun even the smallest Jewish settlement, and this despite the lack of military experience and weapons among the Jews.

Mention has been made of the support, both propagandistic and financial, given to the insurgents by Axis powers; some leaders of the rebellion, including the Mufti of Jerusalem himself, settled in Germany during the war. But the same was true of nationalist rebels from other parts of the globe, such as, for instance, Subhas Chandra Bose. It would be mistaken to exaggerate the significance of such collaboration with the Axis countries. Nazi Germany and Fascist Italy were the natural allies for these rebels because they were "anti-imperialist," meaning anti-British, and because it was widely believed that the days of the Western democracies were numbered. Hitler and Mussolini were popular figures among nationalist rebels not alone in Palestine, just as, after the defeat of the Axis, Stalin and Mao were to have so much appeal. Nationalist movements were primarily concerned with their own cause; whether they turned "right" or "left" depended on the general political constellation. Authoritarian regimes had of course greater attraction as models than the democracies because, their "anti-imperialism" quite apart, they seemed far more dynamic and purposeful.

LATIN AMERICA

The most colorful incidents of guerrilla warfare at a time when small wars seemed to have gone out of fashion took place in Latin America. Among them were Pancho Villa's and Emiliano Zapata's operations in Mexico, Luis Carlos Prestes's "long march" in Brazil, and the Sandino rebellion in Nicaragua. Of these movements, only the last was guerrilla in the strict sense of the word, although it bears repeating that to apply purist standards with regard to guerrilla warfare is as misleading as the indiscriminate use of the term in general.

Porfirio Díaz, who had started his career as a partisan leader in the struggle against Emperor Maximilian and who had subsequently ruled Mexico for thirty-five years, was overthrown in 1911. A decade of civil war and anarchy ensued and it took another decade until central state power reasserted itself. In 1911, too, the Manchu dynasty was overthrown in China with similar results. But whereas China for the next twenty years was ruled by warlords, the Mexican situation was different inasmuch as there were more horse thieves in Mexico at the time than soldiers, which made for a warlordism of another sort. (The effective strength of the Mexican army in 1911 was eighteen thousand troops, quite insufficient to keep order in Mexico's many provinces.) Most of the Mexican *caudillos* to emerge in the interregnum were not military men by training but local chiefs who imposed their leadership by force of personality.

To review the main developments of these years, the ever-changing alliances and frequent betrayals, the campaigns and the intrigues, or even simply to list the names of the main protagonists, would be to write the history of that chaotic decade. Zapata and Pancho Villa, the two most important guerrilla leaders, had their bases in the south and north respectively; Zapata's "Liberation Army of the South" in his native Morelos, Puebla and Guerrero, Villa's "Division of the North" in Chihuahua and Durango. Villa had been a popular bandit, his politics, in as far as they went, vaguely populist. He was a local hero, a crude and frequently cruel man, brave, a patriot, and in his way a radical. Zapata, a peasant leader, thirty years old at the time of the revolution, made his name in the struggle against the *hacendados* who had illegally acquired land belonging to small farmers. "The land free for all, land without overseers and masters, this is the war cry of the revolution." He sponsored an agrarian reform program (the "Plan of Ayala") that was subsequently adopted in its essentials by his rivals, and he was also the author of several memorable phrases such as "Men of the South, it is better to die on your feet than live on your knees," and "Seek justice from tyrannical governments not with your hat in your hands but with a rifle in your fist."[80] However, the struggle in Mexico was complex, it was not a clear-cut confrontation between the forces of reaction and the party of the revolution. Once the Díaz regime had been overthrown and Victoriano Huerta had been exiled, all the chief protagonists in the conflict were men of the left, or in any case left of center. A good many vested interests were involved in the struggle for power and it was not always readily obvious who were the most consistent and radical revolutionaries. Zapata distrusted, not unjustly, the urban leaders and civil servants who, he suspected, would sabotage, or at least water down, agrarian reform. But Zapata's urban critics claimed, again not quite unfairly, that Carranza, Zapata's major foe, was also committed to agrarian reform. In contrast to Zapata, Huerta had the support of sections of the urban working class. Furthermore, the acts of brigandage committed by Zapata's men made orderly agrarian reform very difficult indeed. Pancho Villa's interest in politics was minimal and capricious; he hated foreigners, especially North Americans and Chinese, he fought for the government and against it, he entered an alliance with Zapata which never really worked, and when a leading Zapatista writer published an article critical of Villa, he had him shot by way of rebuttal.

Pancho Villa was the more spectacular guerrilla fighter. Within six months his little army swelled from eight to eighty thousand; this figure included the raiders' women, who frequently participated in the fighting. Villa's "Division of the North" defeated the government forces in several battles at Torreon in the summer of 1914. It was the most important achievement of his military career; in later years he was to put even larger armies into the field, and seized (and lost) countless cities, but he could never hold his gains for any length of time. He was in substance an audacious buccaneer and master of the ambush and hit-and-run attack who vainly sought to excel as a regular army general. His successes in open field battles were largely thanks to the advice he received from Felipe Angeles, a French-trained general who was his artillery commander. Villa was at his most potent when he could play his old guerrilla game; he eluded General Pershing's expeditionary force which had been sent to Mexico to punish Villa for raiding Columbus, New Mexico, and murdering American civilians. When he chose to fight an able Mexican general such as Obregón at Celaya in 1915, he suffered heavy losses and this despite numerical superiority. The Villistas were better equipped than the Zapatistas; Villa usually did not lack money and he liberally nationalized (and resold) horses and cattle wherever he went. But for all his astonishing tenacity and ability to reassemble new bands after each defeat, he never quite recovered his strength after 1915. After finally making peace with the government in June 1920, he was given an estate of twenty-five thousand acres, and his seven hundred followers were also offered land, a time-honored Latin American manner of settling a dispute in the case of a draw.

Zapata, a *mestizo* like Villa, began his career as the head of the defense committee in his native village, and emerged during the last year of Díaz's rule as the supreme revolutionary chief in the state of Morelos. His army may have been the poorest in the Mexican civil war, suffering from a chronic lack of money, arms, ammunition and supplies, yet it was also the one most adept at employing guerrilla tactics. Whenever the government forces attacked in strength, as in 1913 and again in 1916, the Zapatistas just melted away in small groups to reassemble in neighboring districts. The government troops would seize the towns and major villages, only to withdraw after a short while on account of severe casualties from malaria, dysentery and, of course, innumerable small ambushes. The government could mobilize an army of forty thousand against Morelos, but it could not permanently station them there. Unlike Pancho Villa, Zapata hardly ever concentrated his troops and was reluctant to fight in open battle. Only once his modest army had greatly expanded did he besiege and occupy towns such as Cuernavaca, Puebla and eventually Mexico City. At the height of his power in 1915, when Zapata withdrew from Mexico City, he had (nominally) some seventy thousand men under his command. A year later their number had dwindled to five thousand. There was much revolutionary enthusiasm, but discipline was lax, officers were unreliable and unpunctual; if the government forces committed horrible excesses in their pacification campaigns, the Zapatistas also burned, raped, plundered and killed civilians and prisoners. From time to time Zapata would express regret about these abuses, but he knew that he could not really restrain his followers. His army was not a centralized body, but consisted of units of several dozen or several hundred men, acting most of the time independently. The composition of these units would constantly change, for the guerrillas would be released to work their fields during the agricultural seasons.[81]

Eventually, the central government reasserted itself and the bands grew weaker. By 1920 the guerrilla war came to an end; the year before, Zapata had been lured into an ambush and assassinated. Felipe Angeles was executed a few months later and Carranza was shot in 1920. Pancho Villa was murdered by private avengers three years after he had retired to his large ranch, and Obregón was killed by a religious fanatic posing as an artist who wanted to draw him.

The Mexican revolution, like others, devoured its children; among the few to escape unscathed was Adolfo de la Huerta who became an opera singer in his North American exile. But the revolution itself was not abortive; there was no return to a Porfirian dictatorship; agrarian reforms were carried out and, in fact, gathered additional momentum in the 1930s. Granted, the guerrillas, whether of the south or the north, could not provide any political leadership for the country; their resistance movements were regional, entirely wanting in organizational ability and the necessary minimum of political sophistication. Villa, for all his populist-radical slogans was, after all, only a bandit-cum-*caudillo,* and the Zapatistas had little support in the towns and could not transform themselves into a broad, national movement. Zapata led his *peons* through the desert, but like a more monumental leader, did not live to witness their arrival in the promised land.

THE PRESTES COLUMN

The military coup which occurred in São Paulo in July 1924 seemed at the time no more than just another coup of which Latin America has seen so many. It collapsed after a few weeks and would hardly be remembered today but for the initiative of a twenty-six-year-old captain, Luis Carlos Prestes, who decided to move with a column into the interior of the country. There he hoped to continue his struggle, shake the country out of its apathy and perhaps trigger off an eventual general insurrection. The attempt failed, but not before the *Coluna Prestes,* made up of about a thousand men, had covered some sixteen thousand miles in a giant raid unprecedented in military history, traversing Brazil from north to south, from east to west (and vice versa), while fighting government troops. When Prestes with six hundred and twenty of his men crossed into Bolivia in February 1927, he was still undefeated. His mounted column (it reportedly used as many as a hundred thousand horses during the campaign) originally consisted of regular army officers and soldiers, but about half of them were killed, wounded or fell ill and were replaced by volunteers. The column thwarted innumerable attempts by government troops to surround and capture it and had to fight, moreover, the *Cangaceiros* of the north, bandit groups which had been given official status as counterguerrilla units by the central government; it found these enemies far more dangerous than the government troops which showed little fighting spirit. The *Coluna Prestes* with its vaguely revolutionary watchwords had the passive support of the populace, but the general insurrection it had hoped for simply did not get off the ground. It was a heroic episode without political effect and all that remained was the folk myth of the *Cavaleiro da Esperanca,* the Knight of Hope and his companions, a symbol of the struggle for a new and better Brazil.

The ideological makeup of the *coluna* presented a picture almost as curious as a map of its raids; it was a mixture of revolutionary nationalism compounded of both extreme left-wing and rightist philosophies. In terms of this, the future paths chosen by its leaders are peculiarly enlightening. Several of them took part in the Vargas coup in 1930 and in this roundabout way rejoined the political-military establishment to become in due course generals and ministers. Prestes, on the other hand, turned to the Brazilian Communist party, serving as its secretary general. But the very man who had shown such exceptional skill as a guerrilla leader became the left's chief opponent of guerrillaism when in the 1950s and 1960s it had a revival.[82] Brazil in the 1960s was in many ways an altogether different country, with great conurbations, a modern industry and a growing working class; a rebellion in the backlands must have appeared even less promising than forty years earlier. But this alone may not be sufficient to explain Prestes's disenchantment with guerrilla tactics; it also reflected the general Communist aversion to this kind of warfare, a subject to which we shall have to return in a separate context.[83]

SANDINO

Whereas Prestes's long march was the unexpected sequel to a traditional Latin American military coup, Augusto César Sandino's guerrilla movement in Nicaragua grew out of an equally traditional civil war, a confrontation between conservatives and liberals. Sandino, a *mestizo* of upper-class origin who had spent some years in Mexico and was strongly influenced by the heady wine of Mexican revolutionism, became after his return the leader of an armed band within the liberal camp. When in the summer of 1927 a compromise was reached between the two sides which gave the liberals most of what they had asked for, they laid down their arms. Only Sandino, the most radical leader among them, declared his intention of continuing the struggle until a truly democratic regime was installed and the American marines, which had intervened in Nicaraguan politics on and off since 1909, were once and for all withdrawn.[84]

Nicaragua, a sparsely populated country with mountains, many forests and few roads, was in many ways admirably suited for guerrilla warfare. The social structure was conservative-traditional and a populist leader was bound to gain the sympathy of the poor villagers. Sandino's campaign lasted for five years, and with a force never exceeding a thousand men he imposed his rule over large sections of the country, establishing to all intents and purposes a countergovernment, even levying its own taxes. He adopted guerrilla warfare only by trial and error; in the beginning there were tactical mistakes for which his force had to pay dearly. But he quickly mastered the guerrilla approach, and once he had done so, all attempts to destroy his scratch little army came to nothing. The government force, the *Guardia Nacional*, was small and ineffective and the U.S. Marines, a few thousand at most, forever failed to catch up with Sandino in the impassable forest. Attempts to bomb him from the air were no more successful. Sandino made peace with the government in 1933 after the last marines had been evacuated and once the liberals had again come to power. He was assassinated by political enemies within the year.

Sandino (*El Liberador*) became, like Prestes, a legend in his own lifetime; his operations were closely studied by later generations of Latin American guerrilla leaders. Not that he was without blemish as such; there was the usual high incidence of professional murderers and robbers in his ranks, the familiar atrocities. Yet service was well rewarded — every sergeant was made a general, or at least a colonel — and there was great emphasis on military pomp. Sandino's social radicalism, while shocking in the eyes of his contemporaries, was exceedingly mild in retrospect; he was not a socialist, just a radical *caudillo* with a populist program. True, hardly ever before in the history of Latin American guerrillaism had the anti-American element been so pronounced. As Castro attracted the Argentinian Guevara, so Sandino was joined by radical militants from Honduras and Guatemala. Lastly, Sandino looked for, and found, some support among the native Indians, who had hitherto been neglected by both the political establishment and opposition alike. But if in its immediate effect the Sandino revolt did not entirely fail, it did in its longer-range purpose; under the dictatorship of the Somoza dynasty Nicaragua became politically one of the most backward countries in Latin America with a small clan monopolizing political and economic power to the detriment of every other section of its society.

THE SPANISH CIVIL WAR

When the Spanish Civil War erupted in July 1936, it was widely expected that large-scale partisan warfare would be one of its chief characteristics both in view of the specific character of the war and because Spain was the country with the richest guerrilla tradition of all. When the war ended, it was clearly apparent that the assumption had been altogether mistaken and there was much mutual recrimination bearing on who carried the blame for this sin of omission. The Communist leader Enrique Lister (Manuel in Malraux's *L'Espoir*) wrote in 1965 that it was the fault of the indecisive Republican government not to have organized a powerful guerrilla movement in Franco's rear. The Anarcho-Marxist Abraham Guillen, on the other hand, who had fought in the civil war to become later an ideological mentor of the Tupamaros, wrote in 1969 that the Russians were the guilty party because they had always pressed for frontal attack.[85] Both Lister and Guillen were right, but neither version provides a full explanation. The Spanish Republican government of the day gave guerrilla warfare low priority because the main task facing it was to create a regular army as a defense against Franco's troops. Since most regular army officers were anti-Republican, this was a formidable undertaking. The problem facing Madrid was very similar to that confronting Lenin and Trotsky in 1918: not to give additional encouragement to the irregulars of whom there were too many anyway, but to weld them into a regular army and to create a central command. The Russian advisers in Spain did indeed press for "confrontation," in accord with their military doctrine, but they also established schools for guerrilla specialists on Spanish soil. These institutions were run by the Soviet secret police who trained their students for acts of sabotage to be carried out by individuals or small units — but *not* for guerrilla warfare. These operatives, more often than not, would be foreigners, figures like Kashkin or Jordan in Hemingway's *For Whom the Bell Tolls,* who could not possibly have played the role of a Mina or an Empecinado among the Spanish peasants. Over a thousand men were trained in the six guerrilla schools and eventually a special unit, the 14th Guerrilla Corps, was established.[86] But all this refers to missions of very short duration, to "diversionist acts," not to the organization of guerrilla units. Robert Jordan's three-day mission to blow up a bridge (described in Hemingway's novel) was quite typical.

Guerrilla units in the truer sense did come sporadically into being in the winter of 1936–1937 in various parts of Spain, particularly in the center and the north.[87] These were usually small bands which did not last long and whose activities left no great imprint. The war was decided in the battles for Madrid, at Guadalajara and the Ebro; partisan warfare made no difference to its course. Even those Spanish leaders who were most predisposed toward partisan warfare — anarchists like Durutti — fought at the front in Barcelona and Madrid; there were not enough soldiers to spare for partisan operations. Even a born guerrilla leader such as the Communist El Campesino was appointed a division commander, in which capacity he showed much less aptitude.

Given the long-standing Spanish propensity for guerrilla warfare, the political support of substantial sections of the population for the Republican cause and the inefficiency revealed by the right-wing regular army commanders, would not the Republicans have been better advised to put stronger emphasis on guerrilla warfare? Even with the benefit of hindsight the question cannot be answered in the affirmative. True enough, the Nationalist army (as Stanley Payne has written) never became a first-rate twentieth-century military machine. "It won because it proved less ineffective than the motley contingents of the Popular Front."[88] But however incompetent, the Nationalists would still have been able to seize the major cities within a short time but for the Republican forces concentrated in their defense. Precisely because they were not really a modern army, the Nationalists were not that dependent on supplies, and damage to their lines of supply and communication would not have been fatally harmful. Partisan units could have been concentrated in the mountainous regions of central and northern Spain, but the military decisions fell in the plains. The presence of active, strong and highly mobile guerrilla units in Franco's rear might have made a difference in the battle for the Basque country; it is most unlikely that they would have influenced the outcome of the fighting in the south.

The Nationalists, too, had many irregulars in their ranks; during the first year of the war regular army units were in a minority. There were many *banderas* — Carlists, Falangists, and other right-wing volunteers, some of them counting a few hundred members, others, such as the *Tercio de Navarra,* ten thousand or more. But they were gradually integrated into Franco's army and there was never any systematic attempt on the part of the Nationalists to wage guerrilla wars in the Republican rear.

GUERRILLA WAR AND THE REGULAR ARMIES

On the eve of World War II, the attitude toward guerrilla warfare that its advocates had noted four decades earlier still held true: the various army general staffs had no interest in it, and the military academies saw no reason to include courses on it in their curricula. Individual officers had gained guerrilla knowledge in the interwar period. Major Wingate's experiences in the Palestine rebellion helped him in the jungle of Burma, doubtful though it is whether General Patton, the tank commander, drew on the inspiration acquired by Lieutenant George Patton in the raid against Pancho Villa. By and large the feeling prevailed that guerrilla warfare, half-brigandage, half-political in genesis, was a messy business best left outside the confines of regular armies and their commanders. One German writer's view was that it could endanger the country's war effort by very reason of its methods being so diametrically opposed to the German way of waging battle.[89] But Arthur Ehrhardt was almost the only German author in the interwar period to concern himself with the prospects of guerrilla warfare in modern conditions. He pointed out that aircraft and motorized columns would make for armies being able to advance far more rapidly than ever before. But this meant that their supply lines would be much more extended and that the advancing units would be infinitely more dependent on supplies, above all of fuel, ammunition and spare parts. Long and vulnerable supply lines would be an obvious target for enemy partisans.[90] Ehrhardt also calculated that the average modern airplane was much too fast to be of help in combating guerrillas and that special aircraft would be needed for this purpose. He envisaged the possibility of enemy partisans landing in the German rear, and of motorized guerrilla units. He even weighed the potential use of chemical warfare by guerrillas, or in the fight against them, but dismissed this as impractical. These, however, were only the views of an outsider, the German military command remained uninterested; among the hundreds of books and the thousands of articles on military topics published in the 1920s or 1930s one looks in vain for any serious discussion of guerrilla warfare. There was some logic in Germany's neglect since she was prepared (and preparing) for a *Blitzkrieg,* and in a war of this kind, if successful, guerrillas could not possibly play a part of any significance. There was less logic in French, British and American ignoring of the subject, none of them having that faith in a *Blitzkrieg.* Yet they nevertheless equally shared the German conviction that guerrilla warfare was unimportant. True, all the British army was asked to prepare itself for in the 1920s was small wars only, for which it was generally assumed that no special provision was required since these would surely be, *grosso modo,* on the pattern of previous colonial wars.[91]

Warnings about impending changes in the character of guerrilla war were so infrequent that they deserve to be singled out. Thus Major B. C. Dening in an essay on "Modern Problems of Guerrilla Warfare," published in 1927, pointed with astonishing foresight to three important contemporary processes favoring the guerrillas. The first, and the most important, was that in view of the increasing influence of public opinion at home, the Great Powers could no longer act with the same "ferocity" as on past occasions. "Otherwise such an outcry would arise as would be certain to bring about either the fall of the government responsible or the intervention of an interested outside power."[92] Secondly, the development of modern weapons favored the guerrilla more than those operating against them. These weapons could be readily concealed and lent themselves to the first principle of guerrilla warfare — rapid concentration and equally rapid dispersion. Last, there was the difficult problem of combating guerrillas in thickly populated areas. "Here the guerrillas have opportunities to make propaganda, to destroy property and to deliver attacks with great ease. The task of the army becomes essentially a police task."[93] Major Dening also suggested that it was quite likely that guerrillas would in future try draining the financial rather than the military resources of a great power, as the Cuban rebels had done with considerable success in 1898. But such predictions, to repeat once again, were the exception, not the rule in the interwar period.

There was an upsurge of nationalist and revolutionary movements in Asia and Africa in the 1920s and 1930s, but their activities were largely political. British and French, Belgian and Dutch colonial administrators were not unmindful of these activities, and from the time of the famous Baku Congress on, there was a tendency in the European capitals to attribute most colonial rebellions to Soviet propaganda and intrigues. Quite frequently it was argued that these operations were carefully prepared and coordinated in Moscow. This was almost certainly untrue at the time or in any event exaggerated; the Comintern supported existing colonial insurgencies, but such support was usually quite limited and often as not refused. The Germans, who had lost their colonies after the war, followed the anti-imperialist struggle with great interest; the geopoliticians, with General Professor Karl Haushofer at their head, were quicker than most others to realize the potential importance of coming national liberation wars. It was another former German general, Max Hoffmann, who predicted that the explosive mixture of Communism and nationalism would result in protracted colonial wars, different in character from those in the past, which the British and the French could not win.[94] Even if the colonial forces were to defeat the enemy, there would be a subacute revolutionary situation which would make it impossible for the British to withdraw their units. This constant combat readiness would wear out the colonial troops, there would be no clear and distinct enemy to combat and, furthermore, it would be impossible to employ native troops. Even if Britain were able to crush an insurgency in Egypt, its position would become untenable should there be simultaneous risings in Bengal, in Iraq and elsewhere — the power and the resources of Britain and France would sooner or later be exhausted in these unending colonial wars.

Such predictions were rejected as too pessimistic at the time. In the 1930s the Communists favored popular or national front policies that excluded the armed struggle. The radical nationalists had no such inhibitions, but the time was not yet ripe for mass campaigns against foreign rulers or domestic enemies, nor was it certain that the guerrilla approach would be the most effective weapon in any such struggles. The development of modern military techniques had seemingly made regular armies well-nigh irresistible. Guerrilla movements could hope to challenge regular armies only in certain exceptional conditions which in the 1930s did not exist. World War II was the turning point in this respect; Europe's (and Japan's) decline was, to paraphrase Wolfe Tone, the guerrillas' opportunity. World War II caused the collapse of the colonial powers, it undermined the confidence of the European ruling classes, led to deep economic and political unrest and created revolutionary situations the world over. In these conditions, following a major shift in the global balance of power, it was again possible for a few determined people to find support for an all-out assault against the established powers by other than political means. Once it had been demonstrated that guerrilla war worked, the example was bound to be emulated.

5

The Twentieth Century (II): Partisans against Hitler

Resistance in the German-occupied countries ranged from a refusal to read Nazi newspapers to organized armed struggle. Resistance fighters gathered and transmitted military intelligence to the Allied commands, printed and disseminated anti-Nazi newspapers and leaflets, sabotaged the German economic war effort. Partisan warfare aimed above all at disrupting lines of supply and communications, at creating a general climate of insecurity, compelling the Axis powers to divert some of their forces from the main theaters of war. The story of the resistance movement in the various countries, its achievements and many setbacks has been amply documented; resistance organization, political views, activities and way of life have been studied in the greatest detail. Nevertheless, the effectiveness of partisan warfare is still in dispute. The claims made by the partisans as to the damage and the losses inflicted on the enemy are often ridiculously high.

There is a tendency in every war to magnify one's successes in the heat of battle; even in so notable a one as the Battle of Britain the official British announcements on enemy aircraft shot down were proved after the war to have been greatly exaggerated. But when it comes to guerrilla warfare, there seems to be, as remarked long ago by Denis Davydov and other outstanding guerrilla leaders of the past, an almost built-in temptation to overstate, and here the possibility for verification is usually all but nonexistent. Rumors or wild estimates are repeated and passed on so often that they eventually enter the history books as the established truth. According to General Bor Komorowski, for instance, the *Armia Krajowa* had four hundred thousand sworn-in members in 1944. There is no reason to doubt this statement; for all one knows, the number of its sympathizers may have been even greater. But judging by all available evidence, only a very small proportion of them were in physical partisan actions at any one given time during the war (one percent in 1943, perhaps five to ten percent in 1944). Soviet partisans in the Orel region claimed after the war to have killed 147,835 Germans.[1] But Western sources give the total number of Axis soldiers killed by partisans in Russia during the war as only about thirty-five thousand; and of these not more than half were Germans. According to the British official history of the war, the Greek Communist guerrillas wounded and captured five thousand Germans in October 1944. Brigadier C. M. Woodhouse, deputy head of the British military mission to Greece at the time, called these figures absurdly inflated. "I myself never saw more than two or three [Germans]. The fact is that the guerrillas' claims were simply copied out from hand to hand, without any attempt to evaluate them, until ten years afterwards they had become part of the official history of the war."[2] The published figures of French Maquis membership are substantial, but to this must be added that the great majority joined only during the last few weeks (or days) before the liberation.

If Allied partisan claims cannot be trusted, German internal reports, if for different reasons, also bloated the strength of these partisans and the importance of their activities. A local German commander would deliberately exaggerate the number of partisans operating in his vicinity, either because of deficient intelligence reports or in the hope of getting reinforcements for the next counterguerrilla operation. Following such operations, the Germans would magnify the extent of their successes. Of the total number of "bandits" killed, wounded or taken prisoner, the bulk would more likely than not be peasants who had no connection whatsoever with the partisans.[3] There are, in short, few reliable facts and figures. Internal accounts of the German Army Railway Command 4 record two thousand cases of railway lines being mined or trains attacked in one sector of the Eastern front during 1943. This is an impressive figure, except that on breaking it down further, one finds that only ninety-four servicemen and railway personnel were killed in these many incidents and only in fifty-two cases were the lines closed for longer than twenty-four hours.[4] An overall assessment of the cost effectiveness of these attacks would also have to take into account the resources invested in producing the mines, in transporting them from the Russian rear to the partisans, the man-hours spent and the losses incurred during the attacks themselves. But such calculations are virtually impossible.

Mining railway lines was the most important activity of Soviet partisans; after the war they declared there had been five hundred thousand such operations whereas the Germans were aware of only a hundred and fifty to two hundred thousand. Partisan avowals were of fifteen thousand locomotives destroyed throughout Europe; again, the real figure seems to have been considerably lower, for Germany had more locomotives at the end of the war than at its beginning. A locomotive which had been derailed might be reported to headquarters by five or six different partisan units;* besides, the partisans had no way of establishing whether a locomotive had been only slightly damaged or permanently disabled. Whether successful or not, these actions undoubtedly involved a great deal of courage and many partisans paid for it with their lives. They merit our admiration, but the historians' assignment is not that of the hagiographer and he cannot uncritically accept their claims.

According to Yugoslav sources, Tito's partisans fought in early 1942 against 616,000 Fascist soldiers; later that year their number is said to have risen to 830,000. Altogether, 450,000 Fascist officers and men were reputed to have lost their lives fighting the partisans. But the total number of troops under the German commander in chief "Southeast" (Balkans) never exceeded 467,000. Many of the men under his command did not fight in Yugoslavia at all, and most of them were not killed.[5] Mikhailovich's *Chetniks* reported military operations against Germans whereas in reality they had concluded a *de facto* armistice with them.

The probable truth is that the political impact of partisan activity was far greater than its military contribution. It helped restore the self-respect of a defeated nation and gave new impetus to the spirit of defiance: *sanguis martyrorum, semen ecclesiae.* It was in any event restricted in its scope, actual guerrilla war being confined almost exclusively to parts of East and Southeast Europe. In Denmark, Holland and Belgium there could be strikes, acts of sabotage and of individual terror, but partisan warfare was ruled out. Toward the end of the war Maquis congregated in certain areas of France (such as the Massif Central) and in upper Italy, but their military

* This point is made by a recent Soviet source in a footnote to the statistics issued by the partisan general staff, i.e., 1.5 m. enemies killed, wounded or taken prisoner, 4,000 tanks and 16,000 locomotives destroyed, etc. *Bolshaya Sovietskaya Entsiklopedia* (1975), vol. 19, 234.

operations were of no great consequence. The Slovak uprising in the latter days was not guerrilla in character, which, incidentally, may have been one of the reasons for its failure. Partisan units existed in Greece but they spent more time fighting each other than the German invaders.

The two major theaters of guerrilla war were Russia and Yugoslavia, but again, if chiefly for geographical reasons, it was contained within certain areas; in Russia the partisan movement was strongest in the central sector, in Yugoslavia it was at its most active in Bosnia, Montenegro and parts of Serbia. Within these regions, the partisans' hold on the countryside was virtually unchallenged, German rule being limited to the towns and the main lines of communication.

In Yugoslavia, as in Russia, the partisan units were still weak in 1941, they gathered strength during 1942, and by late 1943 reached the height of their power — following Italy's collapse and the German defeats on the Eastern front. This gradual development of partisan ascendancy was perhaps only natural, but it also points up the tenuous effect the partisan movement had on the course of the war; the partisans were at their weakest when they were needed most and might have rendered the greatest service — before Stalingrad. They could cause serious harassment to the Germans only after the tide of the war had already turned.

In many European countries the Communists were the leading force in the armed resistance. They were the only political party organizationally prepared to operate in conditions of illegality. To be a Communist involved an unqualified measure of discipline and commitment, it meant not merely paying one's dues, it meant fighting for the cause and, if necessary, giving one's life for it. Opposition to Nazism was widespread in all social classes and most political parties, but it was almost solely the Communists who had the machinery to channel resistance into armed struggle.

Relations between Communist and non-Communist partisan units were always strained. Non-Communists were usually suspicious of Communist intentions and reluctant to cooperate with them; the Communists tried wherever they could to isolate and destroy their rivals. From time to time, under the pressure of the Allied powers, Communists and non-Communists would collaborate on some specific operations, but sooner or later hostilities would again break out between the two sides. This applies above all to Greece, Albania and Yugoslavia. In Poland the Communists were in lame shape; their party had been dissolved following the execution of most of its leaders in the Moscow purges. It was re-established in 1942 but remained much weaker than the *Armia Krajowa*. There was tension between Gaullists and Communists in occupied France, but no armed clashes.

The structure of the partisan movements and the leeway for their operations varied from country to country. After an initial period of confusion and disorder, the Soviet partisan units were in close contact with the partisan general staff in Moscow, and all major units were in radio communication with the "center." They had airports of their own, received arms and supplies at frequent intervals; wounded partisans were evacuated by air, there was a steady stream of new commanders, political commissars, demolition experts and important visitors. Some Soviet partisan units grew within a year from a few dozen members to a few thousand (those, for instance, headed by Kovpak, Melnik, Saburov and Naumov) and from time to time several units would combine for a large-scale operation, but on the whole their assignment was to act as individual units, not to become an army in the enemy's rear. Some units were dispatched on long-distance raids, but only very rarely would they try to occupy cities. The few attempts that were made (by Kovpak) failed. The dispersal of forces was in keeping with the overall strategy of the Soviet High Command; the establishment in the German rear of large infantry units lacking armor would have exposed the partisans to dangerous counterattack.

German forces in the Balkans, on the other hand, were not strong and the Yugoslav partisans concentrated their units into an "Army of National Liberation" subdivided into divisions as early as November 1942.[6] (A partisan division, however, numbered no more than three thousand men at the time, frequently fewer.) This partisan army fought large-scale battles against Axis forces in 1943 at the Neretva and the Sutjeska rivers. It occupied, and held, cities for long periods. In 1943 the German High Command admitted outright that it no longer faced "bandit" (i.e., partisan) warfare, but that a new front had been opened in Southeast Europe. Thus, Yugoslav strategy followed the Chinese pattern, even though there is no evidence that Tito and his comrades were familiar with developments in China. According to Yugoslav sources, Tito's army numbered eight hundred thousand men at the end of the war, whereas Soviet partisans never exceeded two hundred and fifty thousand; the quality of the forces the Yugoslav partisans were up against was, however, distinctly inferior to the antiguerrilla ones on the Russian front.[7] Still, it must be borne in mind that the Yugoslav partisans received no supplies at all from outside during the first years of the war; a massive Allied airlift was organized only in 1944. The partisan movements in other parts of Europe were considerably smaller and the operations against them, until the last year of the war, were, with rare exceptions, carried out by special police units.

THE SOVIET PARTISANS

Guerrilla warfare in the Soviet Union was officially initiated with Stalin's appeal on 3 July 1941, calling Soviet patriots to establish infantry and cavalry partisan units everywhere in the enemy's rear, to mine bridges, to cut his communications and supply routes; in short, "to make life intolerable for the invader." Mention has been made of the fact that before the war the NKVD, the secret police, had been responsible for diversionary action. With the outbreak of the war the Communist party apparatus and the Soviet army also became prominently involved in the enterprise; a central partisan staff was formed under the command of P. K. Ponomarenko, former party secretary in White Russia, and Marshal Voroshilov. The major partisan commanders (Naumov, Saburov, Medvedev) were usually either NKVD men or leading party officials (Begma and Fyodorov, former first secretary of the Chernigov party district).[8] Later on, regular army officers were attached to the units as chiefs of staff or as their commanders. The chief of the Ukrainian general staff was T. Strokach, formerly head of the Ukrainian NKVD; he was later replaced by V. Andreyev, a partisan commander. There were some notable exceptions; Sidor Kovpak, a veteran of the civil war, was fifty-four years of age when the war broke out. He had been mayor of the city of Putivl; his deputy, P. Vershigora, who later became commander of another major unit, had been a movie producer. The party officials who became partisan leaders showed enterprise and courage well above what could be expected of the average bureaucrat. Few of them would have collaborated with the Germans in any case, even if it had not been Nazi policy to execute all leading Communists.

No exact data exist about the social composition of the partisan units; according to one Soviet source, of twenty-five thousand partisans in the Orel districts, thirty-eight percent were workers, thirty-one percent peasants, and thirty percent belonged to the "intelligentsia." Of sixty-two thousand Ukrainian partisans, thirty-six percent were workers, forty-seven percent peasants, and seventeen percent "employees."[9] Such statistics are, however, of dubious value, for during the early period of the war the partisan units consisted for the most part of Red Army stragglers who in the general retreat had been cut off from their units. Later on, the composition of the units changed rapidly as young villagers were recruited in the Nazi-occupied areas.

The attitude of the population during the first months of the German occupation was one of *attentisme,* especially in the Ukraine and the Baltic countries. However, this mood did not last. It was official Nazi policy to treat the Slavs as *Untermenschen;* the Germans engaged in wholesale requisitions, they employed forced labor and carried out mass executions. The partisan leaders would have found it much more difficult to attract recruits had the Germans treated the populace decently, but this would have been quite incompatible both with the character of the Nazi leaders, their doctrine, and their aims.

The initial defeats of the Red Army had come as a shock to the inhabitants of the occupied territories, but with the failure of the Germans to take Moscow and at the very last with the battle of Stalingrad, the belief in a German victory waned. As a result, the partisan movement continued to grow; if there had been something like thirty thousand partisans by the end of 1941, their number had risen to a hundred and fifty thousand during the second half of 1942, and reached its peak, about two hundred to two hundred and fifty thousand, in the summer of 1943.[10] At first partisan units had been organized on a territorial basis; platoons and companies were mobilized by the local party secretaries, the Komsomol officials and the NKVD. Later, as the units congregated in wooded or marshy areas far away from their original base, the territorial system was given up. There were relatively few guerrilla units in the northern sector of the front and south of Kharkov; the Ukrainian steppe and the lowlands of the north offered little cover and the partisan units operating there endured heavy losses. The main concentrations were around Smolensk and Minsk, the forests of Bryansk, the Pripet marshes, and White Russia generally.[11]

The area occupied by the Germans during the first fifteen months of the war was several times that of Germany proper, the population also exceeded that of Germany itself, and supply lines had become a logistic nightmare. Troops could not be spared for operations in the rear and the German presence, as in Yugoslavia, was necessarily confined to the towns and the main traffic lines. There were large areas which throughout the occupation remained altogether outside German control. Thus, a region of several thousand square miles southeast of Minsk was in partisan hands without interruption from the summer of 1942 onward and there were a number of other such little partisan republics dotted about elsewhere. The Germans were fully aware of their existence but could not divert sufficient forces to destroy them. Even the front line was not continuous; certain gaps existed, such as the Vitebsk corridor, through which the partisans maintained contact with the *bolshaya zemlya*, that is, Soviet territory.

The vast spaces and the lack of manpower thwarted the German attempts to suppress the guerrillas. The German army group "Center," the one most exposed to partisan attacks, had at its disposal in late 1941 no more than four regiments and one SS brigade to police an area the size of England. Throughout the spring and summer of 1942 the German armies largely ignored the partisans, with something of fitful actions against them launched only in the autumn of that year, and during the following spring. Meanwhile, however, control of antipartisan operations, hitherto left mainly to the local commanders, was coordinated and non-German units, including Russian collaborators, Latvians and Lithuanians, were set up to fight the "bandits." In the end, the number of Russians serving in assorted German auxiliary units far outstripped that of the partisans. Hitler objected to the use of Russian volunteers; shortly after the outbreak of the war he had even welcomed the existence of Soviet partisan units, for this, so he said with his own unfathomable brand of illogic, would make it easier to recognize the enemy and to destroy him. His few instructions about antibandit warfare left little to anyone's imagination; there was to be no misplaced chivalry in this struggle — the enemy had to be exterminated.[12] Villages were destroyed and their inhabitants killed, or at the very least rendered homeless. In the countryside there had been no overwhelming enthusiasm for Soviet power; the circumstances of collectivization were still fresh in the memory. But indiscriminate arson and murder soon made the Germans as much hated in the villages as in the cities. In the later phases of the war, the German command switched its tactics and made all kinds of promises to partisans who surrendered. But of three major antipartisan operations, only one (*Zigeunerbaron*), produced a sizable number of deserters — 869 — whereas *Nachbarhilfe* netted a mere twenty-four, and *Freischütz* only five, a poor response to half a million leaflets that had this time been distributed.[13]

During the first months of the war the partisan units were building up their organizations, training for action, gathering weapons and establishing contact with the "center." Alongside this, and on a more political level, they circulated propaganda broadsheets, assassinated individual collaborators and tried, by persuasion if possible, to enlist support among the civilian population, appealing to the villagers' patriotism (Communist slogans were dropped during the war), but also not hesitant to use intimidation.

Fearing for their life and property, many mayors and policemen appointed by the Germans opted for collaboration with the partisans. True, the German military command had published countless warnings that all those who gave cover or supplies to partisans would be executed. But the next German police post was usually far away, whereas the partisan was the man with the gun in the doorway. In the circumstances, the decision was not difficult to make.

Another major partisan task was to spoil and destroy the agricultural crops, preventing their shipment to Germany. Since the main wheat-growing areas were a long way from the main partisan concentrations, they succeeded only partially in this in 1942; in 1943 the results in this respect were more impressive. They almost never managed to interfere with the extraction of minerals and oil in the occupied areas; the mining centers were located in the south, beyond the partisans' reach. They did better in the battle for the railway lines; to hamper German transport had been one of their primary missions from the start, but in 1943 it was given absolute priority. The Smolensk-Bryansk-Orel and Minsk-Gomel-Bryansk railways which were of vital importance to the German central army group were temporarily paralyzed during a decisive phase of the war. In late July 1943 new plastic mines, against which metal detectors were ineffective, were used for the first time and in their greatest single operation of the war, on 2 August 1943, more than 100,000 partisans planted 8,422 of these mines on the railway tracks. This sabotage coincided with the Soviet offensive following the battle of Kursk.[14]

Occasionally an especially large partisan body would be given instruction to carry out penetration raids deep in the German rear; the intention was usually to relieve pressure on the Soviet army.[15] During the winter of 1942/43 several major partisan units transferred their activities to the Western Ukraine far beyond the German lines; the Saburov and Bogatyr units marched some four hundred miles through the German rear, crossing five rivers in the process. From his hideout in the Bryansk forests, Kovpak moved to the same target area, west of the Dnieper, where by May 1943 a partisan concentration of twenty-two thousand men had assembled. One White Russian unit went on a six-hundred-mile raid, but the most spectacular operation was Kovpak's in 1943 which took him to the Carpathian Mountains and the Slovak border.[16] Sections of his unit remained there and participated in the Slovak rising of 1944. Later, Kovpak ran into trouble and his unit was almost wiped out. The partisans were unfamiliar with the area and the population was frequently hostile. The partisans had to cope not only with the Germans but also with Ukrainian nationalist irregulars who were both anti-German and anti-Soviet.

After May 1943 operational cooperation between the partisans and the Soviet army was near total, the partisans coming under direct army command. During this period partisans became more active in the northern sector of the front where they had previously been weak. Having already been of assistance in preparation for the Soviet offensive in 1943, they also helped to pursue the fleeing Germans, while the great Soviet summer offensive of June 1944 was again preceded by massive partisan attacks on railway lines, with more than ten thousand minings taking place two nights before its start.[17]

As the Soviet army crossed into Poland and Germany, the partisan units were gradually disbanded. Their general staff had already been dissolved earlier, in January 1944, for reasons which are not entirely clear; that of the Ukrainians, in contrast, continued to function up to the end of the war. Perhaps some influential party leaders such as Khrushchev and Korotchenko took a personal interest in the Ukrainian partisans.

As more and more territory was liberated by the Soviet army, the local partisans would be absorbed into regular units; their former commanders, awarded some military decoration and promotion, returned to their old positions in government, political parties and the secret police. Their subsequent personal fates fluctuated with the postwar purges and shifts in the Soviet party leadership. Among those who were to rise to high rank was Mazurov, the partisan Komsomol secretary in White Russia, who eventually became a member of the Politburo.

The place of the partisan movement in Soviet historiography and literature underwent certain ups and downs in the postwar period. In professional military literature there was a tendency to downgrade the importance of the partisan units; they were honorably mentioned and due tribute was paid to their heroism, but there was a reaction against the wartime disposition to exaggerate their part in the victory over Nazi Germany. Some of the firsthand accounts written while the war was still on, or immediately afterward, had to be rewritten, because, as highly placed critics argued, the authors had paid insufficient attention to the "leading role of the Communist party in organizing and guiding the partisan units."

Conditions of partisan life varied inevitably from place to place, but they improved everywhere as the war continued. During the first winter, partisans underwent great hardships; a typical account relates that members of one of the larger White Russian units were given sugar — a great luxury — in their tea and salt with their bread only on special occasions.[18] Most partisans were forced to hibernate in their hideouts during the first winter of the war because they had no suitable clothing or equipment, and furthermore their traces in the snow would have betrayed them.[19] Eighteen months later the situation had improved so much that when a temporary food shortage occurred following a German antiguerrilla operation, the problem was solved by flying in supplies from behind the Russian front. Altogether, partisan life on the whole was neither as dangerous nor as strenuous as was generally imagined. It was far more risky to engage in anti-German activities in a town; the illegal party leadership in the cities lasted, as a rule, no longer than six months, whereas most leading partisan commanders survived the war. (One notable exception was Rudnev, Korpak's chief of staff, who fell in action in 1943.) There are no detailed statistics, but it would appear that the chances for survival among the partisans were no worse than among Russian front-line units.

One of the reasons for the partisans' relative safety has already been mentioned. The Germans never had sufficient manpower during the war to cope with them, nor could they concentrate enough armor and aircraft for such operations. The partisans, on the other hand, had radio contact with their general staff and could ask for air support. Radio communications were of great psychological importance; there was no feeling of isolation, one of the commonest drawbacks of partisan warfare throughout history. Add to this that the partisans received warnings of impending German attacks and their offensive operations were effectively coordinated.

Arms and supplies were flown in by air.[20] This included medical supplies and personnel; in one partisan unit, seven doctors took care of five hundred men. During the antiguerrilla operation *Zigeunerbaron,* the Germans found to their consternation that the partisans had not only heavy guns but even a few tanks. (The Germans seldom used tanks against partisans, partly because they could not spare them, partly because the terrain was unsuitable.)[21] Some larger partisan units had been supplied with 45 and 76 mm guns. Although the Germans used aircraft against partisans, it was chiefly for reconnaissance, on only rare occasions for tactical support. The Soviet partisans, in brief, were in almost every respect in a more advantageous position than partisans in other European theaters of war.

Partisans helped the Soviet war effort in many supplementary ways. They collected taxes for the Soviet government, recruited soldiers for the Red Army and, in the Leningrad region, transported food from the Pskov *kolkhozes* into blockaded Leningrad.[22] They acted as the long arm of the Soviet government disseminating propaganda and policing the countryside. In all this, however, their function was almost entirely political and administrative. Militarily, their activities, as has been noted, were by far the most frequent and forceful in central Russia, affecting the operations of the German Army Group Center. Years later, a military historian analyzing the Battle of Kursk, the turning point in the history of the war on the Eastern front, mentioned the many partisan operations against railway lines, and the fact that, as a result of them, agricultural deliveries decreased by two-thirds. But for all that, he concluded that until the spring of 1943 partisan activities did not seriously influence the operative planning of the German army.[23] It was only during the German retreat from Russia that the partisans became more than a nuisance. During 1942 and 1943 many German generals grumbled about the intolerable situation in the rear, and there were similar complaints emanating from German headquarters. But Himmler, in a secret speech to his *Gauleiters* in October 1943, summarized the situation in the rear in saying that it was nonsense to make so much of a mere inconvenience; the soldiers at the front were not dying of hunger, nor were they short of supplies and reserves, which arrived as planned.[24] It was only in May 1944, when the German armies were about to evacuate the last parts of the occupied Russian territory, that they realized that they might have saved themselves much inconvenience if they had treated the partisans differently. An official handbook on antipartisan warfare published in May 1944 decreed that partisans should not be shot but treated as prisoners of war unless they were caught in German uniforms.[25] Such recognition of past mistakes was by that time of purely academic interest.

A balance sheet of partisan operations in Russia, then, would have to be based on an equation including many incommensurate, and immeasurable, factors.[26] From a purely military point of view it is not certain whether the effects (and the losses incurred) were worth the effort.[27] The same resources, used in a different way, might have produced greater gains. Nor does one know whether the intelligence provided to the Soviet general staff by the partisans was of crucial importance. Partisan brigades were not needed for the collection and transmission of military intelligence; this could be done by individuals.

The decision to wage partisan warfare in the German rear was only partly motivated by military considerations; political reasons were assuredly more telling. Stalin's views on its value in general had changed more than once; having been one of its early advocates during the civil war, he later took a dim view of its military significance. In the 1930s the civil war partisans were systematically denigrated, together with other old Bolsheviks, and many came to grief in the great purges. Still, Stalin believed that if Russia were attacked, partisan warfare could do no possible harm and might do some good. Perhaps he was already concerned with the more distant future, the postwar period and the restoration of the image of omnipotence and the omnipresence of Soviet state and party organs.[28] And in this respect the partisan movement could play a very important part indeed.

YUGOSLAVIA

To the Russians, the creation of partisan units was an auxiliary weapon of the regular army to carry out certain tasks behind enemy lines; to the Yugoslavs the partisans were the army.[29] The Yugoslav partisans fought alone and their achievements earned them the admiration of friends and enemies alike. "I wish we had in Germany a few dozen Titos," Himmler said in one of his secret speeches in 1944, "a man with such a strong heart and such good nerves; he has really earned the title of marshal."[30] The fight of the Yugoslav partisans is indeed in many respects unique; a mere handful of dedicated Communists, with little experience of tactics and none of strategy, they succeeded against all odds in building up a military force of considerable potency. During the critical period of their struggle they received no outside help. The Russians provided advice of doubtful value, but nothing else; the Western Allies sent substantial military assistance, but only after the partisans had emerged as the leading resistance force in Southeast Europe.

Partisan warfare in Yugoslavia has been fully documented, yet to some of the main questions there still are no ready answers.[31] Outwardly, Communism had been little in evidence in Yugoslavia in the interwar period, but there was a fairly strong Communist tradition going back to the early 1920s. The Communist party, illegal for most of the time, had been steeled in the underground struggle; hundreds of its members had fought in the Spanish Civil War. The Yugoslav establishment, on the other hand, the monarchy and its political supporters, had been discredited by the military defeat of April 1941. Yugoslavia was a house divided against itself—between Croat chauvinists, Serbian nationalists and Montenegrin fanatics, between Roman Catholics, Orthodox Christians and the Muslims. The Communists, paradoxically, were almost the only political force which could provide a platform for all nationalities. The cadres of the party were almost entirely urban, with a heavy preponderance of intellectuals and students; Tito and Rankovic apart, there were hardly any leaders of working-class origin.

How, one wonders, could urban intellectuals not only make common cause with the villagers, but transform themselves into highly effective mountain fighters, the most militant in the struggle against the invader? It is true that many of them were only one generation removed from village life, that they were young and enthusiastic. It is also true that in Serbia and Montenegro there was a strong pro-Russian tradition. The partisans benefited from the systematic extermination of Serbs by Ante Pavelic, the *Poglavnik* (leader) of the new Croatian state; of those who survived, many fled and joined Tito's forces. All these facts help to explain the partisans' success, but they do not provide a conclusive answer. The partisans made mistakes as well, more perhaps in the political than in the military field. The Russians were horrified by the political extremism of their overzealous Yugoslav comrades. In Montenegro and Slovenia, in the middle of the war, partisan leaders gave orders to assassinate local patriots because their attitude to the Communist party was not sufficiently enthusiastic.[32] This policy was discontinued after Tito had reprimanded them.

The achievements of the Yugoslav partisans cannot begin to be satisfactorily explained without reference to the men who led them. In Tito they had a great political and military leader, imperturbable, a man of iron will, a true believer yet not a fanatic, a civilian with an uncanny military instinct. Yet Tito alone would have been able to accomplish little but for the presence of younger men of great capacity, a Kardelj and a Djilas, a Ribar and a Popovic, willing to accept his authority, yet able to act independently. For once intellectuals were also men of action; the partisans were not only more intelligent than their opponents, they were also tougher.

The Yugoslav revolt began with a call for a general insurrection by the Communist party in July 1941; the appeal had some effect in Serbia and Montenegro, none at all in Croatia. But the Germans and their local collaborators suppressed the revolt without difficulty. On 16 September Tito left his hiding place in Belgrade and went to the mountains to assume leadership of a more intensive struggle; a decision had been taken previously by the party executive to convert itself into the general staff of a guerrilla army and to create operational bases in certain parts of the country from which the enemy would have to be evicted.[33]

By September 1941 Tito's supporters numbered about fifteen thousand but many of them were without arms; Colonel Mikhailovich, who had organized a resistance group in the mountains even before the Nazi invasion of Russia, had some five thousand followers at the time. The menace of the "bandits" was thought by the Germans to be sufficiently serious by then to warrant a major offensive against them; in a matter of days in late November 1941 three German divisions cleared Serbia, and the partisans had to escape to Bosnia-Herzegovina. In Yugoslav partisan history, as in China, the enemy offensives are the great milestones; the Yugoslavs counted seven, the Chinese six (up to the Sian incident in 1937). The second German offensive was of no great import, but the third, carried out mainly by Italians and Croats, compelled Tito to withdraw even farther to the south. Montenegro had the great advantage of being virtually inaccessible, but it was also exceedingly poor and supplies were almost impossible to obtain and in June 1942 Tito decided to march north again, where he occupied a fairly large area in the heart of Yugoslavia, including the towns of Jajce and Bihac. The fourth and fifth enemy offensives (operation "White," January–March 1943, and operation "Black," May–June 1943) were successful inasmuch as Tito lost about half of his troops, but, as on previous occasions, he once more broke out from the enemy encirclement. When Italy surrendered in September 1943, his forces were back in Bosnia, disarmed some ten Italian divisions and seized great quantities of arms and supplies. By the end of 1943 Tito's partisan army numbered almost three hundred thousand according to Yugoslav sources; two hundred thousand according to others. These forces were not, however, all concentrated in one region; besides Tito's own army, major partisan units were fighting from Slovenia in the north to Macedonia and Montenegro in the south.

Early on in the war Tito had realized that the strength of the partisan movement lay in its dispersal, that the establishment of one compact front would be more than dangerous.[34] He had told his comrades in Montenegro in November 1941 that given the conditions, partisan warfare was the best means of getting a popular uprising underway, aware of course as long ago as then that the tactics of guerilla war alone would not be suitable for large-scale offensive operations aimed at the liberation of vast stretches of territory; hence his later decision to form larger mobile units (brigades) which were not tied to any specific locality. But these did not replace the guerilla units, which continued to operate in Serbia, Montenegro and Macedonia. Even in November 1942, after fifty thousand square kilometers had already been liberated, Tito insisted that the guerrilla tactics used previously — meaning the harassment of the enemy's supply lines, the destruction of his bases and so on — must still remain integral to the overall struggle of the Popular Army.[35] By the late summer of 1942 Tito's partisans had become an "operative problem" for the Germans.[36]

One year later, in November 1943, the German High Command determined that a "Soviet State" had come into being behind its lines, and new major antipartisan operation was decided upon, with the object of clearing eastern Bosnia, and above all the Dalmatian coast.[37] This was a matter of utmost consequence to the Germans, because the partisans' presence could have opened the road to Allied landings. So operations *Kugelblitz* and *Adler* were set in motion. *Adler* succeeded, but *Kugelblitz* was a failure; Tito again broke through the enemy lines without undue hazard. Paradoxically, the last German offensive in May 1944 (*Rösselsprung*), in which relatively small German units were parachuted near Tito's headquarters, was the most nearly destructive; a German battalion almost seized Tito and his staff. Six thousand partisans were killed at the cost of only a few Germans. But by that time the Allies were in a position to provide more effective air support to the partisans, and the German columns had to withdraw, their basic mission, save for the casualties inflicted, in no way fulfilled.

Mikhailovich's *Chetniks* continued to exist but played no significant role in the war against the Germans, he having meantime, through various intermediaries including the Belgrade puppet government, concluded an armistice with the Germans and the Italians. The Mikhailovich tragedy was, to some extent, the fault of the partisans. Mikhailovich certainly was no collaborator or traitor in the strictly accepted sense of either word; he could, after all, have stayed in Belgrade in 1941 in the first place instead of leading a far less comfortable existence in the mountains. In November–December 1941 the two factions — his and Tito's — negotiated the coordination of their activities; Tito claims that he even offered Mikhailovich the supreme command. But Mikhailovich prepared himself for a long war; he thought partisan activism misplaced and in any case deeply distrusted the Communists. His attitude was shared by other Serbian officers; it is one of the ironies of history that Kosta Pecanac, who had been the principal guerrilla leader behind the German lines in World War I, and who served for a time in 1941 under Mikhailovich, was among the first *Chetniks* to make peace with the Germans. A Serb, a conservative and a regular army officer, Mikhailovich was suspicious of all non-Serbian Yugoslavs; the Croat Tito was in his opinion a mere "jailbird." At his trial he said that he was not a politician but a military man; he certainly lacked the Communist ability to mobilize the masses. The Royal Yugoslav government in exile had appointed him minister of war in 1942, but as his forces disintegrated, he lost his official position (June 1944). He managed to evade the Communists for about a year after the war had ended, but was captured in March 1946 and sentenced to death.[38]

While they forged their guerrilla units into an army, the partisans made similar progress on the political front. At a meeting in Bihac in November 1942, the Anti-Fascist Council of Yugoslavia (AVNOJ) was established, superseding the Movement for National Liberation ("We are now setting up something like a government," Tito wrote to Dimitrov, the general secretary of the Comintern in Moscow).[39] A few non-Communist politicians belonged to AVNOJ but this was mere window-dressing, for all effective power was in the hands of the Communists. Partisan discipline had been strict from the very beginning; with the establishment of a quasi-state, their units increasingly resembled those of a regular army. Ranks and decorations were formally introduced, several generals were appointed, and in November 1944 Tito was made Marshal of Yugoslavia.

The partisans did not lack money during their campaigns; in the early days of the war they had commandeered considerable sums from provincial banks. It was far more difficult to get arms and for a long while the partisans depended on those taken from the enemy or manufactured in their own small factories. The first Allied emissaries visited Tito's headquarters in 1942, but a permanent British mission was not installed until 1943 and there were no Russians in the partisan headquarters until late in the war.

At the Teheran Conference it was decided that the Allies should give Tito full support and, from then on, the partisans received more or less what they needed from British and American bases. This help was important, but it was no longer decisive. To the great disappointment of the partisans, no Russian supplies were forthcoming until the very end of the war.

Yugoslavia is one of the few cases in history in which a partisan movement liberated a country and seized power largely without outside help. It is, of course, true that but for their military involvement in Russia and on other fronts, the Germans could have crushed the partisans with the greatest of ease, just as the Chinese Red Armies could not have won their war but for Japan's many preoccupations elsewhere. But this does not detract from the partisans' achievements. As in China, it was essentially a peasant army led by middle-class rebels, mainly intellectuals.[40] But it was not a peasant war, land was not redistributed while the fighting continued. Again as in China, the government which emerged after the war was the wartime general staff of the partisan movement. The partisan experience enormously strengthened the self-confidence of the Yugoslav leaders; unlike the other Communist governments, they had not been imposed by the Russians but had attained power through their own efforts. They were not a satellite and this made for growing strains in their relations with the Soviet Union, culminating in Tito's break with Stalin in 1948. Seen in a wider perspective, it could be argued that Communism would have prevailed in Yugoslavia even had there been no partisan movement, because it had been decided between the Allies that this was to be part of the Soviet sphere of influence. It is quite likely, furthermore, that, sooner or later, Yugoslavia would have opted for national Communism; Rumania after all did so, despite having been "liberated" by a foreign army. Inside Yugoslavia the partisans emerged as the "new class" described by Djilas in his books; their mental makeup and their intellectual outlook differed in some respects from that of the Communist elites in other East European countries and this has had its impact on Yugoslav domestic and foreign politics since 1945. Much of the partisan tradition has worn off with the years, but some still persists, and for this, among all the other reasons, Yugoslavia remains a "special case."

THE SLOVAK RISING

The two major East European risings outside the Soviet Union and Yugoslavia ended in disaster. The initiative for the Slovak revolt came from Lieutenant-Colonel Golian who acted both as chief of staff of the local puppet government and as the representative of the Czechoslovak government in exile. He had conspired with fellow officers during the spring and early summer of 1944; in one rapid action the whole Slovak army was to open the road to the Soviet army. He coordinated his plan with the Communists who had at the time around fifteen hundred partisans in the Slovak mountains. Golian and the regular army officers intended to carry out the operation in late September or October to coincide with the assumed date of the Soviet army's new offensive. The Slovak Communist leadership agreed in principle with this timing, sharing the fear that should the rising start prematurely, it would be quickly smothered by a few German divisions. The Czech Communists in Moscow and the Soviet High Command, on the other hand, preferred an earlier date and were anyway more interested in partisan activities than in regular army participation. Substantial Ukrainian partisan units were parachuted into Slovakia in late July; in August they launched a general insurrection, thus forcing Golian's hand. It cannot be established unequivocally whether the partisan leaders acted on their own initiative, or whether they had explicit instructions from Moscow to pre-empt a non-Communist rising.[41] At first the Germans hesitated; they were not as yet certain about the extent of the revolt and could ill afford to divert troops from the Eastern front. However, on 28 August a German military mission was stopped in transit in the north of Slovakia and its members killed. The day after, Tiso, the puppet president of Slovakia, announced that German forces had been "invited" to suppress the partisans; within twenty-four hours the first *Wehrmacht* units entered Slovak territory.

At the beginning of the rising, Golian had at his disposal twenty thousand regular soldiers in central and eastern Slovakia; in the western part of the country his appeal had found little echo. There were also some twenty-five hundred partisans; within the next few weeks their number rose to seven thousand. The German advance did not go too well during the first ten days of fighting; but by mid-September the tide turned in its favor, and while operations lasted until late October, the outcome was no longer in doubt. The insurgents received small quantities of arms from both the Western Allies and the Soviet Union. But they had expected the arrival of the Soviet army, or at least Czech units fighting with the Soviet army, and neither of these joined them in time to avert the disaster. The most they were given was occasional air support by Czech pilots operating from Soviet airfields. During October the Soviet army tried to force the Dukla Pass, which would have allowed of rapid advance into Slovakia. But they encountered unexpectedly heavy German resistance which they overcame only after the Slovak rising had ended. As the rising collapsed, several thousand Slovak soldiers joined the partisans in the mountains, but they no longer constituted a military danger for the Germans.

The story of the Slovak rising has been written and rewritten several times since the end of the war. Communist historiography in the Stalin period has it that the Slovak rising had been led by the Communists but had failed owing to the incompetence and the intrigues of the bourgeois elements which had been involved. The party line was modified in the 1960s and a collection of documents published which showed fairly accurately what had actually happened in 1944.[42] Seen in perspective, it was a case of bungling and bad timing; there was no deliberate attempt to sabotage the rising. The Soviet command could not have known that it would face such determined German resistance in the Carpathian Mountains, and that the advance into Slovakia would take so long. The Communist partisans' decision to launch the insurrection prematurely for political reasons certainly did little good to their own cause. They would have helped the Russians more by intensifying partisan warfare; it is doubtful that this would have provoked a German invasion. In the fighting against the Germans the partisans played only a secondary role; the Slovak army units bore the brunt of the German attack. Had they employed partisan tactics, occupying the mountain heights rather than trying to defend a front of some hundred and thirty miles, they would certainly have held out longer, although it is less guaranteed that they would have been able to resist the Germans for the further seven months until the Soviet troops at last arrived.

WARSAW

The Warsaw rising in 1944 was the one major urban insurrection of World War II. The resolve of the Polish Home Army (AK) to accelerate the struggle against the German occupiers reaches back to late 1943. As the Germans retreated from the eastern regions of Poland, the AK appeared more or less openly in the countryside and in some cities such as Vilna. For the Polish leadership this operation, *Burza* (the Storm), was of the greatest political significance. For centuries Russia, the occupying power, had been the enemy *par excellence,* and to the Polish eye the attitude of the Soviet Union differed in no essential respect from that of Tsarist Russia. As the Soviet troops advanced into Poland, the people's fear was that they were come not to liberate the country, but to occupy it. The arrests of Polish Home Army commanders and the execution of some of them by the Russians proved that these fears were not groundless.

The decision to mount the Warsaw rising was rooted in the assumption that the Soviet army would very soon be at the gates of the Polish capital; the Home Army would forestall it, thus creating a political fact. There was no unanimity about this decision; Sosnkowski, the Polish minister of defense, strenuously opposed it; one of his chief reasons was that the insurgents could not count on Allied help. Warsaw was still outside the effective range of Allied aircraft, and the Soviet Union refused to give the Allies landing rights for supply missions to Warsaw. The rising began on 31 July 1944 with an attack of between twenty-five hundred and three thousand Poles against German strongpoints; of the twelve thousand members of the Home Army in Warsaw, only every fourth one was armed.[43] The city was defended by five thousand German soldiers who were supplemented inside a week. There was no moment of surprise, the Germans had foreseen what was coming. "The expected rising has started," ran their army report.[44] The insurgents' intelligence and communications system was totally inadequate; orders scarcely ever reached the units at the right time. The Germans repelled attacks against almost every strategic point.

In effect, the rising was defeated on the very first day, when the Poles lost one-half of their manpower and failed to seize a single bridge or the airfield. Yet against all odds the struggle continued for another fifty days. Fighting courageously, the Poles received reinforcements from other towns and they still hoped that the Russians would perhaps after all come to their rescue or that the Allies would somehow assist them. The Germans, who for all their more professional equipment had nonetheless too few forces to quell the rising entirely within their predicted day or two, resorted to calling in the most notorious cutthroats such as the Dirlewanger brigade and the Kaminski Russian volunteer units for antipartisan operations. In one Warsaw region alone, Vola, ten thousand civilians were killed. As a result of these atrocities, the Polish will to resist stiffened all the more.

Meanwhile the Russians were camping just outside Warsaw, on the opposite shore of the Vistula. They had not stirred the inhabitants of the Polish capital to rise (as some of the Poles later claimed) in order to watch them being killed by the Germans. But equally, they had not the slightest intention of coming to their aid. Stalin told Churchill and Roosevelt that the whole enterprise was a "contemptible adventure," the Home Army was not an army; it could do no more than hide in the forests, and was quite incapable of challenging the German army. Even earlier, Soviet spokesmen had declared that there was no fighting in Warsaw; it was all an invention by the Polish government in exile in London.

The fight for Warsaw ended in early October. A hundred and fifty thousand Poles were killed in the rising, among them some sixteen thousand members of the Home Army. German casualties were eleven thousand including two thousand killed. The German conditions of surrender were curiously and surprisingly magnanimous; prisoners were not to be shot but to be accorded combatant status. The Germans had of course every incentive to bring the struggle to a quick end for they were aware that the Russians might at any moment resume their offensive. Hitler's instructions were that Warsaw was to be razed to the ground, to "disappear from the face of the earth." Soon after it was all over, the Soviet army entered a ghost city.

It is only too easy in retrospect to conclude that the rising was doomed from the start. Even had the insurrection succeeded on the very first day, and even had the Russians come to its help, a Stalinist regime would have been imposed on Poland. The Poles lost thousands of victims and their capital was destroyed, but even a military victory would not have affected the political outcome; there was no hope of Poland regaining its independence. In purely military terms, however, the rising proved once again that failing the element of surprise and sufficient arms and ammunition, an insurrection cannot succeed. (The insurgents had initially ammunition for five days only.) But it also demonstrated simultaneously that, given equal determination on both sides, it is very hard indeed, short of overwhelming military superiority, for regular army units to reimpose their control on a major city.

ALBANIA

The development of the Albanian resistance movement during the war much resembles that in Yugoslavia. The Communists were the strongest group and they revealed not only greater military ability but also considerable political acumen. They outmaneuvered, isolated and ultimately destroyed their enemies of the *Balli Kombetar.* There was in Albania a guerrilla-banditry heritage dating back to the Middle Ages which had never been altogether stamped out. The topography of the country made it all but impossible for any invader to establish effective control unless he had almost unlimited manpower at his disposal. Since the country was neither rich nor strategically important, it had usually escaped the worst effects of foreign domination.

Italian and German control during the war was always limited to the major towns and lines of communication. In large parts of the country the partisans could operate virtually without hindrance. The Communists were at first undeniably handicapped by their lack of military experience, and they had great difficulty in acquiring arms and supplies. The existence of deep ethnic conflicts made it, in addition, anything but easy for the Albanian Communists (as for their Yugoslav comrades) to establish a unified guerrilla command. Initially they based themselves mainly on the landless peasants among the Tosks; their primary foe, Abas Kupi (the Albanian Mikhailovich), belonged to the Gheg, the rival tribe. Born in the same little town as Skanderbeg, he had fought in World War I with a little guerrilla band behind the Austrian lines.

The Yugoslav Communists already had an active party organization to draw upon when the war broke out, not to mention two decades of political and conspiratorial experience. The Albanian Communist party, on the other hand, came into being only during the war. Its backbone was two to three thousand young urban intellectuals; as in China, they constituted the leadership which mobilized the peasants.[45]

Julian Amery, who was a British liaison officer with the Albanian resistance, has pointed to the social and political roots of the partisan movement in Albania; a new class of officials and merchants had emerged in the 1920s and 1930s whose children had received a European education, either in Albania or abroad (frequently in Paris).

> These young men had no roots in landed property or among the tribes and could find no outlet for their energies within the narrow limits of independent Albania such as the Ottoman Empire had offered earlier generations. They were thus peculiarly susceptible to the influence of revolutionary ideas. In other countries such young men often inclined to Fascism, but in Albania Fascism was the creed of the foreign overlord and, in their search for faith and discipline, they therefore turned to the Communists.[46]

Operating under the cover of a "National Liberation Movement," founded at a conference in Peza in September 1942, the Communists gradually subverted and took over the other bands — a considerable achievement by any standards. Their contribution to the Allied war effort in 1941–1942 was negligible apart from such projects as cutting telegraph lines in July 1942. Boasts by Albanian spokesmen that they had detained a hundred thousand Italians and seventy thousand Germans can hardly be taken seriously. Some Italian and German troops were kept in Albania as a precaution against an Allied invasion, not to fight the partisans.

The great hour of the Communist partisans came in the summer of 1943 when Italy capitulated to the Allies, and much of Albania, including some of the major cities, passed into the partisans' hands. Paradoxically, they came under much greater pressure toward the end of the war, for whereas the Italians had not been that eager to fight, the Germans, with relatively small forces, launched a counteroffensive in November 1943 which almost proved fatal for the partisans. They were pushed back into the mountains where they had to spend a most uncomfortable winter. But by 1944 Germany was no longer in a position to squander its men and resources in so very minor a combat area; its troops were gradually withdrawn and by November 1944 no more German units were left in the country. Some of the non-Communist guerrillas escaped abroad, others were seized and shot, a few continued their struggle in the mountains. To all intents and purposes the victory of the Communists was complete by late 1944.

The Albanian Communists had received both guidance and support from Yugoslavs during the war, but this did not prevent their turning violently against Tito in later years. In the Khrushchev era their country became the last bastion of Stalinism in Europe. This was not altogether surprising in view of the cultural level of development of the country, but the spirit of defiance cannot perhaps be explained entirely in terms of political, cultural and social backwardness. The case of Albania is yet another to illustrate that a partisan movement coming to power mainly through its own efforts will still remain stubbornly independent and strongly nationalist in inspiration quite irrespective of its internationalist slogans.

GREECE

As in other European countries, the resistance against the Axis powers in Greece was split; most of the guerrilla fighting was done, in the event, after the war had ended. Great claims were later made with regard to the Greek contribution to the Allied war effort. But the German War Diary noted on 5 November 1943 that "nationalist and communist bands, altogether some 12–15,000 men, oppose each other; British officers in both camps have been unable to bring about unity of action. So far these operations have been of little significance."[47] The Greek Communists (ELAS) were the sturdiest of the partisan movements; their party had been in existence since the early 1920s, and they were among the first to take to the mountains after the Russians had entered the war. Their closest rivals were EDES under General Napolean Zervas, the "National Band" of General Serafis, and EKKA commanded by Colonel Psaros. Serafis was taken prisoner by the Communists and joined them subsequently, Psaros was killed by ELAS. Only the units commanded by Zervas survived the war more or less intact, primarily owing to British support. Zervas and the Communists collaborated from time to time, as in the first major action of guerrilla warfare — the mining of the Gorgopotamos railway viaduct in October 1942 under the direction of a party of British parachutists. But fighting each other was their more frequent occupation.[48] The Germans later maintained that even the mining of Gorgopotamos was of no military consequence, because the British assumption that this was the main supply line to the *Afrika Korps* no longer held good; Rommel had already retreated from El Alamein.

According to Colonel Woodhouse, second in command of the British military mission among the partisans at the time, the value of the guerrilla operations was not inconsiderable in 1943; they created the impression that an Allied landing was about to take place in Greece rather than in Sicily, and thus drew into southern Greece a German armored division which the Germans could not withdraw in time when it was needed elsewhere. But in 1944, again to cite Woodhouse, partisan activities were not important in scale, and this despite the guerrillas having grown substantially in strength and their seizure of great quantities of Italian arms and ammunition.[49] The Communists were being more farsighted than their rivals and the British, whose overriding concern was to win the war as quickly as possible. The Communists realized that the decisive contention for power would take place only with the end of the war. Their fundamental task was therefore to increase their strength and, if possible, to destroy their own internal antagonists before the war should be over. Although the Communists fought the Italians and the Germans only on rare occasions, they still were more active than other Greek resistance groups; since the British command appears to have been oblivious of the Communists' post-war ambitions, it seemed only natural to send them more supplies than the other factions.[50]

As in Albania, partisan life was neither particularly strenuous nor risky under Italian occupation. The situation changed radically after the Italian surrender; during the last months of 1943 the Germans launched an offensive in Thessalia, Epirus and the Peloponnese which involved the partisans for the first time in heavy fighting and compelled them to retreat to distant mountain hide-outs. But the overextended German forces were no longer in a position to sustain a prolonged campaign and the partisan movement survived this difficult winter.

The Communists feared, not without reason, that the British would support the restoration of the monarchy after the war; they also suspected, quite mistakenly, that the British intended to destroy Communism. In fact, the British merely insisted on a plebiscite to decide the future of the monarchy; they were quite ready to accept its results. What the British were not willing to accept was a Communist coup; it was this fatal misreading of the situation that led first to the Communist refusal to disarm and subsequently to the insurrection in Athens in December 1944 with its tragic consequences.[51] Years later, the Communists were to regret their precipitate action; Siantos, who had led the insurrection, was denounced by his colleagues as a traitor, enemy agent and British stool pigeon.[52]

The partisan experience in Greece during the war reveals a picture similar to that in other European countries. Relatively weak before the war, if with a sounder nucleus than the rest, the Communists emerged as the most dynamic party, the best prepared psychologically and organizationally to operate in illegal conditions. The prestige of the party grew by leaps and bounds during the war; its military contributions to the victory had been small, but it had still done more than the rival groups. Since the Axis forces were not numerous enough to occupy the whole of Greece, the Communists stepped into the vacuum and eventually dominated about half of the mountain regions. During the war years new cadres were trained for the postwar struggle and, by war's end, the Communists had become a powerful military force. For three years they were to engage the Greek army in a bloody and costly guerrilla war.

FRANCE

Frenchmen resisted the German occupation by collecting information of military value for the Allies, by acts of sabotage and individual violence. Guerrilla warfare played only a minor role in the French resistance, except in the weeks following the Allied landings in 1944 and again during the liberation of France. On 6 June 1944 when it received its orders to move to Normandy, the 2nd SS Panzer Division (*Das Reich*) was stationed near Toulouse. Owing to resistance harassment, railway sabotage and RAF attacks, it reached the battle zone only sixteen days later. But the price that had to be paid when it did arrive was high; it was this division which destroyed Oradour-sur-Glane and killed its inhabitants. It has been argued that it was not a very good division anyway and its presence at the front did not make much difference.[53] While giving their blessing to the various Maquis, most of which had sprouted spontaneously, both the Gaullist leadership and the Communists had reservations about the military value of large partisan concentrations. These doubts were only too justified. The north of France was, with some exceptions (the Ardennes and the Vosges), unsuitable for guerrilla operations, quite apart from the fact that many German divisions were stationed in this area. Certain sections of the south, on the other hand, were thinly populated and the rough terrain favored the defender. This was true of the Massif Central, for instance, and the Vercors, an Alpine plateau to the south of

Grenoble. But while the Maquis were relatively safe in these mountainous areas, their ability to strike from there at the main lines of communication was limited.

The first Maquis had come into being in late 1942; by early 1943 there were so many that it was already impossible to list them in full.[54] Most of these, however, were very small groups, only rarely existing for more than a matter of weeks or a few months. Many of their members were *réfractaires*, young Frenchmen escaping forced labor and deportation to Germany, and more anxious to hide in the woods and mountains than to indulge in military operations. The idea of establishing a major concentration of Maquis in the Vercors originated with Pierre Dalloz, the secretary of the French Alpinist Association, and Yves Forge, a resistance leader in Lyons. Another plan aimed at concentrating some ten to fifteen thousand partisans in the Mont Mouchet region near Clermont-Ferrand. Both projects were undertaken in the belief that the local Maquis would be joined by French paratroops and that heavy equipment would be provided. (War materiel was dropped by British planes on many occasions in 1943 but consisted mostly of light weapons and explosives.) The Maquis congregated but neither the paratroops nor the heavy arms arrived. Furthermore, the basic concept of a "mountain fortress" violated the most fundamental guerrilla principles.[55] The Germans stormed Mont Mouchet on 11 June 1944; most of the defenders were fortunate enough to escape. The main battle for Vercors began on 13 June 1944, the Maquis having struck prematurely. Following the failure of the first German assault, the local Maquis had some four weeks' respite, but eventually the superior numbers of the Germans and their heavier equipment told and most of the Maquis perished in the battles of July 21–23. There were bitter complaints that the Maquis had not received the promised help, but the hard-pressed Allies could not divert forces to a military sideshow, hundreds of miles from the places where more imperative battles were being fought. Nor could the responsibility be so easily shifted; the original notion of using guerrilla forces for conventional warfare was itself of course at fault.

These two major partisan efforts apart, there were countless minor operations, the story of which has been described in exhaustive detail.[56] When General Marshall wrote in 1946 that the resistance in Normandy had assured the success of Allied landings by delaying the arrival of German reinforcements and in preventing the regrouping of enemy divisions in the interior, he referred not to one major, spectacular engagement, but to hundreds of small acts of sabotage, especially on the part of transport workers. Generally speaking, there is reason to believe that the railway workers of Europe contributed more to the Allied victory than the partisans.

Between 1942 and 1944 the French resistance suffered more losses as the result of betrayal within its own ranks than through German action. By early 1944 it had to some degree recovered and it played a certain role in the liberation of French towns and villages. But in the last stage of the liberation of France, military operations were dictated by the overt scramble for power. The Communist (FTP) decision to launch an insurrection in Paris while the Allied columns headed for the capital were yet on their way is the best-known example. Elsewhere the Gaullists tried to pre-empt the Communists. One circumstance, however, that perhaps distinguishes the French resistance from others of its kind at the time is that although it was internally no less divided than any politically, the conflicts between Communist and non-Communist groups brought no ostensible armed clashes as long as the Germans were the common foe. It was only in the interregnum between their retreat and before full civilian control was established that a free-for-all took place. Most of the participants in these postwar struggles had never even seen action against the Germans.

ITALY

The Italian military resistance differed in some immediate respects from partisan warfare elsewhere in Europe. In the first place it began only after Mussolini's downfall, when the final outcome of the war was scarcely any longer in doubt. Secondly, relations between Communists and other anti-Fascist forces, despite occasional strains, were more harmonious than was the usual rule. And further, if it produced no giant world figure, its partisan movement was headed by men such as Longo, Parri, Pajetta, who were to play an important part in Italian politics for years to come. Activist almost by definition, it ranged itself deliberately against the policy advocated by the less militant anti-Fascist leaders. As the movement saw it, Italy had to redeem itself before the world and to regain its self-respect after submitting for two decades to Fascist domination. The Italian resistance, to put it somewhat crudely, was the return ticket to democratic, anti-Fascist respectability. The Rome government, along with the Allied command, was less enthusiastic about partisan operations. Italy, like France, was save for a few exceptions not ideal guerrilla country: it could well be that the partisans might hinder more than help the Allied campaign. The Italian government tried to bring the partisan movement under its direct control, but as in France, the exercise was not wholly satisfactory. The Rome government appointed General Cadorna as the partisan chief of staff, whereupon the resistance (CLNAI) named two of its leaders, Longo and Parri, to keep a watchful eye on him.

The resistance claimed that the lack of Allied encouragement stemmed from political reasons, a not altogether unfounded charge. The Western governments were aware that at least a third of the partisan movement was under Communist administration. The second largest contingent was that of the Action party — the former *Giustizia e Libertà*. It is undeniable both that the tactics of the Italian resistance were not very well thought out and also that its strategy was politically motivated. A striking instance of this was the decision (against Allied advice) to establish large rather than small fighting units. Granted all the undoubted heroism, there was also a bombastic element in the heady invoking of *il secondo risorgimento.*[57] True, the difficulties facing the early partisans were formidable; they fought not only the Germans, but the diehards of the Duce's Fascist regime which still had its stalwarts throughout the German-occupied areas. Kesselring, the German commander, estimated that some thirteen thousand soldiers were killed and the same number wounded as the result of Italian partisan action; transport between Italy and the south of France was interrupted for ten days in October 1944 and some major Italian cities were temporarily cut off because of mined bridges, roads and railway lines.[58] But the partisans suffered many more losses — some estimates range as high as sixty-five thousand killed and wounded, and up to ten thousand civilians were executed during German and Fascist reprisals.[59]

Some partisan leaders had acquired military experience in the Spanish Civil War but this was of no great help in such different circumstances. They committed the same mistake as the French Maquis, attempting to establish liberated areas at a time when German military power was yet unbroken. For a while small partisan republics existed near the Swiss frontier (Ossola, Monferrato, Carnio), but at the end of six weeks, in late October 1944, they were destroyed by the Germans who did not even have to employ strong forces for the purpose. The partisans found themselves in the worst possible situation that could face a guerrilla — having to defend a static line without real fortifications, armor and artillery. It is only fair to add that Longo and other partisan commanders had warned against this strategy.[60] Following these disasters, General Alexander, the commander in chief of the Allied forces in Italy, appealed to the partisans on 13 November 1944 to stop large-scale operations; he did not tell them in so many words to go home, but advised them to save arms and ammunition for a better day and more propitious circumstances. From the partisans' point of view such advice was highly demoralizing; they regarded his call as an underhand trick, giving a new boost to bourgeois *attentisme*. But in truth there was no other strategy the partisans could have adopted once they had been defeated; it was only commonsense to prevent further unnecessary losses.

The partisan movement recovered its momentum the following spring. During the winter the organized bands numbered no more than a few thousand members, whereas on 1 March there were again eighty thousand men in the partisan army. By mid-April that figure had swollen to a hundred and thirty thousand, and at the time of the general insurrection there were two hundred and fifty to three hundred thousand.[61] But "general insurrection" usually meant not much more than stepping into the vacuum created by the German retreat, ousting the Fascist bureaucracy, and turning the Fascist *casa del popolo* into the headquarters of the Communist party.

The Italian partisans were from the start a political movement above all, their military activities were subordinated throughout to their political aims. In military terms their operations were of no great consequence, but their political impact was indubitable; but for them the country might have remained a monarchy — at least for far longer than it did.[62] As in France, the resistance movement in Italy was the cradle of many idealistic schemes for a better postwar world, of far-reaching internal changes, social justice, industrial democracy. Even in defeat these ideas continued to be cherished. And through its unfulfilled dreams the spirit of the resistance was to materialize as a distinctly tangible factor in the country's postwar political history.

POSTWAR REFLECTIONS

When the war broke out in 1939 no one thought that guerrilla operations would play any material part at all in the critical years ahead.

And it was perhaps by very virtue of this revival of partisan war being so unexpected in the first place that the pendulum later swung to the other extreme and its importance exaggerated. The impression is sometimes even created that it was the guerrillas who in fact won the war with occasional help of this or that regular army. These sources have it that there were two million partisans in the USSR, 50,000 in France, 462,000 in Italy and 250,000 in Bulgaria.[63] The partisans claimed to have killed millions of enemy officers and soldiers, not to mention local traitors; they allegedly destroyed fifteen thousand locomotives, fifteen thousand bridges, four thousand tanks and a thousand aircraft.[64] They claim to have diverted enemy forces amounting to between two hundred and fifty and three hundred Axis divisions, forty divisions in the USSR, fifty in Poland, fifty-five in Yugoslavia, more than thirty divisions in France, twenty in Greece, ten to fifteen in Czechoslavakia, eight in Albania and so on down the line. These figures certainly do not err on the side of understatement.

The real number of partisans is virtually impossible to establish. Much, of course, depends on the definition of the term. If one applies it liberally and includes men and women who were ready to hide partisans for a night or who expressed sympathy with them, there may have been millions of them. The dangers involved should not be belittled; it took a brave person to give shelter to a partisan for it could mean execution. If, on the other hand, one counts only those who actually participated in the armed struggle against the enemy, the number is much smaller; the less resistance there was, the taller quite often the claims. Furthermore, there are great discrepancies between the figures given by Soviet, Yugoslav, French and other sources at various times and in various contexts. Recent Soviet sources quote the total number of people involved at one time or another in the partisan movement as seven hundred thousand; Western sources put it at five hundred thousand. But the rate of attrition was high; many lost their lives, others were wounded or captured, some deserted or were sent back to join the regular army. The maximum strength of the Soviet partisan movement at any given time was two hundred to two hundred and fifty thousand men and women.[65]

The discrepancies are even greater when it comes to the results of the partisans' operations. The Yugoslav partisans alone, it was noted, claim to have killed and wounded almost a million "enemies." If this figure is correct, most of the victims must have been Yugoslavs. The same applies to other European countries; a great many people perished in Europe during and after the war in civil wars, purges and the settlement of all manner of accounts, but this did not necessarily weaken the German war effort. If two to three hundred divisions had been diverted by the partisans, the war would have been over by 1943 at the lastest. Again, much depends on the interpretation of the term "diverting enemy forces." A French author rightly observed that in the German operations against partisans in the Soviet Union far more forces were diverted than in the entire North African campaign which involved a mere twelve divisions.[66] But with equal justice it could be claimed that the number of German troops stationed in Norway in 1944 (372,000 men) exceeded that of German antiguerrilla forces in Russia and Yugoslavia, and this despite the fact that there was no guerrilla warfare in Norway. Such uncritical comparisons are, of course, absurd; occupying armies have to station some divisions in their rear quite irrespective of the incidence of guerrilla warfare. The larger the territory occupied, the greater the number of forces that have to be deployed as garrisons, to police it and to safeguard supplies and communications. Considering that the Germans occupied vast territories with a total population of more than two hundred million, the forces stationed in the rear of the German armies were few, and the aggregate of tanks, aircraft and heavy artillery diverted for antiguerrilla warfare was insignificant.

Some of the reasons making for the partisans' exaggerations have already been gone into. There were other motives, conscious and unconscious, such as the compulsion in some of the occupied countries to wash away with braggart tales of gallant resistance exploits the shame of the defeat and of collaboration; the majority of the population, so it was avowed, had been actively anti-Fascist all along. There were numerous such brave men and women who fought the occupiers from the very beginning, but for every one of them there were a hundred (or perhaps a thousand) last-minute resistance fighters who would put on a red armband or don some fancy uniform to join the partisans in the victory parade. The genuine partisans were contemptuous of these late arrivals but they still needed manpower, so welcomed them in their ranks nonetheless. Not that these tardy recruits were necessarily all cowards or opportunists; many of them had perhaps for long sympathized with the resistance. But not everyone is born to be a hero, and besides, in some countries it may indeed have been physically impossible to join a partisan group earlier.

The very fact that the military experts had prematurely announced the demise of guerrilla warfare, or ignored the subject altogether before 1939, acted as a spur in itself to further the tendency after the war to overrate it. Now it became the fashion to proclaim Marx and Engels, Lenin and Stalin as the great strategists of guerrilla warfare.[67] German writers who had taken part in anti-guerrilla operations discovered belatedly that partisans were not just bandits, and that the problem could not be solved by military repression alone.[68] To which one must hasten to add in due fairness that in the documents and memoirs of the supreme warlords, of Hitler and Stalin, of Churchill and de Gaulle, there are but few references to guerrilla warfare. Liddell Hart, who had been so enthusiastic about Lawrence's accomplishments in the First World War, had strong reservations about the efficacy of partisan warfare in the second.*

The real causes of the proliferation of guerrilla warfare are not shrouded in mystery; they were the same, broadly speaking, as those that provoked resistance against Napoleon in Spain, Russia, the Tyrol and elsewhere. Both Napoleon and Hitler had occupied many lands and dispatched their armies to faraway countries. They had both overextended their supply lines and spread their forces very thin. But whereas Napoleon scarcely intervened in the internal affairs of occupied countries, except perhaps by appointing a relative to rule it, the Germans interfered brutally and on a massive scale, and this was bound to intensify the struggle against them. Terror produced counterterror, and given the heavy demands on their manpower, the Germans lacked the soldiery to destroy the partisans if these operated in favorable conditions.

The partisan experience during World War II again demonstrated the paramount importance of geography. But it also pointed up the significance of psychological factors; but for the dedication of capable cadres, and rallied political sympathies among the local populations, there could have been no guerrilla warfare. A partisan

* Liddell Hart's enthusiasm for guerrilla war waned because he found it both ineffective and politically counterproductive. He wrote in 1954 that the partisans in the Second World War had rarely had more than nuisance value except when their operations coincided with the imminent threat of a powerful offensive absorbing the enemy's main attention: "At other times they were less effective than widespread passive resistance — and brought far more harm to the people in their own country. They provoked reprisals much more severe than the injury inflicted on the enemy. They afforded his troops the opportunity for violent action that is always a relief to the nerves of a garrison in an unfriendly country." Liddell Hart recalls a meeting with Wingate, then serving as a captain in Palestine, in the late 1930s. He was beginning to have doubts about the long term consequences of guerrilla war — the political and moral ill effects which would inevitably continue after the invaders had gone. (B. H. Liddell Hart, *Strategy* [New York, 1967], 368–370.)

movement needed space to maneuver, and areas in which it could hide. This ruled out most of Western Europe; with German concentrations in all the major cities, urban guerrilla operations were also impossible there. If the local garrisons had not sufficed to subdue the partisans, the German air force would have bombed them out of existence. Allied intelligence agents and individual resistance fighters could hide in a city, major partisan units could not. In the German scale of priorities, Western Europe mattered far more than the Balkans. The presence of small guerrilla bands in the mountains of Albania was a mere pinprick, it did not threaten any vital military or political German interests. If, on the other hand, the Maquis had been able to gain a firm foothold anywhere in France, the German military leadership would have had to destroy it at any price because this would have constituted a direct military threat. But even if geographical conditions had been more propitious, it is doubtful whether a major partisan movement would have emerged in Western Europe. Partisan war, *guerre à l'outrance,* was not in the tradition of civilized nations; as Engels had noted many years before, Western Europe was no longer conditioned for a war of this kind. There was opposition to foreign rule, but not necessarily the willingness to sacrifice life and property, to risk the destruction of cities and perhaps the entire nation. Furthermore, Nazi terror was much less in evidence in Western than in Eastern Europe. The West, as far as the Nazis were concerned, was merely decadent, whereas Eastern Europe was racially inferior; the fewer Slavs, the better.

The attitude of the civilian population was, of course, of crucial importance. The partisans could knock on many a door and expect help, or at the very least be certain that they would not be betrayed. The strength of the partisan movement's appeal lay in its patriotic character; revolutionary slogans would have been wholly ineffective and the Communists were the first to acknowledge this.

Conversely, it was this very strength of patriotic feeling that prevented the emergence of a resistance movement directed against the government inside Nazi Germany and inside the Soviet Union. The German and the Soviet political police had no difficulty in putting down any manifestation of political dissent and obstructing any organized resistance. The majority of the population in these countries either supported the regime or was cowed into submission. It is revealing that of the German Communists parachuted into German territory by the Soviets to organize cells of resistance and to collect information, only one was not caught within a week or two, and this lone survivor, significantly, infiltrated toward the end of the war, was an agent of Polish extraction who hid among Polish friends in Upper Silesia. (One or two German socialist parachutists also survived, but this was in the last phase of the war when the activities of the Gestapo were already disrupted as a result of saturation air raids.) The Germans tried on various occasions to send agents to the Soviet rear, but there is reason to believe that most of them were arrested before very long. The only trenchant resistance could have come from inside the German army, but the oppositionist generals and colonels were not at all sure whether their troops would follow them.

From what has been said so far it would appear that the partisans in Europe were not strikingly successful in their paramount task, namely, to inflict decisive damage on the Axis forces. One of the foremost historians of the resistance, who can scarcely be suspected of lack of sympathy, later wrote that "the vast majority of German units never came into direct contact with the guerrillas."[69] The only expression of real anger and concern about partisan warfare on Hitler's part was his directive No. 46, dated 18 August 1942. ("In recent months banditry in the East has assumed intolerable proportions.") But in 1943–1944, when "banditry" had assumed far greater proportions, there was no such outburst. As the German position deteriorated, the German military leaders were totally preoccupied with the regular forces facing them.

Conditions differed from one country to another and political perspectives were subject to change. The Soviet partisans certainly had no doubt that their chief assignment was to wreak the worst possible havoc on the Germans. Elsewhere in Europe the situation was far more complex; by the time the Balkan partisans reached the peak of their strength — in the autumn of 1943 — the German summer offensive in Russia had failed and Italy had surrendered. It was obvious by now that the Germans were going to lose the war and this affected the political perspective of Communists and non-Communists alike. Seen from the Communist point of view, it was far more important to destroy their domestic rivals than engage in costly battles against the Germans who were doomed anyway. The anti-Communist partisans also stopped operations against the Germans, unless they were attacked, because survival had become their top priority. True, some of the heaviest fighting took place in the Balkans in late 1943, but this was invariably the result of German antiguerrilla campaigns, not of partisan offensives. As in China, priorities had changed; if at the start everything (or almost everything) had been subordinated to the anti-Fascist war effort, from 1943 on at the latest, the anti-Fascist fight was subjugated to winning the struggle for power inside each country.

Whether one considers the record of the European partisan movements a story of success or failure or something in between depends entirely on the yardstick employed. Insofar as winning the war is concerned, Alan Milward correctly observes that, on the evidence, the history of the resistance should have a far lesser place in the history of World War II. "As an individual act resistance was liberating, satisfying and necessary; on a coordinated level it seems to have been seldom effective, sometimes stultifying, frequently dangerous and almost always too costly."[70] But then, the task of the Communist resistance movements was, of course, not only to help win the war.

Seen in this light, their operations were not, of course, a failure but an outstanding success; the guerrilla movement was an excellent school for the mobilization of masses and the training of cadres, it was the most effective tool for the seizure of power. There was widespread feeling in the west of Europe as well as in the south that the old elites had failed — hence the defeats in the early days of the war. It was widely held that the old system had been corrupt, that out of the ashes of destruction a new order would emerge — just, more humane and effective, under the leadership of the progressive forces headed by the Communist parties.

The record shows that in Yugoslavia and Albania the Communists destroyed the rival forces before the end of the war, in Greece they almost succeeded, in France and Italy they were in a very strong position but failed. Wherever Communist partisans did not succeed it was principally because the presence of Allied troops made an armed takeover impossible. Once parliamentary democracy was restored, the majority was bound to reassert itself, and the "vanguard" had missed its opportunity. Irrespective of the existence of partisan movements, Communist regimes would of course have been imposed in Eastern Europe, but this does not detract from the overall achievement of the partisans. The political repercussions of the resistance were for long to echo — still echo resoundingly — through the west, south and east of Europe.

6

The Twentieth Century (III): China and Vietnam

Of all guerrilla wars in modern history those in China and Vietnam have been the most important, and all others have been in comparison of regional significance only. The war in China resulted in the victory of a new social and political order in the most populous nation in the world; the Vietnamese war caused a deep crisis in the United States, and it is too early to assess its impact on the global balance of power. The wars in China and Vietnam were won by political parties inspired by a mixture of Communism and nationalism whose elites came to correctly understand and apply the ideas of peasant support, political organization and propaganda. By trial and error they chose the military strategy most likely to succeed in local conditions. The Communists won because the Japanese occupation of China and Vietnam had discredited these states' former rulers and had created, with Japan's defeat, a power vacuum. The Japanese were not defeated by the Chinese Communists, and in Vietnam there was hardly any anti-Japanese resistance; but in both countries, the Japanese acted as the catalyst for the victory of Communism. The Communists prevailed in the struggle for power that followed the Japanese defeat because they were the only modern political party that appeared both as the "party of the poor" and at the same time as the leading patriotic resistance movement.

For many years the Chinese Communists received virtually no help from outside. Their tactics were traditionally Chinese; foreign examples, such as the Russian revolution, were of little relevance to them. New elements for the Communists were the political techniques, and the subordination of the military struggle to political aims. The Vietnamese Communists were in a more fortunate position inasmuch as they received foreign assistance on a massive scale from a very early stage of their war. On the other hand they had to do more fighting than the Chinese Communists, and their room for maneuvering was more limited; they also had to cope with an enemy using superior equipment. For both the Chinese and the Vietnamese Communists guerrilla warfare was a transitional stage in the struggle: they were fully aware that guerrilla tactics could not possibly win the war and that sooner or later these would have to be supplemented by and ultimately subordinated to regular war. The Chinese Communists had used guerrilla tactics on and off from the very beginning but they were employed, on the whole, in the framework of regular warfare; it is quite likely that the Communists would have been defeated by Chiang's fifth encirclement campaign anyway, but the defeat was certainly precipitated by their engaging in positional rather than guerrilla warfare. In view of the numbers involved in it, the Long March was not, and could not possibly have been, a guerrilla operation, though in the skirmishes guerrilla tactics were of course frequently used. Guerrilla warfare was an essential stage in the war against Japan, but the Communist commanders sounded sometimes almost apologetic: thus Peng Te-huai said in 1938, "The growing partisan units in North China will very rapidly be transformed into a regular army, and this new army will be better than the present one . . ."[1]

In fact, the transformation took much longer than expected, and it was only after the surrender of the Japanese that the "new armies" entered the fighting. As usual reality was far more complex than doctrine: up to 1945 there were guerrilla bands that had the strength of a division, and on the other hand there were regular army units that used guerrilla tactics; the dividing line was far more fluid in practice than in theory.

In Vietnam the Communists engaged in regular army operations early on in their war against the French, and suffered some major defeats. By trial and error they came to combine guerrilla and conventional warfare and, as in China between 1928 and 1935, they constantly discussed to which form of military operations preference should be given. Eventually their emphasis also shifted to regular warfare.

The facts about guerrilla warfare in China and Vietnam are not easy to unravel since they have become the object of hagiography rather than of critical study. Those who fought the Communists, having originally underrated their tenacity, fighting spirit and inventiveness, frequently ended by regarding the victors as a race of supermen, thus exaggerating the originality and general applicability of the guerrilla strategy of Mao and Lin Piao, of Ho and Giap. They came to accept "revolutionary warfare" as an irresistible force, as the greatest revolution in modern politics as well as in modern warfare.[2] Thus a recent historian has written: "For the first time in a century the armed might of an industrial power was fought to a standstill, its dreams of empire crushed, by a people in arms."[3] It is certainly true that Japan in 1937 was an "industrial power" to the extent that it had an industry. But the military importance of the textile industry — its chief industry at the time — was strictly limited, and its heavy industry was still in its infancy. Its share in world industrial production was only about three percent. Japan had started to produce motor vehicles only in the early 1930s, and by the end of the decade the number of Japanese trucks and cars was less than ten per cent of Britain's. Furthermore, Japan's dreams of empire were shattered not in China but on other fronts; the number of Japanese casualties in the Philippines alone far exceeded those suffered at the hands of the Chinese Communists. If, towards the end of the war, the Japanese evacuated much of China this was not the result of fighting in mainland China but of the defeats suffered in the Pacific, mainly a consequence of the American victory at Okinawa. The Russians, needless to say, have not joined the general chorus of admiration for obvious political reasons: poor Mao had always been mistaken, had engaged in either right- or left-wing deviations; the general line had somehow always eluded him.[4]

In the search for historical truth one has to proceed beyond mythology and political polemics: despite all the tenacity and courage displayed by the Chinese Communists, they were on more than one occasion exceedingly lucky. They operated in near ideal conditions: there was no strong central authority in China even before the Japanese invasion. Once the war had started in 1937 the Communists enjoyed virtual immunity in their bases in northern China. For the Japanese, the Communist guerrillas were not a serious danger, and this despite that the Japanese occupation army was small by any standards. Indeed, the Chinese Communists did little fighting against the Japanese after 1940, though Chiang Kai-chek's troops did even less. Mao's policy was to devote seventy percent of the Communists' effort to expansion, twenty percent to coping with the Kuomintang government and ten percent to fighting the Japanese. The Vietnamese Communists' war of liberation against the

Japanese belongs almost entirely to the realm of mythology, for their military organization came into being only towards the end of the Second World War. After the Japanese surrender and before the French returned in strength, there was an interval during which the Communists obtained a great many arms and consolidated their political and military power base. After that they faced the troops of the French Fourth Republic, a regime quite incapable of waging and sustaining a "dirty colonial war." By the time American troops became involved in Vietnam a major Communist army had already come into existence that was better trained and more experienced, if much less well equipped, than its enemy. During their struggle the Chinese and Vietnamese Communists displayed a great many sterling qualities and, of course, greater political foresight. But given the vast size of China's territory, the weak base of Japanese militarism and the chaotic state of Chinese domestic politics, it cannot fairly be argued that the victory of the Chinese Communists was an unprecedented, superhuman achievement. Seen in historical perspective the military odds were heavier against the Vietnamese Communists, but the political factors involved in the struggle for Southeast Asia favored them to a great extent, and these, in the end, proved to be of decisive importance. The stories of the guerrilla wars in China and Vietnam have been told countless times; the following account concentrates on a discussion of some of their essential features.

CHINA

Guerrilla warfare in China developed against the background of peasant unrest, the breakdown of the central authority and the presence of a Communist elite which, by trial and error, evolved a more effective strategy than did its rivals for coping with China's problems. The decisive turning point was the Japanese invasion, which aggravated the general discontent, discredited and paralyzed the central government and thus created the pre-conditions for the Communists to spearhead the armed struggle in northern China.

Agrarian unrest had been endemic in China for many centuries; furthermore the peasant wars had been on a much greater scale than those in Europe. In the T'ai Ping rebellion some three million insurgents had fought, and millions of civilians had been killed. The peasantry suffered from natural calamities and rising rents, as well as the requisitions and pillaging of warlords and their soldiers. Local industry and handicraft were undermined by cheap foreign imports.[5] Most of this had happened before in Chinese history; the young peasants would "take to the mountains," a euphemism for banditry, and, generally speaking, *éléments déclassés* proliferated. Those remaining in the villages would establish self-defense societies to struggle against bandits, warlords, tax collectors and other plagues. The secret societies such as "Red Spear" and "Great Knife" had supporters in many parts of the country due to widespread unemployment in the villages and increasing militancy among a peasantry wanting land.

Since the fall of the Manchu Dynasty in 1911, there had been no effective central authority in China. Chiang Kai-chek's attempts to reunite China in the late 1920s were only partly successful; the warlords were integrated into the system as subcontractors who collaborated, or refused to do so, as they saw fit. Their loyalty could never be fully trusted.

The Communist Party of China was founded in 1921. The main stages of its growth during its first decade are well known and need not be reiterated. Its influence spread among sections of the intelligentsia, particularly the students and the urban younger generation. It also had some support among the working class. The party established a united front with the Kuomintang but it was defeated in the struggle for power by Chiang, its erstwhile ally, and its urban insurrections failed totally. The Communists had never entirely neglected the revolutionary potential of the countryside, but they had given it low priority; according to Marxist-Leninist doctrine only a mass movement led by the industrial proletariat could be truly revolutionary and so prevail in the end. From time to time resolutions would be passed by the Comintern (as in 1922 and again in 1926) that the revolutionary movement in the backward countries could not be successful unless it relied on the broad peasant masses. But the magic wand to mobilize the peasants and to combine their struggle with that of the industrial proletariat had not yet been discovered. The Chinese Communists began to reveal greater interest in the countryside only in 1927, the year of their great defeats in the cities. The two processes of revolution were not, of course, unconnected. When Mao, head of the peasant department of the party, returned from an inspection tour in Hunan he wrote that the upsurge of the peasant movement was a colossal event, that several hundred million peasants would rise and that this would be a hurricane, a force so swift and violent that no power could hold it back.[6] Mao was, of course, not the first to discover the revolutionary potential of the peasantry. The tactics he suggested at the time were, in fact, less far-reaching than the Comintern line which advocated confiscation of the land. But Mao was among the first to make use of the peasants' potential and to move beyond extreme but noncommittal slogans. There had been among the precursors of Communism in China a strong pro-agrarian, populist bias mingled with antiurban sentiments and nationalist overtones: the young intellectuals (it was argued) should go back to the villages to free themselves from the corruption and the vices of town life. There was also a great belief in the spontaneous energies of the people and in revolutionary voluntarism.[7] The idea that people, not things, are of paramount importance was not entirely new; the theory and practice of Maoism did not develop in a vacuum. Mao's great merit was to open a new vista to the party at the very time when its fortunes were at a low ebb.

From August 1927 to the "Autumn Harvest Uprising" and the establishment of a first Red Army to the beginning of the Long March in October 1934, Mao and his supporters, with Chu-Teh and Peng Teh-huai as the chief military commanders, engaged in armed operations in the Hunan-Kiangsi-Fukien area and built their first revolutionary bases. But the great hurricane Mao had predicted did not yet break: there were many military defeats, and he came under strong attack in his own party. In fact, in November 1927 he was ousted from the Politburo. He made a temporary comeback, but from 1932 to 1934 his influence was again on the wane. It was only in January 1935 that he was elected chairman of the Military Affairs Committee and became the *de facto* leader of the party. The story of these seven years was basically one of an almost uninterrupted struggle for survival. By trial and error the Communists gradually developed in the Kiangsi Special Border Area a system of guerrilla warfare that in some respects followed traditional Chinese patterns. In later years Mao summarized the experience that had been gathered in the fighting:

> Divide our forces to arouse the masses, concentrate our forces to deal with the enemy. The enemy advances, we retreat; the enemy camps, we harass; the enemy tires, we attack; the enemy retreats, we pursue. To extend stable base areas, employ the policy of advancing in waves; when pursued by a powerful enemy, employ the policy of circling around. Arouse the largest number of the masses in the shortest possible time and the best possible methods. These tactics are just like casting a net; at any moment we should be able to cast it or

draw it in. We cast it wide to win over the masses and draw it in to deal with the enemy. . . .[8]

Mao claimed that these tactics "are indeed different from any other tactics, ancient or modern, Chinese or foreign." But neither for Sertorius nor for Viriatus two thousand years earlier would these ideas have been startling revelations — nor for guerrilla leaders all over the globe throughout history. Although neither the strategy nor the tactics were novel, their use in the framework of a political doctrine was. The idea to combine them had occurred to others before but they had never been applied on such a scale, or ultimately with such effect.

Mao and his comrades-in-arms faced overwhelming odds in the early years of the fighting. The enemy was weak, but they were even weaker. They had no supplies, no arms, no money except what they could seize from the enemy. The human material they were dealing with was unpromising: bandits, vagrants, various *éléments déclassés*, deserters from the enemy camp who had not been paid their wages — a veritable riffraff. They had, as a historian of the Chinese army put it, "few bricks and little straw, but they had a clearly defined goal and the determination and perseverance needed to attain it."[9] The Communists were not just another party or clique of warlords; they had an ideology that at one and the same time provided an explanation of the world and a guide to action for changing it. It generated among the younger generation enthusiasm and the willingness to sacrifice. They were far more highly motivated than their rivals and they realized that it was necessary to establish a much closer relationship between the soldiers and their commanders and between the army and the people; they had a method to mobilize the masses that was more effective, more in line with Chinese realities than that of the Kuomintang or the warlords. It has been noted that there was no great difference between the social background of the military leaders of the Communists and their enemies;[10] some Communist commanders (such as Nieh Jung-chen, Ho Lung and Chu Teh) hailed from poor peasant familes, and there were also officers of similar background in Chiang's armies. Communist and Nationalist commanders had studied in the famous Whampoa military academy; their careers were parallel in some respects, and they were heirs in many ways to the same tradition. The main difference was that those who had embraced Communism were willing to share the life of the common people and its misfortunes and deprivations. The Communist party curbed its officers' individualistic traits, whereas the Nationalist officers were not subject to such discipline; they lacked the feeling of serving a common cause and were incapable of cooperating.

During the early years of its mountain warfare the Red Army was engaged in guerrilla operations; these, however, were on the whole subordinate to regular army activities. Mao was not in charge of operations during much of the time and it is in retrospect almost impossible to say with any degree of accuracy when and in what circumstances guerrilla doctrine developed. In 1928 Mao called for strict discipline because "guerrillaism" ("the tendency to destroy cities and kill, burn and rob purposelessly") had to be extirpated. This (he argued) was merely a manifestation of the *Lumpenproletariat* and peasant mentality that might hamper development of the party among the peasant masses.[11] In 1929–1930 Mao bitterly denounced the idea of "pure guerrilla warfare," the "aimless and ineffectual raids of vagabond elements." The Red troops should not be dispersed but instead should establish and consolidate revolutionary bases.[12] Communist experience with guerrilla warfare frequently had been negative; several guerrilla leaders had even deserted. Some of the leading military experts such as Chu Teh talked about the "guerrilla quagmire" into which the party should not sink; guerrilla tactics were useless against the blockhouse system that Chiang's troops applied in accord with the advice given by foreign military specialists.[13] It was only after the massive defeat of the Red Army at Kuangch'ang (April 1934) that guerrilla tactics were widely employed, and that the Tsunyi Conference passed a resolution that constituted a decisive switch from positional to mobile warfare. Preference to real guerrilla warfare was given only after the Japanese invasion in September 1937.

The acrimonious disputes about what was the correct strategy lasted throughout the early 1930s. Mao and his supporters saw the main task to be the establishing of "Red Areas" (the "highest form of the peasant struggle"), the foundation and development of a Red Army and the building up of a political and military power that would eventually lead to the encirclement of the cities by the countryside. Li Li-san, Mao's chief antagonist during that period, maintained that the center of gravity was still in the cities and that protracted warfare was "boxing tactics"; he for one was not willing to wait for victory until his hair had turned gray.[14]

The Maoist line in Chinese Communist strategy prevailed only after the struggle for power in the party had been settled. Only after it appeared that Communism had no future in the cities was Li Li-san's approach discredited and did the various emissaries from Moscow lose their influence. The Communists, to be sure, had not done that well in the countryside either, but at least they had preserved their strength. Chiang's first two encirclement campaigns against the "Communist bandits" ended in failure: the government troops were not familiar with the terrain, their units were dispersed and they had no intelligence about Communist movements. Above all, they thought the campaigns would be a walkover and did not take the enemy seriously. The Communists captured thousands of prisoners and seized considerable booty. In the third encirclement campaign Chiang himself took command and employed some of his best troops. The Communists were on the brink of defeat when the Mukden incident (the occupation of Manchuria by the Japanese in October 1931) compelled Chiang to call off his campaign. This campaign had shown that in the long run the position of the Communists was untenable.

In the fifth encirclement campaign (October 1933) the Communists' area was ringed with blockhouses, their supply lines were cut and their sources of intelligence dried up. With the Nationalists' evacuation of part of the population and the organization of local militias against them the Communists' survival was no longer certain. The Communists had used the wrong tactics — "halting the enemy in front of the gate," that is, fighting for every inch of territory. This was, of course, exactly what Chiang had hoped for because the elusive enemy had at last become easily identifiable. In later years the blame would be put on one Li Te (Otto Braun), a German Communist who had been delegated as military adviser. But the Communists would have lost even if they had reverted to guerrilla tactics. In this desperate situation the decision was taken to transfer the Communist forces to some other part of China, and on 16 October 1934 some hundred thousand men and women set out on the famous Long March. No one knew at the time where it would lead to.

The Long March has entered the annals of Chinese and world history as an achievement without precedent. Fourteen months later, after the Red Army had arrived in Shansi, Mao claimed that the Long March was a "manifesto, an agitation corps and a seeding machine." For twelve months (he related) the Communists had been under daily surveillance and bombing from the air; they had been encircled, pursued, obstructed and intercepted by hundreds of thousands of enemy soldiers; they had crossed eighteen mountain ranges, twenty-four rivers, had taken sixty-two cities. They had engaged in two hundred and thirty-five day marches and eighteen night marches and fought almost one skirmish a day against government troops and provincial warlords. They had covered a distance of over six thousand miles.[15]

As so often, it is not easy to differentiate between fact and fiction. That the Long March was not a picnic goes without saying; for to move units such distances involved great stamina even if there had not been an enemy to harass and obstruct them. On the other hand enemy attacks were not as frequent and fierce as the Communists claimed in later years: according to the official Communist version the First Front army averaged twenty-four miles a day over a 368 day period — including 44 days of rest in Szechuan. From a practical point of view no army involved in constant combat, even to a limited degree, could have maintained this pace.[16] There was no unified effort to attack the Red Armies: Chiang Kai-chek's units were only rarely in contact with them and the local warlords were only too eager to leave the Communists unmolested, provided that they did not bother them. Some of the hardest fighting went on not against the Nationalist armies but against the (non-Chinese) Lolo tribesmen in southwest China. The main obstacles on the march were in fact natural ones such as the eternal glaciers of the Great Snow Mountains, the marshy grasslands of Ching hai and the savage Tatu River.

These obstacles caused the most casualties. Of the hundred thousand who had set out on the march, only about a third arrived at Shensi in October 1935. This threadbare band represented at best only a marginal element in Chinese political life, but "sustained principally by discipline, hope and political formulae it had . . . several hidden assets which were later to prove of major significance" (Howard L. Boorman). In retrospect it has been maintained that the operation was a stroke of genius meticulously planned from the beginning. In fact, there is every reason to believe that the Communists were simply improvising; for a long time they had no clear idea about their eventual destination. On at least one occasion, Mao noted that the whole march had been unnecessary and that the original base could have been kept.

The Long March was not a major victory but a great retreat. But the Communists turned military defeat into a propagandistic victory, for Chiang had after all failed to destroy them; their forces seemed invincible. Above all, the march had given them outstanding training, far better than any military academy could provide. Those who had been steeled in battle during this *annus mira-*

bilis would stick together in the stormy years ahead until they became the masters of a new China. (After the victory, to be sure, human nature and the vicissitudes of politics would reassert themselves, and the veterans of the Tatu River would find themselves in opposite camps in the struggle for power.) The choice of northern Shensi as the main base was an astute one and for the next ten years Yenan was the capital of Chinese Communism — partly perhaps by accident. This remote, exceedingly poor and backward part of China offered little temptation to warlords and was more or less ignored by the Chinese central government and the Japanese. On the other hand, there was already a Communist guerrilla tradition in this area adjacent to the north China plain. Within less than three years of their arrival the Communist forces counted a hundred and fifty thousand soldiers; by 1940, with the expansion of their bases, their number had risen to some four hundred thousand. Almost unnoticed the Communists had re-emerged as a major factor in Chinese politics.

The Long March was a great test of endurance but otherwise it offers little that is new to the student of military history, for it proceeded, roughly speaking, like all long marches in history from the *anabasis* onwards. The Communists occasionally used guerrilla tactics, but the retreat of so big a force had to proceed largely on orthodox lines. At times the Red units presented a "snake" fifty miles long. Chiang Kai-chek himself would fly reconnaissance missions: there was never a secret about the location of the Red Armies. It was just another maneuver, albeit one of historical importance, in the general course of the Chinese civil war. If the politics of the Communists were revolutionary, their military strategy was fairly orthodox, and it changed only to a certain extent during the Yenan period.[17]

The wilderness of northern Shensi was, as a Communist leader told Edgar Snow, culturally one of the darkest places on earth: "We have to start everything from the beginning."[18] But some politico-military spade work had already been done in the area; for almost a decade previously banditry had gradually changed to viable guerrilla operations. In the course of eight years of setbacks, local partisans had independently developed a set of military and political postulates appropriate to survival and revolutionary growth in the northern Shensi area.[19] The Shensi disputes about whether and in what way guerrilla tactics should be used were almost identical to the discussions among the Communist leadership in the Kiangsi Soviet. The issue was decided in favor of the "guerrillaists" (such as Kao Kang) only upon the arrival of Mao and his companions in 1935.

Having made its way to northern China the Red Army began to reorganize and expand its units and to consolidate its political base. There was hardly any outside interference. In 1936 Chiang had instructed the Manchurian army to attack the Communists, but its attacks showed little determination: to Chinese patriots the Japanese, not the Communists, were already the main enemy. A second united front between the Communists and the Kuomintang came into being after the Sian incident (December 1936), and hostilities ceased for a number of years. The understanding between the two sides was rather vague, however. It certainly did not prevent the Communists from expanding the areas under their control nor Chiang's troops (in December 1939) from taking some of these areas back and imposing a blockade on the Soviet regions. In January 1941 Kuomintang forces moved against the New Fourth (Communist) Army which was based south of the Yangtse and which had made inroads in what the KMT considered its own territory. But the blockade apart, there was little fighting in northern China. Having lost their major cities and lines of communication and having to bear the brunt of the Japanese offensive, the KMT could no longer launch a major campaign against the Communists. It was reduced to adopting a "negative policy" of restricting as much as possible the areas of Communist control and suppressing Communist activities in areas under its control.[20] Thus, for the first time in their existence, the Red Armies were no longer fighting for mere survival; they could invest their energies in the expansion of their military units and bases. Russia's and America's entry into the Second World War offered further protection, for the Japanese could no longer concentrate their military efforts in China.

Communist political activities in Yenan, such as land reform, political organization and indoctrination, were, of course, closely connected with military strategy; they have been described and analyzed in great detail and need not be retold here. We should instead turn to the military doctrine as formulated in the late 1930s by Mao, Lin Piao and others and to the specific place of guerrilla fighting within the general concept of revolutionary warfare. It has been noted that Mao's three major essays on guerrilla warfare were all written during the first half of 1938;[21] Lin Piao's one major essay on the topic was also apparently written in 1938.[22] Mao's ideas on revolution in China rest on three basic assumptions: that military forces will play a decisive role and that the fight will be carried on by the villages, which will eventually "encircle" the cities — hence the importance of rural base areas. Lastly there is the concept of "protracted struggle." These ideas took about a decade to evolve, the concept of protracted war being the last of the three to be pronounced and in some ways the most difficult to accept, for it is the natural inclination of soldiers, as of politicians, to prefer a short war to a long one. On guerrilla warfare Mao usually took a "centrist" position between those who advocated "pure guerrillaism" and dismissed regular warfare altogether and the military (or party) experts who took the opposite view. Mao always claimed that, in the last resort, the correct choice of an approach depended on the circumstances. He argued that positional warfare was ruled out as long as the Red Armies were not strong, lacked reserves, supplies and ammunition. Mobile warfare was the answer since in China as in the Soviet Union in 1918–1919 or in other revolutionary wars there could be no fixed battle lines. "Fluidity of battle lines" meant that the base areas, too, had to be fluid, constantly expanding and contracting, "as often as one base area falls another rises."[23] In a rare flash of irony Mao notes that the very comrades who had been the strongest advocates of regular warfare in 1933–1934 "managing affairs as if they were rulers of a big state," had in fact caused extraordinary and immense fluidity — the "Long March."

Mao noted that while military doctrine in general favored both "moving" and "fighting," few people did as much moving as the Chinese Communists: "We generally spend more time in moving than in fighting and would be doing well if we fought an average of one sizeable battle a month."[24] As Mao saw it, the rejection of guerrilla warfare and fluidity had been practiced on a large scale. But upon the arrival of the Communist army in northern China and the establishment of base areas, conditions had again changed. Guerrillaism was part of the infancy of the Red Army; this involved irregularity, decentralization, lack of uniformity, absence of strict discipline and, generally speaking, simple methods of work. These were the negative aspects, but there were others that were still valid: the principle of mobile warfare, the guerrilla character of certain strategic and tactical operations, the inevitable fluidity of base areas and the rejection of premature regularization in building the Red Army.[25]

The task, as Mao saw it in 1936, was to end those practices of the Chinese guerrilla struggle that had become obsolete and to prepare for a new stage in which battle lines would become more stable and positional warfare more frequent. At the same time, it was vital not to discard the positive and still applicable lessons of the guerrilla past and not to "rush blindly" into something new, simply due to its newness.

In 1937, China, "a large and weak country," was attacked by a "small and strong country." Mao was optimistic about the outcome. The Chinese regular armies had been beaten but the Japanese did not have sufficient manpower to occupy the whole of China, and many gaps were left. In these circumstances Communist strategy should be to engage in independent ("exterior line") operations rather than in interior line actions in support of regular troops.

Mao's prescriptions for the war against Japan were simple and straightforward, and can be summarized as follows: the basic aim in war is to preserve one's strength and destroy that of the enemy. In revolutionary war this principle is directly linked with basic political aims — to drive out the Japanese and to build an independent, free and happy (that is, Communist) China. The correct approach during the early phase of the war is the strategic defensive or, to be precise, the frequent and effective use of the tactical offensive within the strategic defensive. Guerrilla warfare involves careful planning and flexibility ("breaking up the whole into parts" and "assembling the parts into a whole"). Pure defense and retreat can play only a temporary role in self-preservation; the offensive is the only means of destroying the enemy, and it is also the principle means of self-preservation. Offensive operations must be well organized and not be launched under pressure.[26]

The constant repetition by Mao of these elementary insights does not detract from their educational value. But there was yet another major problem: how to coordinate guerrilla with regular warfare. Such coordination should exist on the level of general strategy — pinning the enemy forces down behind the front line, disrupting his supply lines, and, generally speaking, spreading demoralization among his troops and providing fresh courage to the Chinese people. But there should also be coordination in specific campaigns and battles. Mao saw the third main problem of the Communists' revolutionary war to be the maintaining of the vital base areas: it would be impossible to sustain guerrilla operations in a protracted war behind the enemy lines without such strategic bases. They constituted the rear in a war without rear; without such bases the guerrillas would be mere "roving rebels" in the ancient Chinese tradition, noisy but ineffectual.[27]

Mao correctly predicted that a long time would pass before the Communists could launch their strategic counteroffensive to recover the lost territories. Meanwhile a great part of China's territory would remain in Japanese hands. Guerrilla warfare behind enemy lines would have to be extended over all this vast area. This would be impossible without bases either in the mountains, the plains or "river-lake-estuary regions." Mao differentiated between guerrilla zones and base areas. A "zone" becomes a base area only after large numbers of enemy troops have been annihilated, the puppet (collaborationist) regime destroyed and the masses roused to activity. The key to establishing a base area is the building up of armed force. Economic and also geographic conditions — mountains being preferable to plains — are of importance, but in view of the vastness of China's territory and the shortage of Japanese troops, guerrilla warfare can be conducted and sustained also in the plains. Propitious geographical conditions are desirable but not vital. Essential on the other hand is coordinating with and mobilizing the masses and organizing self-defense corps among the villagers to assist the hard core guerrillas. Existing base areas should be consolidated so as to enable them to withstand enemy attacks. At the same time they should be constantly expanded: conservatism in guerrilla warfare will not achieve the target, namely the confinement of the enemy to a few cities.

How are guerrillas to cope with enemy attempts to conquer the base areas? Mao advised guerrilla commanders to smash the enemy's converging attacks upon them. They should contemplate giving up their base area only after having failed repeatedly to achieve this objective. Mao (wrongly) predicted that the Japanese would not be able to use the blockhouse system because they were too few in number to make it work effectively. Once the enemy offensive had been halted the guerrillas could go over to the offensive — not by attacking the main enemy forces entrenched in defensive positions, but by driving out or destroying small enemy detachments and by expanding their base areas.

Mao maintained that since the war would be protracted and ruthless, the Japanese invaders could be defeated only if the guerrillas gradually transformed themselves into a regular army and adopted a system of mobile warfare. In certain mountain regions the elements of mobile warfare had existed from the beginning of the war and could simply be expanded. This involved increasing numbers and improving military and political quality.[28] Lastly was the problem of command. There cannot possibly be a high degree of centralization in guerrilla operations, otherwise their flexibility would be restricted and their vitality sapped. On the other hand, the centralization of command is essential for coordinating guerrilla with regular warfare. If absolute centralization is harmful and absolute decentralization ineffectual, the answer is to centralize the strategic command for the overall planning and directing of a guerrilla war and to decentralize it for campaigns and battles. In other words the central command should not interfere in details such as the specific dispositions for a battle.

Ten months after the outbreak of the war with Japan Mao outlined in detail his famous concept of protracted war.[29] There was no ground for pessimism (he wrote), for China would prevail in the end. Nor was there reason for excessive impetuosity; victory would take a long time. Even before the outbreak of the war Mao had stressed the importance of mobile warfare in combination with the operations of a great number of guerrilla units. The war of national liberation would gain support both at home and abroad. In view of China's vast territory, rich resources and large population, it could sustain a long war despite its being a semicolonial, semifeudal country. Japan's war, on the other hand, was reactionary and barbarous, its manpower and material resources inadequate and its international position unfavorable. Thus everything depended on perseverance. Mao assumed that the war would pass through three stages: at first the Japanese would take the strategic offensive. Later on the enemy would engage in strategic consolidation and the Chinese would prepare for the counteroffensive ("strategic stalemate"). Lastly the Chinese would pass over to the strategic counteroffensive. From inferiority China would move to parity and eventually to superiority. Mao assumed that the first stage would last a long time because the Chinese forces would not soon have the technical equipment for a massive counteroffensive. Chinese guerrilla forces would operate in the enemy's rear in large numbers, and the country would suffer devastation. Mao predicted a very painful period that eventually, with the change in the balance of forces in China and the changes in the international situation, would be overcome. In the last stage of the war mobile warfare would still be conducted but positional warfare would be of growing importance.

Mao attributes the greatest importance to man's dynamic role in war. His emphasis upon the decisiveness of men, not arms, explains the central role of political mobilization in his thinking. Guerrilla operations *per se* cannot win a war. The strategic role of guerrilla troops is twofold: to support the regular armies and to transform themselves into regular army units. In the second stage of the war against Japan the guerrillas will fulfill the first role, but in the third stage mobile warfare will become paramount and be conducted by former guerrilla forces.

Mao was preoccupied with problems of grand strategy and political analysis, but he also provided guidance on basic tactics. His lectures provide a great deal of sound advice: avoid attacking strong enemy positions and fighting hard battles; take precautions when on a march or a halt; stage concealed attacks from ambushes; and try "to cause an uproar in the east, and to strike in the west."[30] Most of this is derived from the writings of Sun Tzu, *The Water Margin*, *The Romance of the Three Kingdoms* and, of course, Mao's own experience since his first exploits in the "Autumn Harvest Uprising."

Other Chinese Communist military leaders were, for all one knows, Mao's equals with respect to understanding the theory and practice of guerrilla warfare, but he was the only one to develop a comprehensive guerrilla doctrine. Lin Piao in 1938 explained why northern China was ideal guerrilla country: "Enemy mechanized units find it difficult to penetrate mountain ranges; large units learn that it is logistically infeasible to encamp in hilly regions, while smaller units are easily annihilated on arrival."[31] The enemy could probably neither use blockhouse tactics nor enjoy local support; he would exist in a vacuum. The main difficulty facing the Communists was to defeat a stubborn enemy imbued with the Samurai tradition who believed that his victory was inevitable. A developed network of railroads and communications facilitated swift movement against the guerrillas by enemy cavalry and mechanized units, but these and other unfavorable conditions would be overcome in time.

Mao emphasized more than once that it was necessary to study what foreign specialists (including, needless to say, the Marxists) had to say about military problems. But he ridiculed those who wanted to apply foreign models in China. This attitude can also be found in Chu Teh's occasional speeches. On the basis of their experience the Chinese Communists had developed a native military science which combined theory with practice and which corresponded with the needs of the Chinese people. According to Chu the three essential points for conducting a battle were to avoid rashness in attack, to avoid conservatism in defense and to resist any tendency to run in panic from the enemy when withdrawing from a point.[32] All this clearly is elementary common sense; all military leaders in history have tried to act according to these principles, even, it must be assumed, the Kuomintang generals. Thus, the key to Communist success can be found neither in Chu Teh's speeches nor even in Mao's writings. The Chinese Communists' theoretical contribution to military science is limited. Their real achievement was a practical one: first, to survive; later on, to consolidate and expand their bases; and lastly, to establish a regular army that eventually defeated the government forces.

Mao's military writings are likely to create a mistaken impression in some respects: that the Chinese Communists were engaged in constant fighting against the Japanese invaders, that the Communists were facing (strategically) an overwhelmingly strong enemy, that most of the Japanese war effort was directed against the Communists and that in the end the Japanese were defeated mainly due to the relentless attacks of the Chinese Red Armies. In reality the Communists devoted more of their time and effort to fighting the Kuomintang (and vice versa) than the Japanese. The whole of occupied China — an area larger than Western Europe — was held by small Japanese forces (about 400,000 in 1939–1940, about a million by the end of the war). Until 1940–1941 the Japanese virtually ignored the Communists, whose pinpricks had hardly any military impact. But even after 1940 only about one-quarter of the Japanese forces in China were operating against the Communists. The Chinese Red Army, on the other hand, counted 400,000 soldiers by 1940; in other words it had overwhelming numerical superiority over the Japanese forces confronting it. The Communists' arms and equipment were vastly inferior, but the supposition that a few courageous guerrillas out-fought forces superior to them in every respect is historically quite untenable. The number of Japanese soldiers tied down by the Chinese Communists was proportionally smaller than the number of German soldiers that were needed to police the occupied European countries, even if one includes the puppet troops whose military value was very nearly nil.

The Japanese troops in China were not defeated by the Communists; the change in the balance of power that led to the Japanese surrender in 1945 had little, if anything, to do with the fighting in China. It seems virtually certain in retrospect that the Japanese troops in China would have surrendered even if neither the Kuomintang nor the Communists had fired a single shot against them between 1937 and 1945. Mao correctly understood, even if he did not directly say so, that the main obstacle on the Communists' road to power was the Kuomintang, not the Japanese. The Japanese forces, as he saw it, would be defeated sooner or later because the material base of Japanese imperialism was simply too narrow to dominate a country such as China for any length of time. Once the United States had entered the war, its outcome, despite the initial setbacks, was virtually certain because the Japanese High Command had to disperse the limited manpower at its disposal over a vast area from Burma to the Philippines, not to mention Manchuria. The anti-Japanese struggle was essential from the Communist point of view. But equally there is no doubt that from 1942 both the Communists and the Kuomintang prepared themselves for the post-war contest and that this became their overriding concern.

During the whole war the Communists fought two major battles against the Japanese — the Battle at P'inghsingkuan in September 1937, in which they inflicted some 5,000 casualties on the enemy, and the offensive of the "hundred regiments" which began in August 1940, lasted for three months and resulted in some 25,000 Japanese casualties and some 20,000 prisoners. As a result of the "hundred regiments" offensive, the Japanese began to take the Communists seriously and initiated the "three-all" counterinsurgency policy — burn all; kill all; destroy all. By the end of 1942 the number of Red Army soldiers had fallen from 400,000 to 300,000 and the population in the base areas had shrunk from 44,000,000 to 25,000,000. The Japanese, however, were too weak to keep up their counteroffensive. After 1942 the Red Armies again expanded; by the spring of 1945 there were almost a million soldiers in the 4th and 8th Communist Route Armies (not to count a militia of about 2,200,000) and the population in the base areas had risen to well over 100,000,000. Wherever the Japanese were really concerned, it must be noted, they managed to stamp out the guerrillas without undue difficulty; this refers above all to Manchuria, China's main industrial center. By staging paramilitary special operations and establishing defense hamlets, and, above all, by following a policy of propaganda and pacification (treating the peasants less brutally than elsewhere) they had liquidated the guerrillas there by 1940. It should be added in fairness that the quality of Japanese forces in Manchuria was superior to that of Japanese troops in other parts of China and that the Manchurian guerrillas had no active sanctuary. During the long severe winters they had nowhere to hide.[33]

The one battle of 1937 and the campaign of 1940 apart, there were no major encounters between the Chinese Communists and the Japanese. The transition to mobile warfare that Mao had demanded did not, in fact, occur until after the war had ended.[34] Until then there was some but not much guerrilla warfare. The occasional mining of a road or a railway line were pinpricks, and not enough to provoke Japanese retaliation on a massive scale. Most of this was "sparrow warfare," carried out by groups of three to five men. The Communists "encircled the cities" in some parts of China because the Japanese did not have sufficient forces to garrison the whole countryside. But the Japanese garrisons were not cut off and it is not even certain that they suffered any serious inconvenience.

The main effort of the Communists was invested not in the military but in the political field, in organization and propaganda. Their prestige inside China and among foreign observers was steadily rising: the Communists had purpose and determination; they were well organized and there was no corruption in Yenan. The Communist victories of 1937 and 1940, even if militarily insignificant, had a considerable psychological effect, whereas the prestige of the Kuomintang, which had to bear the brunt of the Japanese attacks up to the end of the war, lacked inspiration and steadily dwindled;[35] they had neither the ability nor the will to fight. In the Japanese offensive of 1944 the Chinese government lost 700,000 troops and areas with a population of sixty million without putting up any serious resistance. The steady rise of Communism has to be viewed against the dismal record of the Nationalists and the decay and disintegration of the state apparatus and the army.

An investigation into the etiology of Communist achievements shows no single pattern of success.[36] Communist influence certainly did not increase as a result of extreme revolutionary politics. If the Communists had learned anything from their unfortunate experience in Kiangsi, it was the need for caution and a moderate line in dealing with the peasantry. While carrying out agrarian reform in the areas under their control in northern China, they acted with great prudence, and at times even refrained altogether from carrying out any changes affecting land tenure. It has been maintained that the Communists' appeal to peasant nationalism was the key to their success, just as in Yugoslavia the partisans prevailed as the patriotic party rather than as a movement of socialist revolution.[37] This interpretation has much to recommend itself: Japanese terror, exploitation and devastation was certainly a very important factor in mobilizing the peasant masses. But, on the other hand, the peasants of the Shensi-Kansu-Ninghsia Border Region, where the Communists had their main base, were not exposed to Japanese brutality. For them the invasion was something fairly remote. And for the Communist enclaves in the Japanese rear, the Red Armies could not offer real protection against Japanese punitive raids. The

Communists' influence in these areas remained limited even though they recruited many soldiers from them. The Red Chinese were the party of the poor in contrast to the Kuomintang, and peasant nationalism did play an important role, but this was only a prerequisite for success; it does not, by itself, explain the Communist victory. They made real headway only after the peasant war became a patriotic war.[38] But conversely, it was only after the Japanese invasion had wrought havoc in China, when the Chiang government was fatally weakened and could no longer fight the Communists, that they could expand their bases.

Thus the two main causes of the Communist victory were the general disruption caused by the Japanese invasion and occupation and the superiority of Sino-Communism as a force rallying the masses. The Japanese victory destroyed the hold of the KMT; the Nationalists had been weak even before: the military defeats accelerated and deepened the process of decomposition. The government lost control over most of China and a power vacuum resulted. The extent of Nationalist incompetence came as a surprise even to the Communists; Mao had written in 1940 that

> Whoever can lead the people in driving out Japanese imperialism and introducing democratic government will be the saviour of the people. If the Chinese bourgeoisie is capable of carrying out this responsibility, no one will be able to refuse it his admiration; but if it is not capable of doing so, the major part of the responsibility will inevitably fall upon the shoulders of the proletariat.[39]

The Chinese "bourgeoisie" (meaning Chiang and his supporters) lacked the faith, the stamina, the resolution and the political know-how to pursue the anti-Japanese war successfully. The Communists had these qualities; furthermore, they were in the enviable position of not having to share responsibility for the grave domestic and military setbacks during the war years.

Marxism-Leninism as it developed in China provided a new world view that inspired its followers with something akin to religious faith and with the readiness to fight and sacrifice. It increasingly appealed, as the Kuomintang never did, to the activist and idealist elements among the younger generation, in whose eyes the government had dismally failed. Traditional Chinese political ideas and practices were outdated, and Western liberalism was quite unsuitable for China; Maoism seemed to provide the answer for most of the problems besetting the country. Old-fashioned authoritarian government divorced from the masses was no longer effective; Marxism-Leninism, suitably modified to Chinese conditions, provided an excellent method by which to reach the masses and to organize them. The idealism of the elite and its ability to mobilize mass support at a time when the central government had all but broken down were the decisive factors in the Communist victory. Thus, the key to Communist success was the Sinification of Marxism by Mao, who did not stick too closely to foreign models but adapted Communist ideology and practice to the needs of China:

> If a Chinese Communist, who is part of the great Chinese people, bound to his people by his very flesh and blood, talks of Marxism apart from Chinese peculiarities, this Marxism is merely an empty abstraction. The Sinification of Marxism — that is to say, making certain that in all of its manifestations it is imbued with Chinese peculiarities, using it according to these peculiarities — becomes a problem that must be understood and solved by the whole party without delay. . . .[40]

This is the reason for Mao's voluntarism (the downgrading of "objective conditions") and his orientation towards the peasantry and other "peculiarities," large and small. It is quite irrelevant in this context that the ideology to which Mao referred as "Marxism" was in fact Leninism, a doctrine that in many essential respects had already deviated a great deal from historical Marxism. Nor does it matter that the new ideology and the new political techniques that emerged as the result of the Sinification of Leninism had about as much to do with Marxism as Boulez with Beethoven (or Kandinsky with Fra Angelico). All that mattered in the final analysis was that the new system worked.

The victory of Communism in China, an event of world historical importance, has induced many students of guerrilla and revolutionary warfare to regard Mao's military writings as the greatest revolution in military thought in modern times. But there have been few, if any, such revolutions in recent centuries; the basic principles of warfare have been known since time immemorial: Clausewitz, Jomini and other military philosophers were not radical innovators but systematizers. In the same way Mao's military writings do not really contain novel ideas. This refers not only to his "basic principles" and the elements of surprise and deceit, but also to his ideas about flexibility, the coordination of guerrilla with regular warfare, and even the concept of protracted war. It applies to his advice about concentration and dispersal, the strategic defensive and offensive, interior and exterior line operations and the war of attrition and annihilation. There was nothing new with regard to his advice about how to treat the civilian population; the concept of political power growing out of the barrel of a gun has been known to warlords in China (and not only there) ever since the invention of guns and, in a modified form, well before. The one new element was perhaps the concept of base areas. The idea had been known and practiced before, but Mao put far greater emphasis on it. Yet precisely this is the most contradictory element in his doctrine, for it led the guerrillas into dangerous situations.

The Chinese Communists were not revolutionaries in the military field for the simple reason that the possibilities and variations of guerrilla warfare (and of warfare in general) are limited. New, as far as China was concerned, was the use of time-honored military strategies in the framework of a political movement. The idea had occurred to others elsewhere and it had even been practiced, but never in a backward country and on such a scale. The Communist victory in China proves that a few determined and highly motivated people, equipped with an ideology of radical change, are by far the strongest contender for power once established authority is breaking down. It could be argued that the existence of an activist elite and internal chaos are essential factors and that the character of the ideology is largely irrelevant. A militant movement propagating different policies might have prevailed in China in the 1940s —but it did not. That Communists were also doing well in other backward countries shows that the choice of the ideology was perhaps not fortuitous and that Asian Marxism had distinct advantages over rival doctrines as a tool for the modernization of underdeveloped societies. The character of the struggle and the doctrine was of importance not just for gaining victory; it also shaped to a large extent the quality of the political regime that subsequently emerged. The military tradition and the regimentation of the war years reinforced the tendency towards dictatorship by a small elite; the strong nationalist trends contained the seeds for the break-up of what should have been a happy family of Communist states. In later years Soviet spokesmen were to argue in their polemics against the Maoists that, while guerrillas began as Communists, they tended to end up as nationalists. They should have added, in fairness, that the Soviet Union had undergone, *mutatis mutandis*, the same process, and this despite the fact that the Communists in Russia had not come to power through guerrilla warfare.

THE VIETNAM WAR

The Vietnamese war was the longest, bloodiest and most spectacular of all those modern wars in which guerrilla operations played an important role. In contrast to the Chinese, the Vietnamese Communists were in a position early on in their war to set up a regular army that was comparatively well equipped even though it did not have an air force. This army saw action from 1949 onwards. Dien Bien Phu, the turning point in the war against the French, was not a guerrilla operation but a classical case of positional warfare, an eighteenth-century-style siege, in which the Vietminh defeated the French because of their superiority in men and artillery, and because, as General Giap put it, "we overcame the French artillery by digging trenches."[41] The guerrilla tactics used by the Communists were fashioned, broadly speaking, on the Chinese pattern; an American observer writing in 1954 thought that the tactics used were not just similar but identical for all practical purposes.[42] Ten years later, by contrast, another American observer noted that revolutionary guerrilla warfare as developed in the early 1960s in South Vietnam was something new, not just in degree but in kind. He referred, however, not so much to military tactics as to the political use of guerrilla warfare.[43] From a military point of view the main interest of the Vietnamese war is in the use that was made of guerrilla operations in combination with regular and semiregular warfare. For all their technological superiority, the French and American expeditionary corps were less effective in counteracting guerrilla tactics than were the Japanese in China.

There were certain important differences in the strategic political context between the guerrilla war in China and that in Vietnam. A country smaller than France, Vietnam's geography nevertheless offered in some respects ideal conditions: many of the guerrilla operations were conducted in marshy rice lands and in the jungle; there were countless bridges to be mined. Furthermore, individual urban terrorism, almost entirely absent in China, played a significant role in Vietnam. The Communists systematically liquidated their political rivals from the left as well as from the right. One of their main targets during the early period were the Trotskyites, who had been fairly strong in the South.[44] While the Chinese were self-reliant and without an outside source of supply, the Vietnamese Communists received arms and supplies from the very beginning from various well-wishers — first, on a small scale, from America and Nationalist China, then, after 1949, from the Chinese Communists and the Soviet Union. Air power was used against the guerrillas to a much greater degree than in China, but the country's geographical conditions reduced its effectivity; most guerrilla operations were anyway carried out by night. In the war against the French, the Vietnamese Communists enjoyed numerical superiority almost from the beginning. True, on paper the French Expeditionary Corps had a superiority of two to one over the Communists, but more than half of these units were tied down in "static duties," defending cities, villages and isolated strong points.[45] In later years, the percentage of noncombatants among the American troops was even higher. (The French troops, in comparison, always outnumbered the Algerian rebels by ten to one and, towards the end, by twenty or more to one.) The Chinese Communists faced a ruthless enemy, who was unfettered by moral scruples or restraints imposed by public opinion at home, and thus was free to apply even the most inhuman measures. Torture and the "three-all" strategy was official policy; those who practiced it were promoted, not court-martialed. French and American public opinion, on the other hand, narrowly circumscribed the scope and choice of measures of antiguerrilla action.

The social origins of the Vietnamese Communist elite which conducted the armed struggle were very similar to those of the Chinese Communist leadership. Ho Chi Minh hailed from a peasant family but his father had become a mandarin of sorts as secretary of the ceremonial office of the imperial palace in Hué; later he became a deputy-prefect.[46] Giap's father was a poor scholar; the son became a journalist and professor of history. Pham Van Dongh was the son of a high nobleman. There were one or two lower-class, uneducated *montagnards* among the leaders, such as Cho Van Tan (at one time minister of defense), but most of the leading cadres were of the intelligentsia, with a heavy prevalence of teachers — and, more often than not, these were the children of minor mandarins. There was little advancement for the holders of degrees under French colonial rule, but no undue importance should be attributed to the lack of social mobility. Indeed, it is a well-known fact that the intelligentsia has been the standard-bearer of all movements of national and social revolt in colonial countries, and not only there. That the national movement had been systematically suppressed and indeed decapitated by the French authorities between 1885 and 1932 (to select two important dates in Vietnamese history) made it all the easier for the Communists, with their greater cohesion and more accomplished methods of conspiracy and organization, to take over the leadership of the national movement after the Japanese surrender in 1945. The Vietnamese Communists put an even stronger emphasis on Popular Front tactics in their policy than did their Chinese comrades; they entered various coalitions during the 1940s and 1950s and, at one stage, even temporarily dissolved the party as a protective measure. That Vietnam, with its various nationalities, sects and religions, was less homogeneous than China created certain difficulties for the Communists. If they recruited followers among a certain sect that had grievances against the government, other sects would turn against them. But the Saigon government was even more deeply ensnared by the tensions and rivalries between Buddhists and Catholics, between the Cao-Dai and Hoa Hoa, not to mention various pirates, semicriminal sects, the ethnic Chinese and other groups. The Vietnamese Communists, like their Chinese comrades, had little influence among the urban working class even though in theory the struggle of the party was conducted "under the leadership of the proletariat." Frequently the impression was gained that the Communists were active in the trade unions more from a sense of duty to the demands of theory than from genuine conviction. The Communists invested considerable energy in tackling agrarian problems but this issue was more acute in the Mekong Delta than in the northern provinces where they ruled. Nevertheless, despite their appeal as agrarian reformers, the Communists' influence in the South always remained limited and the drastic agrarian reforms in North Vietnam that involved a great deal of terror did them more harm than good.

Students of Vietnamese history have been puzzled by the sources of Communist appeal, despite the fact that the war in Vietnam has been documented in absorbing detail. It can be shown that the effects of colonialism on traditional society was a factor in the rise of Communism, but these effects were more political and psychological than economic. Mention has already been made of the limited significance of agrarian problems. Generally speaking, immediate socioeconomic grievances were in the final analysis not of decisive importance. All that mattered was that the Japanese surrender had caused a vacuum even more complete than that in China, and that the Communists, however few, were the only group with the political will, internal cohesion and organizational know-how capable of filling it. There was no Vietnamese Kuomintang which would have threatened their progress. The old ruling stratum had disappeared and when the French (and later the Americans) tried to build up a counterelite, the Communists were already firmly entrenched in wide parts of the country. Without foreign intervention they would probably have seized the whole country within a few weeks; against French and American opposition they were to take thirty years.

When the Second World War broke out most leaders of the Vietnamese Communist party found themselves in China and, following Russia's entry in the war, they collaborated with the Kuomintang. They received a small subsidy to start a guerrilla war in the Japanese rear, but their operations were limited to the collection of military intelligence. A small guerrilla unit was established under Vo Nguyen Giap in October 1944 in the remote north of Vietnam among the Ho tribes, near the Chinese border. (This region had been throughout history an area of piracy, banditry, insurrection and partisan warfare.) The unit consisted of thirty-four men who had between them two revolvers, seventeen rifles and one light machine gun. On 22 December 1944 it established itself as the "Army of Propaganda and Liberation." There were hardly any Japanese or French units stationed in the region and the only recorded military operation was an attack against a police post at the mountain resort of Tam Dao on 17 July 1945, in which eight Japanese gendarmes were killed.[47] But the political activity of the partisans was far more intensive and spread rapidly: by June 1945 some six mountain provinces were largely under their rule. And three months later the Japanese had surrendered, a general insurgency had been proclaimed, the Vietcong had entered Hanoi and the whole of North Vietnam had fallen into their hands almost without a shot being fired. They seized great quantities of arms (including some Russian rifles that had been used in the war against Japan in 1905). Some modern weapons they had bought from Nationalist Chinese troops and others they had received by way of American airdrops before VJ Day.

Some ten months were to pass before French forces would return to Indochina in strength; in this time the Communists built up a regular army of considerable strength and organized a countrywide guerrilla network. Their position was weaker in southern Vietnam because the earlier return of the French restricted their activities to certain enclaves in the countryside. Throughout 1946 negotiations continued between the French authorities and the

Communists; on 19 December 1946 full-scale war started when Vietminh units attacked French garrisons. In the beginning the war was mainly guerrilla in character; the Communists, with some exceptions, were as yet reluctant to risk their regular army in open combat. Since the French forces were not numerous enough to occupy in strength the whole of northern and central Vietnam, they had to limit their occupation to the main towns and certain select regions. The guerrillas could not stop the French but they could block their routes:

> They blew up bridges, built road blocks, ambushed patrols and convoys, assassinated collaborators, and attacked and eliminated numerous watchtowers and small road posts set up by the French in the hope of keeping the enemy out and their own lines of communication open. The Vietminh fighters did their work chiefly by night; as a result, many regions controlled by the French during the day became Vietminh territory after darkness fell.[48]

The French expeditionary corps was militarily and psychologically quite unprepared for such unorthodox warfare. As the French did not have sufficient forces to engage in systematic antiguerrilla operations, the Communists gradually and with relatively little disturbance built up a "counter-state"; they levied taxes, collected rice, recruited soldiers and disseminated their propaganda. Occasional French attempts to move against the guerrillas were futile:

> The French as a rule conquered only empty spaces, of which there was enough in the marshes, jungles and mountains to allow the Vietminh to become invisible. They dispersed only to reassemble again at a base five or ten miles away, and they would repeat this manoeuvre if the French continued their pursuit. Sooner or later, the French found themselves too far away from their own bases, out of supplies and ammunition, and had no choice but to return, usually followed closely and harassed continuously by the reappearing guerrillas.[49]

In February 1950 major units of the Communist regular army first entered battle; their aim during the next two years was first to remove the French garrisons from the Chinese border and then to drive the enemy out of Hanoi and Haiphong and to seize the Red River Delta. But Giap's forces were defeated on more than one occasion; the French suffered some seven thousand casualties but in the end the Communist "human wave" attacks failed against superior French firepower. As a result of these setbacks, the Vietminh again adopted mobile and guerrilla warfare that could be carried out by smaller units. Against this, the French forces under Navarre, Salan and Lattre de Tassigny had no answer. Dispersed over the country they lost more and more control until, by the fall of Dien Bien Phu (May 1954), only about one-quarter of northern Vietnam remained in their hands. Meanwhile (April 1953) Vietminh forces had entered Laos and Cambodia and given a strong uplift to the local insurgents, the *Pathet Lao* and the *Khmer Rouge*. The French government was no longer able to shoulder the mounting cost of the war so the Americans, somewhat reluctantly, decided to pay for the French war effort, but to avoid direct intervention. However, growing war weariness inside France made a continuation of the war impossible. From an economic and strategic point of view, Indochina was of no importance for France. At the Geneva Conference (April–July 1954), an armistice was concluded and the seventeenth parallel became the border between two sovereign states.

The outcome of the first Vietnamese war was a triumphant justification of the Communists' use of the guerrilla strategy that they had learned from the Chinese. The first to provide a more or less systematic outline of Indochinese guerrilla doctrine was Truong Chih, the secretary general of the party in 1946–1947,[50] and the chief spokesman of neorevolutionary guerrilla warfare. When only twenty-two, he had been the head of an abortive peasant rebellion in 1931; in the late 1950s he became the scapegoat for the "excesses" in the agrarian reform program and temporarily fell into disgrace. Truong Chih envisaged a protracted war but this, he said, was nothing new in Vietnamese history; under the Tran Dynasty their forefathers had fought the Mongols for thirty-one years. It would be a war without fronts, carried out simultaneously by guerrilla, militia and regular army units. The people were the water, the people's army the fish; partisans and small army units would disguise themselves as civilians and thus become invisible. As he saw it, there would be four separate phases of the war. In the first stage guerrilla warfare, by tying down enemy forces, would be of decisive importance.[51] But with a major effort and good leadership it might be possible to win the war during the second stage by using guerrilla and paramilitary forces without a big regular army.[52] Food, money and shelter could be commandeered from the villagers; it was desirable to have the goodwill of the population but it was not a *conditio sine qua non*.

The views of Truong Chih's chief rival, Giap, gradually veered away from guerrilla warfare. Earlier on, Giap had been its major practitioner. His doctrine did not differ essentially from Mao's; a people's war in backward, colonial countries is "essentially a peasant war under the leadership of the working class." Giap meant the working class in the abstract because there were hardly any working-class cadres in the top echelons of the Communist political and military leadership. Guerrilla warfare was needed especially at the outset of the struggle because it could be practiced in the mountains as well as in the Delta and it could be waged with mediocre as well as good material. If the enemy were strong, contact with him had to be avoided; if he were weak, one ought to attack him. Losses must not be incurred at this stage even if it involved losing ground.[53] Giap fully accepted the Maoist concept of protracted warfare; writing as late as 1967, the victor of Dien Bien Phu expected the war to last "five, ten, twenty or more years."[54] He also agreed with Mao that political activities were more important than military operations and that fighting was less important than propaganda.

The second stage of "people's war" (to follow Giap) would retain certain characteristics of guerrilla war but would put greater emphasis on big raids, including attacks against fortified positions. The principles of regular warfare would become more and more important even though there would still be many ambuscades and surprise attacks. It was imperative to make the transition from guerrilla to mobile warfare, otherwise the strategic task of annihilating the enemy's manpower would not be fulfilled. If the insurgents failed to make the transition, they would find it difficult to maintain and extend guerrilla warfare.[55]

In retrospect, Giap regarded Dien Bien Phu as a model of coordination between mobile and guerrilla warfare, a "dialectical connection and interlacement."[56] Eventually, of course, the main tool in a people's war would be the regular army. Like other Vietnamese Communists, Giap correctly analyzed the main American weakness: unlike the Japanese in China, the Americans had the potential to win the war. But they could not use it because they had set definite limits on their military objectives. Neither could they afford to have the war interfere with political, economic and social life in the United States and with American foreign policy in other parts of the world.[57] It was unrealistic to expect military victory over the Americans. But sooner or later the moment would be reached when, for domestic and foreign political reasons, America would no longer be able to afford to continue the war. The American public would tire of the war, and since America, in contrast to Japan in 1938, was a democracy, public opinion would prevail over the government. The war in Vietnam was, in short, a struggle for American public opinion.

Giap's works are not a major theoretical contribution to military thought. His main achievement, it has been noted, was not to commit many mistakes.[58] He was no great strategist even though he did show mastery of the strategic situation in practice. His decisions in 1952–1953 to move his forces into the remote area of Tonkin and later on to give battle at Dien Bien Phu were his main successes. He learned from his costly mistakes of 1950, when he launched frontal attacks against the French army with forces then unprepared for major operations. His experience explains his ambivalent attitude during the internal disputes on Vietcong strategy in the 1960s: he was perhaps more aware than anyone else that guerrillas could not win the war. But, on the other hand, he feared that the comrades in South Vietnam would give up the guerrilla approach too soon. This was the background to his controversy with General Thanh (the commander in chief of the Vietcong up to his death in mid-1967) and General Truong Son. The Vietcong commanders attacked "rightist conservative thoughts" in the North, that is, the tendency to overestimate the strength of the enemy. They thought that the enemy could be defeated in an all-out assault. But the Vietcong failed to follow up their successes of 1964–1965 in the dry season of 1966–1967. Giap advised them to keep up large-scale operations but to regard the improvement and the expansion of the guerrilla forces as their main strategic task. In the words of Le Bao (probably one of Giap's pen names) the Vietcong had been successful as a result of guerrilla attacks that had disrupted American attempts to establish base areas, lines of communication, supply depots and had, moreover, "thwarted the whole pacification program."[59]

These debates belong, however, to a later period. Between 1954 and 1960 there were six uneasy years of peace. South Vietnam was in a state of near anarchy; there was only the shell of a government; the civil service was incompetent and the army far from trustworthy.[60] The position of the Saigon government was further complicated by the arrival of nearly a million refugees from the North, who had to be resettled. It was widely assumed that the Southern regime was not viable and that it would collapse within a few years, if not sooner. But against expectations, Ngo Dinh Diem, Prime Minister of South Vietnam, coped better with the enormous problems facing him than could reasonably have been expected. In North Vietnam the Communists needed longer than expected to consolidate their position and, while doing so, committed various doctrinaire "deviations"; they were guided by the principle that it was better to kill ten innocent people than let one enemy escape.

Guerrilla war in the South never ceased entirely; there was little activity between 1954 and 1957 but, after that time, it gradually increased. Following the Geneva accord some 75,000 South Vietnamese Communists had been evacuated to the North; most of them filtered back after the Vietminh Communists had decided (in 1959) to step up guerrilla warfare. Up to 1964 there were few Northerners among those dispatched to the South to fight and do political work. Later on there was open intervention by the North though, according to the official Communist version, the insurgency in the South was spontaneous, entirely self-sufficient and nationalist rather than Communist in character.

By late 1958 the Vietcong bands constituted a serious threat to the Diem government, which had lost its initial impetus and no longer provided effective leadership. Diem came to rely more and more on members of his family, excluding from power influential groups inside the country. His administration consisted more often than not of corrupt civil servants. Land reform, admirable in principle, was "notably weak in execution and frequently operated to the benefit of absentee landlords rather than of those who actually tilled the soil."[61] Diem's police tried to weed out the Communists who had stayed behind in the villages after the Geneva agreements. Such repression was carried out in an indiscriminate fashion; it was neither as brutal nor as effective as the Communists' campaign against resistance to their regime in the North. There was no attempt to create any feeling of identity between the population and those governing it. Thus the Diem regime neither terrorized the villagers into submission nor attempted to gain their support and goodwill by friendly persuasion. It was maintained that Diem's "excesses" were the fault of local officials and the regional security forces. But the effect, as far as the population was concerned, was all the same: active and inactive Vietminh cadres were indiscriminately lumped together and private accounts were settled by incompetent, arrogant and venal local chieftains:

> The brutality, petty thieving, and disorderliness of which they [the South Vietnam security forces] were frequently guilty was a source of great annoyance to local inhabitants and the Vietminh cadres who

promised to eliminate the security forces and local officials responsible for these indignities found many sympathetic listeners.[62]

This is not to say that the Vietcong behaved like early Christian martyrs. They had already engaged in individual terror on a massive scale in the first phase of the fighting (1949–1954). Systematic assassination of village leaders, local teachers and other "dangerous elements" played a more important role in Vietnam than in other Asian guerrilla wars; the contrast with China has already been noted. Bernard Fall relates that he returned to Vietnam in 1957 after the war had been over for two years and was told by everyone that the situation was fine. He was bothered, however, by the many obituaries in the press of village chiefs who had been killed by "unknown elements" and "bandits." Upon investigation he found that these attacks were clustered in certain areas and that there was a purpose behind them. But hardly anyone paid attention at the time; these activities were considered a minor nuisance, not a military danger. The attacks increased in scale until President Kennedy announced in his State of the Union message in May 1961 that, during the past year alone, the Communists had killed some four thousand small officials in Vietnam. Altogether the Communists had "liquidated" by that time about ten thousand village chiefs in a country with about sixteen thousand hamlets and thus had methodically eliminated all opposition. South Vietnam was subverted, not in the sense that it was out-fought but in the sense that it was "out-administered."[63]

To counteract this danger, the South Vietnamese government, with American help, established strategic hamlets. This limited the Vietcong's freedom of maneuver, but only temporarily. The Communists still had the original bases that they had administered ever since the expulsion of the Japanese. On the other hand, there were the hostile villages, those, for instance, in which Catholic influence was strong. Most of the Vietcong's terrorist activities were concentrated against the third and most numerous group of villages, those in which there was (in Duncanson's words) "symbiotic insurgency," in which both the government and the Communists levied taxes, collected food and recruited young men for their armies.[64]

The Vietcong put great stress on political propaganda and indoctrination. Most Western observers tended to underrate these factors in the early phase of the war, and some tended to exaggerate their importance later on. It was asserted that only men and women with the strongest political convictions and firmly implanted revolutionary attitudes could possibly continue so long a struggle and bear such suffering.[65]

A high degree of political awareness and sophistication was attributed to the people of Vietnam by pro- and anti-Communist observers alike. Thus, in the words of one sympathetic observer: "When I was in North Vietnam during the winter of 1966–67, I got the feeling that every Vietnamese was indeed an internationalist."[66] On the other hand, there were the political science advisers of the American government who installed political reeducation centers for Vietcong prisoners only to realize much later that their protégés were altogether untouched by the intricacies of Marxism-Leninism let alone proletarian internationalism.

It is, of course, perfectly true that the leadership of the Vietcong was more deeply politically motivated than its enemies and that what has been said about the Chinese Communists in comparison with the Kuomintang applies *a fortiori* to the situation in Vietnam. But despite all the indoctrination of the rank and file, there is every reason to believe that politics played a lesser role than was generally assumed. This is not a novel phenomenon: the fighting qualities of the German soldiers in the Second World War were second to none, yet subsequent research has established that their political convictions were only involved to a slight extent.[67] The history of guerrilla warfare, too, is replete with examples showing that men fight for many years and face great hardships with little apparent political motivation. Throughout history it has been strong leadership, the personal example of the commander, the ethos and the *esprit de corps* which have kept guerrilla movements going and not just ideological motivation. When the shooting was over, a French observer watching the Khmer Rouge enter Phnom Penh in May 1975 noted:

> The common soldier did not appear to be very concerned about politics, Cambodia's future or other ideological questions. I had the impression that a lot of them didn't know which group they belonged to. I felt that their fighting spirit and ability came more from the rough discipline rather than from convictions. I rather admired them, but they often seemed like animals being led into the field by the master.[67]

Of decisive importance to the conduct of the struggle was the small elite which formed the backbone of Vietnamese Communism in both the North and South. It was this cadre which was told: "On your shoulders rests the entire burden of the revolution."[68] Through its courage, persistency and frugal habits this cadre made the Vietcong a supreme fighting force.

The guerrillas were poor, which made it all the more important to share what little they had between themselves, and also with the peasants.

> While there was abundance in South Vietnam, the Viet Cong had to impose austerity on its members: in a psychological environment conditioned by Western materialism versus Eastern spirituality, social levelling made the burdens of "protracted war" a little lighter to bear. The less one has to lose the less hardship one will feel. There is no immediate hope of laying down one's personal burden anyway. . . .[69]

The Vietcong cadre came to the village "barefooted and dressed in black like every other peasant. He made tax demands, but they were not excessive . . . he did not talk Communism or Marxism, but exploited local grievances." He had the habit instilled in him to keep away from the corrupting influence of the cities; he had replaced the Confucian mandarins — but was closer to the people than the mandarins had ever been.[70]

The official Communist version of the war rested on a twofold fiction: that it was a spontaneous revolt, which received no outside aid, and that it was led, not by the Communists, but by a leadership that was in the hands of a coalition, the NLF (National Front for the Liberation of South Vietnam). Like the Chinese Communists the Vietcong deliberately played down their basic ideological tenets and put heavy emphasis on their patriotic inspiration. It was widely argued at the time that this was just a case of political camouflage. But it is also true that, with all the invocations of proletarian internationalism, nationalist feeling was very strong in Indochinese Communism (and apparently even stronger in Laos and Cambodia) from the beginning and it became even more pronounced the longer the war lasted. The rift in world Communism further strengthened the trend towards national Communism, and what was in the beginning, at least partly, mere make-believe, became in the end the *alter ego*.

The assassination campaign against government officials in 1957–1958 was the prelude to the second Indochinese war. Even before that date, however, the Diem regime had engaged in anti-Communist repression and the North Vietnamese government had applied similar measures to stamp out domestic opposition, real or imaginary. But, while the South would have been quite incapable of stirring up trouble in the North, even if it had tried, the Communists had a foothold in the South — about 6,000 armed Vietminh, who had stayed behind after the ceasefire. These "stay-behind cadres" were the hard core of the insurrection. Up to 1960 the Vietcong had enlarged its forces mostly on a local basis. Infiltration on a massive scale from the North via the Ho Chi Minh trail began only in 1960 (3,500 men came south in 1960, 10,000 in 1961 and altogether some 50,000 between 1960 and 1965). By early 1966 this force had grown to some 55,000 full-time, main-force soldiers engaged in mobile warfare. In addition there were about 125,000 guerrillas operating on a regional basis, and also some 45,000 political instructors and administrators. The number of North Vietnamese army regulars who had been dispatched to the South by 1967 was estimated at 70,000. These forces faced more than half a million Americans, 700,000 Republic of Vietnam soldiers and some 60,000 troops from Korea, Australia and New Zealand. Most of the foreign soldiers arrived, however, only after June 1965; up to that date, the Americans had been active almost exclusively as advisers. In 1962 helicopters were first used against the Vietcong on a large scale, and for a little while it seemed as if the tide was turning against the insurgents. But the battle (January 1963) of Ap Bac, a village in the Mekong Delta some forty miles from Saigon, showed that the South Vietnamese army was incapable of coping with guerrilla forces in open battle, however great its technical superiority. On this occasion it had 2,500 men in armored amphibious personnel carriers supported by bombers and helicopters, yet it failed to defeat a group of 200 Vietcong.[71] Militarily, the battle was of little significance, but the outcome was ominous because it showed all too clearly that the Vietnamese army was badly led and had little fighting spirit. Following the mission of General Maxwell Taylor it was decided to establish "special forces" that would collaborate in antiguerrilla warfare with irregular Vietnamese units (*Operation Sunrise* 1962, *Operation Hop Tac* 1964–1965, *Strategic New-Life Hamlets* 1966). But the attempts to pacify the countryside by applying certain Vietcong techniques within a totally different political framework were of no avail and, as a result of this failure, there was an ever-increasing demand for regular American forces.

The basic concept of the Vietcong until roughly 1964 was that it would come to power following a general uprising (*Khoi Nghia*). It was expected that following the intensification of the political struggle, as well as of acts of terror and guerrilla fighting, the whole country would explode, the army disintegrate, the soldiers join the people; no one would be left but a few imperialist lackeys in Saigon.[72] There was, of course, a great deal of guerrilla fighting prior to 1964, but with the arrival of the northern cadres it was very much stepped up. The Northerners advised their comrades in the South that the build-up of military forces was the commandment of the hour. With the Tet offensive of spring 1968 that led to the occupation of Hue and parts of Saigon, the transition from guerrilla to regular warfare was made, although later on the emphasis was again put on guerrilla operations. In the third Vietnamese war (1975) guerrilla operations no longer played any significant role; the conquest of South Vietnam was carried out by regular units of the North Vietnamese army.

The military lessons of the Vietnamese war have been discussed in the most minute detail. Some Western observers had misgivings about the war from the very beginning. Experience elsewhere had shown that to suppress insurgency effectively, great superiority in manpower was needed; far more men are necessary for guard duty on a bridge than for blowing it up. General Maxwell Taylor thought in 1965 that a superiority of twenty-five to one was essential; at the time other experts were more optimistic and believed that a ratio of ten to one would be enough. Yet, in actual fact, the forces fighting Communist insurgency had not even a five to one superiority. To make good its lack of manpower, the American command tried to exploit its technological superiority. But the air attacks against selected targets in North Vietnam did not have much effect on guerrilla operations in the South and the systematic bombing of the North was ruled out for political reasons. Air strikes against the Vietcong were not decisive either, because the enemy was usually not visible. Helicopters, on the other hand, while a real danger to the partisans, also offered an easy target to Vietcong machine guns; a disproportionately high number of them were lost in combat. As Colonel John Vann put it: "In a political war the worst weapon to kill is an airplane, the next worst — artillery."[73] There was not much scope for the use of tanks in the marshes of the delta or in the mountains. This was ideal country for staging ambushes, as General Dayan noted during a visit. The main problem facing American forces was to discover enemy positions in a country with many natural places for concealment. Dayan predicted that technological devices would not seal the Ho Chih Minh trail forever: if the Vietcong opted for guerrilla warfare, it could not be subdued.[74] This is not to say that the Vietcong could not have been defeated by a determined adversary who, unhampered by political considerations, would have been willing to employ its military power. The United States, for obvious reasons, was not in a position to do so. The main danger facing the Vietcong all along was the loss of its cadres. The fact that some six thousand of them dropped out or were killed or taken prisoner between 1960 and 1965 was far more serious than the killing of ten or twenty times that number of Vietminh rank and file. The cadres were irreplaceable, except, of course, in the long run. The Communist high command was all along aware that military victory against the Americans was ruled out; the strategic aim was therefore to make the war so costly for the United States that it would tire and withdraw. Once the Americans were out of the country, victory over the Saigon government was just a question of time, and probably not much time and effort at that.

The Vietnamese war was unpopular in America from the very beginning, and it became more so the longer it lasted. Furthermore, American intervention in Indochina had been based on three assumptions, two of which had been proved manifestly wrong by 1971. The domino theory included a larger element of truth than its critics wanted to admit. But by 1971 no one could argue any more with much conviction that the United States had to fight Communism in Vietnam in order to stop Chinese expansionism. For, meanwhile, relations between America and China had been normalized and the continuation of the war in Southeast Asia became a serious obstacle on the road to a further improvement in relations with China and the Soviet Union and thus an embarrassment to the architects of détente. The Vietnamese Communists gravitated more to Moscow than to Peking but they did not want to be a satellite of either. Lastly, it had become perfectly obvious by 1971 that the United States was neither willing nor able "to pay any price, to bear any burden, to meet any hardship" in order to assure the survival and success of liberty in Southeast Asia; the burden and the commitments had become too heavy, the country had become disillusioned and retrenchment was the order of the hour.

The deeper reasons for the Communist victory in Indochina are not shrouded in mystery. The Vietcong fought exceedingly well, but there was nothing novel about its military strategy and tactics. Nor was there any secret political weapon; the Algerian nationalists and the Cypriot EOKA won their wars with much smaller forces against even heavier enemy superiority than the Vietcong faced; they achieved victory in geographical conditions less propitious than those in Vietnam and without the benefit of Communist ideology. The reason for their success was simply that a foreign power could have destroyed the insurgents only by applying a strategy that would have been unacceptable to a democratic society. Above all, no native elite existed which equaled the Communists in enthusiasm, determination and dedication; the Americans could offer guidance, money and arms, but they could not provide the qualities most needed to win the war.

7

National Liberation and Revolutionary War

With the profound global shifts in the post–World War II balance of power, guerrilla warfare received a galvanic fresh impetus. Very much weakened, the European colonial powers could no longer resist the rising tides of nationalism in both Asia and Africa. By 1960 most former colonies had attained independence, the majority without recourse to armed struggle. The breaking of the "colonial yoke" did not, however, inaugurate a new era of peace and stability, for there were many contenders for dominance in the newly established countries. Radicals fought conservatives, national minorities pursued separatist policies, and the conflicts frequently took the form of guerrilla, or quasi-guerrilla war. Of these many wars no two were alike. Some, as in Palestine, predated World War II in origin, some were given a fillip by it, with continuing resistance merely switching its focus — as in Greece, Malaya and the Philippines, for instance — once the territories concerned were no longer occupied by the wartime invader. Some of these wars were short, others protracted, some ended with the victory of the insurgents, others with their total defeat. The Greek and the Malayan insurrections were Communist-inspired and led, the Mau Mau rebellion in Kenya was in the time-honored tradition of anticolonial uprisings. In Malaya, Palestine and Cyprus the wars were further complicated because they took place within a multinational society. In the military sphere, too, the patterns were infinitely variable. In Indonesia the rudiments of a regular army had come into being during the war; in Palestine and Cyprus the accent was for the most part on urban terrorism, in Greece and in

Indochina the Communists transformed their guerrilla groups into militias and even regular army units of brigade and division strength. The Greek and the Indochinese Communists received key support from neighboring Communist countries whereas the Huks in the Philippines were given no such assistance. In Greece most of the fighting took place in the mountains, in Southeast Asia, on the contrary, in jungles and forests.

A periodization of guerrilla warfare is possible only in very general terms. By and large, the first phase was over by the mid-1950s with the end of the Malayan insurgency, the lull in Indochina, the defeat of the Huks and the Mau Mau. But it was just at this moment that the Algerian rebellion began, with Castro's landing in Cuba coming close on its heels. In the 1960s, following the victory of the Algerians and the Cuban rebels, the principal scene of guerrilla operations shifted to Vietnam and Latin America, although there was also some fighting in Africa south of the Sahara. By the late sixties the rural guerrillas in Latin America had been subdued, to be replaced by urban terrorists. Simultaneously there was an escalation in the Middle East, and the war in Indochina reached its climax. But the Indochinese war had meanwhile become increasingly conventional, with guerrilla operations as only a supplementary weapon, while the Palestinians used techniques which were no longer "guerrilla" in the familiar sense.

In some guerrilla wars there was direct superpower involvement, in others help was extended obliquely, and in yet others there was no interference at all. Nor may it be ignored that in addition to the major wars that have been mentioned, guerrilla war was endemic in certain parts of the globe — in Kurdistan and Burma, to name but two. The political character of these more minor wars was in turn so complex as to defy generalization. Some were Communist in inspiration, but with the gradual erosion of the Communist bloc the general trend was towards a nationalist socialism or a socialist nationalism. Some gravitated to Moscow, others to Peking, and they all tried to get support from both. This, however, did not necessarily mean that they were willing to toe either the Soviet or Chinese line; the one common denominator was that each country insisted on its independence. Most Latin American guerrilla movements and the Palestinians were split along political lines, whereas in Africa the divisions derived usually from the tribal or confessional (ELF) background. Sometimes the factions would join forces against the mutual enemy, but more often than not they would be at odds with each other. The differences, in short, were altogether more pronounced than the similarities and any attempt to classify these guerrilla movements according to their ideology, their geographical location or their eventual achievement (victory or defeat) is at best an arbitrary exercise. And yet, for all that, it is only by comparing and juxtaposing the individual wars that something of a clearer picture ultimately emerges.

PALESTINE

The British decision in 1947 to evacuate Palestine and to hand over the thorny problem to the United Nations came after three years of military and political feuding. Jewish resistance was splintered. There was the *Hagana*, a militialike self-defense organization that had been tolerated although never officially recognized by the British Mandatory authorities. During World War II its members had been voluntarily mobilized for the war effort against Nazi Germany. The IZL (*Irgun Zvai Leumi*) had been founded in the 1930s by the right-wing Revisionist party in protest against the purely "defensist" line taken by the Hagana against Arab insurgents. With the outbreak of war the IZL, like the Hagana, declared a truce. But its attitude changed as the danger of a Nazi victory passed, as the full extent of the holocaust in Europe became known and as the British government persisted nevertheless in its opposition to Jewish immigration to Palestine. Thus the IZL renewed its activities in February 1944 with attacks directed against police stations and other government buildings. It was IZL policy at this early stage to avoid, if possible, causing loss of life. The avowed objective of the movement was to expel the British from Palestine and to create a Jewish state.[1] The third resistance group was LEHI (*Lohame Herut Israel* — the "Stern Gang"), an offshoot of the IZL. Abraham Stern, its leader, had been shot early in the war while allegedly resisting arrest, and most of its members had been detained. In November 1943 some twenty of them broke jail and almost immediately reactivated their organization. Their program was a curious mixture of extreme right-wing and revolutionary elements; the enemy was British imperialism, the ally every anti-imperialist force including the Soviet Union and "progressive" Arabs. The great historical model for both the IZL and LEHI was the Irish struggle for independence and, to a lesser extent, the Risorgimento. LEHI had no qualms about political murder and fashioned itself after the classical terrorist organizations reaching back through the ages. An attempt in August 1944 to assassinate the High Commissioner of Palestine was unsuccessful, but in November of that year two of their members killed Lord Moyne, British minister for Middle Eastern affairs, in Cairo.

The Hagana had collaborated with the British police in hunting down members of both the IZL and LEHI because they regarded their activities as detrimental both to the anti-Nazi war effort and the Zionist cause. Nor were they willing to put up with acts of defiance against their own official underground army, representing the majority, the elected institutions of Palestine Jewry. But a few weeks after VE Day collaboration between the Hagana and the British (*saison*) came to an end. In late October 1945 the IZL and LEHI joined with the Hagana in sinking three British naval craft and wrecking the railway lines at a number of points. Throughout 1946 and 1947 the IZL and LEHI continued their operations, directed for the most part against British troops in the major cities. The Hagana's actions were far fewer but on a larger scale, concerned chiefly with the sabotaging of lines of communication. The most spectacular terrorist operation (carried out by the IZL) was the mining of a wing of the King David Hotel in Jerusalem in which several government departments were at the time located, with the loss of more than ninety British, Jewish and Arab lives. Terrorist acts were suspended, however, with the outbreak of the Arab-Jewish war in December 1947.[2]

The dissident organizations nonetheless continued their separate existence until shortly after the end of the war of independence (summer 1948), although while it was yet in progress Ben Gurion, himself an activist second to none, was firmly resolved that the Hagana — or the Israeli army, rather, of which it had become both nucleus and backbone — should impose its authority on the "dissidents," even at the risk of a civil war within the shadow of the wider one being fought; the *Altalena*, a ship chartered by the IZL with badly needed ammunition and provisions, was shelled and sunk off Tel Aviv midway through the war since the dissidents were not willing to hand it over to the government. Eventually the IZL and LEHI were dissolved and their members incorporated in the body of the Israeli army but still not before the Deir Yassin massacre had been perpetrated. And it was members of LEHI who after the war assassinated Count Bernadotte, the Swedish mediator appointed by the U.N. Both dissident groups later went into politics. Members of the IZL established the right-wing *Herut* party, while members of LEHI were involved in the foundation of a short-lived national Communist party.

LEHI in its heyday consisted of no more than a few hundred activists; the IZL had a few thousand members and active adherents. The Hagana was a much bigger, but also much looser organization with perhaps between sixty and eighty thousand members of whom, however, only a small number saw action in the anti-British operations of 1945–1947. The command structure of the IZL envisaged three divisions: the "Army of Revolution" (which somehow never came into existence); "Shock Units"; and a "Revolutionary Propaganda Force." Like the Hagana and LEHI, it had a small, mobile broadcasting station. The IZL and LEHI had only light arms and explosives; for a long time they could not get automatic weapons. But Hagana had no artillery either prior to 1948.

The political effect of the terrorist operations has been hotly debated and has remained a matter of bitter controversy to this day. Some Zionist leaders have argued that without the Irgun the state of Israel would not have come into being; Menahem Begin, commander of the Irgun, has claimed that "we succeeded in bringing about the collapse of the occupation regime."[3] Other authorities maintain that the "dissidents" did more harm than good to the cause. The international auspices were at the time more than favorable from the Zionist standpoint. With delayed realization of the great disaster that had overtaken the Jewish people in Europe during the Hitler era, there was much sympathy for its aspirations to establish a Jewish state. Notably weakened by the war, Britain found the administration of the Mandate a thankless task. From a strategic and economic point of view Palestine was of no great importance. British antiguerrilla operations, though far from ruthless, had a bad press the world over and were not popular at home. In the circumstances a minimum of force was needed to precipitate the British exodus; the dissidents played a certain part in the process but not a determining one.

THE GREEK CIVIL WAR

On 16 October 1949 the Greek Communist radio transmitter situated somewhere in Eastern Europe announced that the Communist army had put a stop to operations in order to "avoid the total destruction of the homeland." The announcement, magnanimous in spirit, came a month after the army had ceased to exist. During the preceding three years it had successfully challenged the Greek government, defeated its armed forces, and a stalemate, if not a Communist victory, had seemed virtually inescapable. The third round in the fight for power in Greece had started with small-scale Communist attacks launched in February–March 1946. Zakhariades, the secretary general of the party, wrote in retrospect that "we all agreed that the situation was ripe, that we should take up arms and fight. . . . The People's Democracies were behind us." But a few British forces were still in Greece and it was not in the Communists' interest to bring about their intervention. The attack, in other words, had to be directed not against the foreign enemy but the domestic foe.[4] It is not entirely clear to this day to what extent the Greek Communists had been encouraged by their mentors abroad; it has been asserted that they were prodded by Tito to start the insurrection, that Stalin was first in favor and then skeptical about the outcome of the venture. Perhaps he had no strong views one way or the other. The ultimate decision had to be taken by the Greek Communists themselves, but it is patent that they would not have gone to war if Stalin had been opposed.

The conditions seemed propitious. The postwar economic crisis had not been overcome, the police force was inefficient and the army in the process of being rebuilt. There was no stable government, let alone strong leadership. Much of Greece consists of mountainous territory which favors guerrilla warfare; the border between Greece and her Communist neighbors, seven hundred miles in length, is almost impossible to secure. Thousands of Communist activists had gained military experience in World War II. The Communists had bases, training grounds and steady sources of supply in Albania, Yugoslavia and Bulgaria. The Greek Communist party, to be sure, was relatively small; it had never polled more than ten percent in general elections, but it was far more tightly knit than the other Greek parties and could rely on a high degree of militancy and discipline. It had supporters both among the urban intelligentsia and the industrial workers; it was especially influential among the Greeks from Asia Minor who as the result of the population transfer after World War I had been repatriated to Greece. Both Markos Vafiades, the commander in chief of the Communist army, and Nikos Zakhariades who later replaced him, had been born in Turkey. The Communist partisans did indeed appear to have the advantage. They were more deeply motivated, their morale was sturdier, they fought better than the government troops.

They were also more ably led, at least until the very last stage of the fighting.

Two reasons have been advanced to account for their defeat nonetheless in 1949. The first is the decision taken in November 1948, apparently against Markos advice, to convert their guerrilla army into one of larger formations (divisions), and to engage in positional warfare in defense of the liberated areas; this, when they had been at their potent best in 1947–1948 operating with units of company strength (fifty to a hundred men) and, at most, in battalions (two to four hundred). The second frequently cited reason is Yugoslavia's rift with the Communist camp which eventually resulted in the closing of the border with Greece. These circumstances had, needless to say, their adverse effect but they were by no means the only causes of the Communist rout. The very decision to embark on an armed offensive had been a mistake, as the Communist leaders themselves later admitted. The country's mood was not as revolutionary as they believed. Greece was, after all, not a dictatorship at the time; the Communists' deliberate boycotting of the first postwar democratic elections was generally interpreted as a confession of weakness.

When they made their fateful choice, the Communists had banked above all on the demoralized state apparatus and an under-mined army, lacking both training and modern equipment. A protracted war, it was figured, would of necessity bring about the collapse of the regime. They overlooked — or perhaps preferred to ignore — the possibility that a long war might precipitate the reorganization of the army. Between 1946 and 1949 the United States supplied Greece with approximately three hundred and fifty million dollars' worth of arms; this undoubtedly contributed to the Communist defeat, but even more telling was the revitalization of the Greek army under Papagos, who became commander in chief in early 1949. As an American observer wrote later: "The army was galvanized into action. Its manpower was not increased, its training was not greatly improved, and there was no significant increase in its equipment. The army was simply made to do what it was capable of doing, and no more than this was then needed to gain the victory."[5] But even if the Greek Communists had not made the mistake of transforming their forces into a semiregular army, if they had continued to operate in small units, they would still have had no chance. The shift to conventional warfare was a fatal error, easy though it is to see why the step was taken. Some Communist leaders felt that time was running out, with Yugoslavia's "defection" the international situation had from their point of view deteriorated, it would be wisest to move against the Greek army while it was still depleted.[6]

Guerrilla tactics would have made it more difficult for General Papagos to defeat the Communists, but without bases inside Greece, which they did not have, the guerrillas could anyway not have existed much longer; their cause in fact lost more of whatever appeal it initially had with each month the war dragged on. The Communists had no popular aim, no obvious, all-embracing slogans such as the overthrow of the tyranny or the reapportionment of the land. They tried to use anti-imperialist slogans, but to little purpose; there were, after all, no more than three thousand British soldiers in Greece at the time, who did not participate in the fighting, and a few hundred American military advisers. On the other hand, the Communists had to defend themselves against charges of treason, since they supported the establishment of an independent Macedonian state. The proposal to surrender Greek territory to the Bulgarians was anything but attractive. Even the Greek Communists were not enthusiastic about this item in their political paraphernalia; perhaps it was part of the price they had to pay for the help they received from their comrades abroad. Their having to press-gang young peasants into their army during the last year of the fighting merely antagonized villagers the more, and at no benefit to their fighting machine either, for there was not time enough to indoctrinate the new recruits.

The Greek Communists were better armed than most postwar guerrilla movements. After the battles in the Vitsi and Grammos mountains, the government forces captured about a hundred pieces of artillery, including anti-aircraft and anti-tank guns, some 650 machine guns, 216 heavy mortars and 142 rocket launchers. Considering that the guerrilla forces had never numbered more than twenty-five thousand soldiers (up to twenty percent of them women), they seem to have had about as much equipment as they could possibly absorb. They had no air force, but the government air force was minute and until the last phase of the fighting played no significant role in the campaign. The war was fought with great bitterness, clemency was rare and atrocities common to both sides. The Greek government forces suffered about sixty thousand casualties, among them sixteen thousand dead and almost five thousand missing. More than four thousand civilians were executed by the Communists. No accurate figures are available about the extent of Communist losses. They were, in all probability, as large as those of the government. Greece, a small and poor country with about seven million inhabitants at the time, needed years to recover from the trauma of the civil war, the material destruction wrought and the loss of life. With the Russian and the Spanish civil wars heading the list in that due order, the Communist insurrection in Greece ranks as third among the major European internal wars of this century.

SOUTHEAST ASIA

As World War II came to its shuddering end, the calls for national independence began to reverberate across Southeast Asia. The easy victories of the Japanese against the European colonial powers had given an enormous boost to the native national movements. In India, Pakistan and Burma, Britain abdicated without a struggle; the course of events in Indochina has been described elsewhere in this study. But in Malaya and the Philippines, guerrilla movements mushroomed, while in Indonesia the Dutch attempts to reimpose their rule also provoked armed resistance.[7] The Indonesian bid for independence was won with relatively little fighting, and this despite the movement there being internally split, with the Communists taking, *grosso modo,* a more militant line than the rest. To compound the confusion, the Communists were themselves divided and all in all there was a real danger that the country might quite literally fall apart. Indonesia's very weakness, however, was its strength, for the Dutch were wary of the chaos whereas the Nationalists and the Communists had no such inhibitions.

There had been no resistance movement in Indonesia during the war; on the contrary, there had been widespread collaboration with the Japanese. Mention has already been made of the fact that under their occupation a small Indonesian army, the *Peta,* had come into being. Furthermore, all the main political parties had their private armies, such as the Masjumi (*Hizbullah*) and the *Darul Islam.* The Peta consisted of fewer than a hundred thousand officers and men, the private armies of somewhere in the neighborhood of double that number. The two Dutch "police actions," in 1946 and 1948, were carried out by much smaller forces, but these were highly trained and well-organized units which had no trouble whatsoever coping with the untrained, ill-disciplined and badly equipped Indonesian troops. But the real problems, as so often in this kind of war, emerged only after the Dutch had seized the key cities and lines of communication. A hundred thousand Dutch soldiers were not sufficient to control the heartland of Java and Sumatra, let alone the other islands. The Dutch army was, in the words of one observer, incapable of occupying an overcrowded area of fifty million people, short perhaps of an outright campaign of terror, for which the Dutch were "temperamentally unsuited."[8] Facing an economy in ruin, the prospect of general turmoil, the condemnation of the United Nations (still a moral force to be reckoned with in those days), facing the strong disapproval of the United States and their other allies, the Dutch opted for withdrawal and Indonesia became a sovereign republic. Weak as the national government was, the Communists were in no position to challenge it for their force had been much reduced in the fighting, notably in the Madiun rebellion. Moreover, by the time they recovered from their internal splits (1952–1953), world Communist policy was no longer that enthusiastic about armed struggle outside the colonial context. So the Indonesian Communists reshaped their strategy to one of political action, demonstrations, strikes, and eventually even of collaboration as the Sukarno government veered towards "anti-imperialism."

Communist guerrilla warfare in Malaya began in 1948, reached its climax in 1950–1952, and petered out in 1956–1958. In the Philippines fighting developed in 1947, continued on and off for about seven years and then gradually died down after the surrender of Luis Taruc, the Communist leader. Both in Malaya and the Philippines the leading cadres of the postwar insurgency were composed of the same men who had organized the wartime resistance. The Chinese Malayans had established a guerrilla force in February 1942, and the following year British officers ("Force 136") landed and cooperated with them; Chin Peng, the commander of the MPAJA (the Malayan resistance forces), which numbered some six thousand fighters, was to be awarded the OBE. The Philippine *Hukbalahap* was founded in March 1942; its relations with the small U.S. guerrilla forces in the islands were, however, anything but cordial, and although the Huks contributed in no small measure to the war effort against the Japanese, they were equally if not more eager to settle scores with their own domestic political enemies. Although the leadership of the Huks was Communist, this fact was not made public at the time. Whereas to all intents and purposes the MPAJA was identical with the Malayan Communist party, the relationship between the Huks and the Philippine Communist party was more complicated. The political situation in Malaya was anyway altogether dissimilar to that in the Philippines. Malaya was still a colony while the Philippines had almost attained independence, even if the Communists would argue that this independence was a mere legal fiction. Which does not alter the fact that the Malayan guerrillas still had to contend with the British army and police, whereas the Huks were by now free to take on their own people. Again, recruits to Communism in Malaya came almost entirely from one community, the Chinese; the membership of the Huks, on the other hand, was not limited to a national minority.

The timing of the insurrections in Southeast Asia was probably not altogether uncoordinated. They all broke out within the space of a few months in 1948 and this has tempted observers to look for a definitive guiding hand behind the eruptions. Attention has been drawn to the Calcutta Conference of the World Federation of Youth and Students in February 1948.[9] The resolutions of the conference attacked the "false independence" of India and Pakistan and called for an intensification of the struggle for true independence, which in the circumstances did not of course mean the concentration of one's efforts on electoral contests. It is most unlikely, however, that the Cominform would have chosen a minor meeting to coordinate its policy in Southeast Asia. Whatever coordination there was had most probably taken place at the highest level. The "general line" of Communist policy, the "two camps" concept, had been defined by Zhdanov and others well before then and the new militancy just happened to coincide neatly with the desires of the Southeast Asian Communists. But this is not to say that the policy was clear-cut, or planned in detail. During World War II the Soviet Union would have taken a dim view of any fraternal party which did not contribute its share toward the defense of the Soviet Union. The world situation after the war, however, was infinitely more complex and there could be no imposing of one rigid universal law. Between 1947 and 1952 the stress was certainly on the armed struggle, at that point peaceful coexistence became the watchword. But just as not everyone everywhere took to arms in those five years, so neither did all armed insurrections cease after 1952. It all depended on local factors, "objective" and "subjective" alike.

Malaya is a small country, four-fifths of its area uncultivated jungle. It was (and is) a major producer of tin and rubber, of its 5.3 million inhabitants (1945), forty-nine percent were Malay and thirty-eight percent Chinese.[10] The Chinese were better educated and had, on the whole, a considerably higher living standard. After the war the Communist party had become legal; its new head was the now twenty-six-year-old Chin Peng, the guerrilla hero who had cooperated with the British. His predecessor, Loi Tak, had been successively a French, Japanese and British agent, a particular learned only later and that anyhow did not affect the party line.[11] At its fourth plenary meeting (May 1948), the MCP decided that "an armed struggle will be inevitable and will constitute the most important form of struggle."[12] This decision coincided with the final victory of Mao's forces in China and was probably not unconnected with it. There is reason to believe that the secretary general of the Australian Communist party, on a visit to Malaya at the time, acted as an emissary and that he advised the local Communists against continuing the constitutional struggle.

Communist strategy, insofar as can be ascertained, was to liberate certain areas near the jungle, to seize plantations and mines, and then to envelop the cities. From the guerrillas' point of view the squatters' villages on the fringe of the jungle were of paramount importance; they relied on them for intelligence, food and supplies. After some hesitation the British authorities decided to resettle the squatters in new villages. This proved an easier task than had been envisaged and it aggravated the guerrillas' supply situation. But it did not solve the authorities' military problem, for with great tenacity the guerrillas continued to fight on in adverse conditions. General Clutterbuck, who was actively involved in counterinsurgency, writes admiringly not only about the organization of the Communist guerrillas but of the "fortitude of tiny bands of guerrillas holding out against the concentrated efforts of twenty or even sixty times their strength of soldiers when the war was already lost— ranking high in the annals of human endurance."[13] It has been pointed out that the main aim of the insurgents at the start of the guerrilla war was to cause maximum disruption to the country's economy and the administration. But in contrast to Vietnam, the Communists never quite succeeded in achieving their objective; the administration continued to function throughout the country, the government collected taxes, the schools were kept open and justice was administered. Unlike their Vietnamese comrades, the Chinese guerrillas in Malaya had no active sanctuary, no secure line of supply from beyond the border. Their staunchest ally was the well-nigh impenetrable jungle in which their camps could be tracked down only with the greatest difficulty. Individual terrorists would be seized only by accident.

The first phase of guerrilla fighting (1948–1949) was not an outstanding success inasmuch as the government failed to collapse. But the guerrillas continued to launch constant attacks from their bases in the jungle and their assault reached its climax in 1951, a year in which the security forces suffered some 1,195 casualties, including 504 killed. (One of the victims was Sir Henry Gurney, the High Commissioner, who was killed in an ambush.) But the toughest year of the emergency was also the period in which the tide turned, though few realized it at the time. Guerrilla casualties in 1951 were over two thousand, including a thousand killed and three hundred and twenty captured or surrendered, an unacceptable number considering that there were no more than ten thousand of them in all, and most of the time only about five thousand.[14]

Looking back on the years of fighting, the Communist leaders acknowledged that certain mistakes had been made. They had subjected the masses to great losses through their acts of destruction and sabotage — "blind and heated foolhardiness" was to be avoided in future, the emphasis was to be on "regulated and moderate methods."[15] This meant among other things no more slashing of rubber trees, and no more indiscriminate assassinations. Internal purges, however, continued; a leading party member was executed for having dared to criticize the top leadership. But instead of having the desired effect, this execution led on the contrary to the defection and surrender of other dissenters. The government tried psychological warfare with some degree of success; once it had been established that the British would not shoot deserters, there was a steady trickle of surrenders, averaging two hundred a year. This damaged the morale of those remaining in the jungle and provided intelligence to the British commanders.

The serious British counteroffensive began in late 1951 and lasted for about two years. By 1953 the security forces were killing or capturing six guerrillas for each of their own men lost.[16] The British had managed to cut the guerrillas off from their regular food supplies, and driven them deeper into the jungle where they lived with the aborigines. They had become far less dangerous, but to flush them out from those tangled depths was an almost fiendishly frustrating exercise. As many as a thousand man-hours could be spent even so much as to encounter a guerrilla. But there were limits, too, to Communist endurance; the guerrillas were aware that they had been isolated and this finally undermined their will to continue the struggle. By the end of 1955 the number of jungle fighters was down to three thousand, by late 1956 only about two thousand were left, the following year the remaining Communist units disintegrated and the "emergency" was virtually over.

One of the major mistakes of the guerrillas (in the opinion of one who fought them) was to adhere too rigidly to Maoist strategy in so altogether different a setting.[17] After 1954 they realized that they had neglected indoctrination and they tried to broaden their mass basis. But several valuable years had been wasted; the British had meanwhile carried out administrative reforms and promised independence. The Communists still found little support outside the Chinese community; their principal bases were the Singapore secondary schools. It has been reasoned with hindsight that the guerrillas might have come closer to success had they engaged in simultaneous urban terror and rural guerrilla operations, or if they had concentrated their attacks against plantations and mines. But they were not strong enough to carry out projects on such a massive scale. Even within their own community they lacked full control; they were not able, for instance, to win over the powerful secret societies. The official name of the guerrilla movement was the MRLA — Malayan Races Liberation Army — but for all that, no determined effort was made to rally Malay supporters, although the tensions between the component nationalities were so palpable that the attempt would probably in any event have proved abortive.

Small guerrilla units continued to exist near the Thai border; the headquarters of the MRLA (later restyled the MNLA) were transferred to southern Thailand as early as 1953. After the collapse of the revolt, the Communist party of Malaya veered toward Peking, to have the decision to give up its active fight criticized later as a "revisionist deviation." But despite the appeals in 1961, in 1963, and again in 1968 to correct the "capitulationist line" and to persist in the armed crusade in rural areas to the very end, Malaya was to remain quiet for almost two decades.

THE PHILIPPINES

In later years Communist writers were to maintain that the guerrilla insurrection in the Philippines was bound to fail because there was no "objective" revolutionary situation.[18] In actual fact the prospects for a successful takeover were better in almost every respect in the Philippines than anywhere else in Southeast Asia. Political and economic power in the islands was in the hands of a small oligarchy which owned all the large farms. The agricultural system was almost entirely feudal in practice, with peonage widespread and an immense landless proletariat. Potentially, the Huks had even greater peasant appeal than had either the Chinese or Vietnamese Communists; they had laid the broad foundations for it during the war on the dual count of fighting the invader and their insistence on a just redistribution of the land. (In China, it will be recalled, the agrarian demands of the Communists were played down while the war continued.) The Philippines had, besides, a long guerrilla heritage dating back to the resistance of the tribes to the Spanish invasion. It had manifested itself again in the struggle of Aguinaldo, the national leader and guerrilla chieftain who had withstood the Americans for two years after the end of the Spanish-American War in 1898.[19] Sixty thousand Americans had fought forty thousand Philippine patriots based chiefly in the north of Luzon, and the U.S. suffered six thousand casualties before they succeeded at long last in surprising and capturing Aguinaldo in his headquarters. Clashes on a smaller scale continued well beyond 1902. (Aguinaldo was still alive when the Huk insurrection broke out in 1947.)

The Huk rebellion reached its climax in the years between 1950 and 1952 when "they were the masters of the countryside and of several cities. . . . The people paid them taxes, fed and sheltered them, gave them valuable information and sometimes rendered military service to them."[20] They numbered then some twenty thousand men with perhaps fifty thousand auxiliaries, and two million people lived in the areas they dominated. Luis Taruc, who had been the commander of the movement when it had first been set up to fight the Japanese, wrote after he had left the party that "errors were made and innocent people died . . . but the common people certainly loved and respected us."[21]

The forces opposing the Huks were weak and inexperienced. The Philippine army consisted altogether at the time of only two fighting battalions, the rest were engaged in service, administration and training and could not be enlisted for active duty.[22] Nevertheless the Huks (whose name had meanwhile been changed to the HMB — People's Liberation Army) could not prevail in their long and bitter war. For all the discontent and the internal tensions, they found it harder to mobilize the peasants against their own masters than against the foreign foe. Nor was there any powerful Communist neighbor to act as a protector, to provide them with a steady supply of arms and ammunition, food and medicines. Further, the

Huk leaders seemed to have no clear notion of how to proceed beyond rural guerrilla warfare. Finally, they had the misfortune to meet up with a gifted opponent in Ramon Magsaysay. Thanks to his initiative, the army and the administration were revitalized, and local government was reformed. Handsome rewards ranging from fifty to seventy-five thousand dollars were given for information bringing about the capture of any of the HMB's leaders. At the same time former members of it received both an amnesty and fifteen to twenty acres of land apiece. This "left hand, right hand policy" produced fairly quick results. In 1953 Magsaysay was elected president; by the time of his death in an airplane accident the war was virtually over. A few hundred Huks remained in their mountain hideouts, but the Communist party was no longer a danger. Jesus Lava, the pro-Soviet secretary general, returned to Manila, abjured the armed struggle and became a loyal oppositionist. The remnants of the Huks engaged in brigandage; in 1969 a new Maoist New People's Army (MNA) came into being. But a more effective threat was the Muslim revolt in the southern islands in 1973.

The Philippine Communists suffered a series of setbacks, some of them self-inflicted. When in 1950 the prospects were at their brightest, half of the Politburo was arrested in a police raid in Manila, leaving the party disorganized. The leadership was ridden by ideological and personal disputes from the start. Above all, as in China, the Communists were entirely on their own. But conditions in the Philippines after the war did not resemble those in China during the Japanese invasion, Taruc and the Lava brothers lacked Mao's qualities, and Magsaysay was infinitely more competent than Chiang Kai-chek.

Since World War II, guerrilla warfare of one kind or another has taken place in every country throughout Southeast Asia. In Burma it is inherent and latter-day Burmese politics have largely been dominated by the struggle between the government and Communist guerrillas of various persuasions, as well as with national minorities such as the Karen. The Indian Communist party engaged in rural guerrilla warfare in Andhra Pradesh in 1948–1951. Later, the Indian army came up against similar warfare in Nagaland. Under Maoist inspiration, the Naxalites in West Bengal organized poor and landless peasants for an armed struggle in 1967. This revolt, which aimed at the physical extermination of the "class enemy," meaning landowners and moneylenders, reached its climax in 1969–1970. Along the way, the Naxalites also killed policemen and teachers, and members of rival political parties — including the pro-Soviet Indian Communist party — and destroyed symbols of enemy rule such as Gandhi and Tagore monuments. The campaign had originally been launched under the umbrella of antifeudalist slogans, but soon the target was redefined as the seizure of power.[23] Following Peking's criticism of their strategy, the Naxalites split into eight factions, and eventually some thirty thousand of their members and adherents, students for the most part, found themselves in Indian jails.[24]

Partisan warfare was conducted in Korea between 1951 and 1954. But none of these campaigns was successful, and though each was different, a detailed analysis would add little to a general understanding of the guerrilla phenomenon.

ALGERIA

Early on in the Algerian war, General de Gaulle had realized that France could not keep Algeria against the wish of the overwhelming majority of its inhabitants. The revolt had started in 1954, by 1956–1957 the FLN thought victory was at hand. Their optimism, however, was premature, for in the following years their units were crushed by the French army. But de Gaulle insisted that there was no solution other than total independence. It would be different if France were still a "mastodon" as it had once been. In present conditions, "only Russia with its Communist methods" could put an end to the rebellion. Having already killed two hundred thousand people (de Gaulle argued), France could certainly continue the war. But where would it lead? The army, seeing no farther ahead than the next *djebel*, did not want to be deprived of its victory, it had only one remedy: to break the bones of the *fellaghas*. But this would merely lead to a new war in five or ten years and by that time the Arabs would be even weightier in numbers.[25]

The French position in Algeria, it goes without saying, was far stronger than in Indochina, quite apart from the fact that the French army had learned from its unfortunate experience in Southeast Asia; a second Dien Bien Phu was ruled out. Algeria was not a colony but part of metropolitan France, the distance from Algiers to Marseille was no greater than from Marseille to Lille. At the beginning of the war there was full support for it in France. Algeria had no jungles or forests in which the rebels could hide; the French air force could easily spot enemy concentrations. One million Frenchmen lived in Algeria and were well acquainted with the country and its inhabitants. The rebels were not members of a monolithic party as with the Communists in Vietnam — there was much less discipline and much more in-fighting; thousands of Algerians were killed before the FLN had defeated its domestic rivals. French army losses were small; during the seven years of the insurrection the average annual number killed was two thousand. For all that, as de Gaulle had predicted, the French army was neither strong nor ruthless enough to win the war. By 1960 half a million troops had been concentrated to police a country several times the size of France; the cost of this amounted to almost a billion dollars a year. Domestically, France passed through the most difficult spell in its postwar history; there was no leadership, no stable government, the crisis in Paris affected the situation in Algeria, and the Algerian war aggravated the French crisis. Most Frenchmen were outraged by the Algerian atrocities: they wanted to keep Algeria but they were no longer willing to fight for it. Gradually the war found decreasing favor at home.[26] To suppress the rebellion effectively the French security forces would have had to use the same means, if not more drastic ones, than the insurgents — indiscriminate assassination, systematic torture — and though the French paras were not plagued by excessive humanitarian scruples, there were in the last resort limits to what means the security forces of a civilized country could apply.

The FLN would still have been routed but for their active sanctuaries in Egypt, Libya, Tunisia and Morocco. Their situation was more advantageous than Abd el-Kader's, for whenever they were hard-pressed they could cross the border, while France, in contrast to a hundred years earlier, could no longer invade Morocco. However much the French generals might rave, they were powerless to pursue the enemy. Even a minor air attack against an FLN base on the Tunisian side of the border (Sakiet Sidi Yusef) provoked a major international scandal; a massive attack was altogether unthinkable since the French government felt it could not commit such an affront to world public opinion.

Algeria had been under French control since the 1830s but native opposition had never been far from the surface. In World War II, France's position in North Africa had become much weaker and in 1945 there was a major insurrection; according to an official estimate, fifteen hundred Algerians were killed, and the nationalists claimed twenty thousand victims, the real figure being perhaps somewhere between five and eight thousand. The fact that Morocco and Tunisia had made greater advances on the road to independence added fresh fuel to Algerian nationalist fervor, so did Nasser's rise to power. Egypt, where the North African liberation committees were located, was the first base of the insurrectionists; only two years later the Algerians shifted their headquarters to Tunisia and Morocco.

The prehistory of the rebellion is still something of an enigma. Officially the coordinating body was the CRUA (*Comité Révolutionnaire d'Unité et d'Action*), an activist group which had split from the MTLD led by Messali Hadj. The nine leaders were Ahmed ben Bella, Belkacem Krim, Mohammed Boudiaf, Mohammed Khider, Mustapha ben Boulaid, Larbi ben M'hidi, Mourad Didouche, Rabah Bitat and Ait Ahmed Hocine. Yet these nine had never actually met before the insurrection started on 1 November 1954.[27] The real kernel was the special organization of the MTLD established in 1948, which engaged in occasional bank robberies, the collection of arms, and sporadic acts of terror. This special organization (OS) was headed by Ahmed ben Bella who had served with distinction in the French army during the war. In 1950 ben Bella was arrested following a robbery at the Oran post office but made a successful getaway from jail. Of the early leaders of the rebellion, hardly any were of peasant background but quite a few had served in the French army. Some had been in politics before: ben Bella had for a short while been deputy mayor of an Algerian town; Khider, who was older than the rest, had been a member of the French parliament. Most had belonged before to the MTLD, a few had been Communists. One observer records that Belkacem Krim, a former French army corporal, had organized a Kabyle *maquis* of his own prior to 1954, a second one puts it that he was a notorious brigand chief.[28] For all their fervent nationalism, most of the FLN leaders were culturally uprooted men; scarcely one of them had a command of literary Arabic.

According to conventional liberal wisdom of the day, the Algerian problem was basically one of poverty, and consequently the solution had to be primarily socioeconomic in character. There was indeed great poverty and social discrepancies were immeasurable. Algeria was still a predominantly agricultural country. Oil had been discovered, but production amounted to only eight million tons in 1960. Ninety percent of Algerian industry was in French hands. Six million Muslims farmed some 4.7 million hectares, whereas a hundred and twenty thousand Europeans had farms of 2.3 hectares. While the urban Muslims had benefited to some extent from the postwar boom, most of the peasants were still desperately poor. There seemed to exist all the makings for a major agrarian rebellion and the FLN leaders stressed in their articles and speeches the importance of land reform once the war was over; in this respect, as in some others, their policy resembled Nasser's. Yet the agrarian issue was far from central to the rebellion and the FLN by no means supported a social revolution. As one of its leaders put it, "The problem is not posed for us as in China. The Chinese carried on both national resistance and social revolution. . . . We have taken up arms for a well-defined aim: national liberation."[29] Some FLN leaders such as ben Bella used more radical phraseology than others, but even most sympathetic observers have noted that much of the ideological verbiage was simply a mask for maneuvers of various groups within the elite which aimed at securing or bolstering their own positions of influence.[30] Thus the bedrock of the struggle against the French was nationalistic, with socialist demands, other than seizing foreign property, little more than scatterings of topsoil dressing. Toward the end of the insurgency there was a shift in FLN orientation in the direction of the Soviet Union, but the motivation was largely pragmatic; the French generals and colonels who claimed that their army was the "first in the world which had agreed to fight on the ground chosen by the Communist revolution to destroy Western civilization" were quite mistaken.[31] The FLN was perhaps more anti-Western than most European Communist parties, but it was certainly not part of a "Communist conspiracy."

With the outbreak of the rebellion CRUA became the FLN, the ALN acting as its military arm. Unlike in China, Cuba or Vietnam, there was no one outstanding figure whose authority was undisputed: some of the founder members were killed in the war, and four leaders, including ben Bella and Khider, were captured by the French in 1957 and spent the succeeding years in prison. The ALN was subdivided into five regions (*wilayat*) under a colonel, with a sixth one (Sahara) added later. Estimates as to the number of Algerian guerrillas vary enormously; at the beginning there were only a few hundred of them, equipped mainly with rifles and some automatic weapons. By 1956 there were forty thousand according to Algerian sources, twenty-five thousand (including auxiliaries) according to the French. After 1955 the rebels were equipped with machine guns, mortars (German 81 mm) and recoilless rifles, and there was no shortage of mines and bangalore torpedoes.

The rebellion had started in the mountainous regions of Kabylia,

Aurès and northern Constantine; during 1955 it spread to other parts of the country. The hit-and-run tactics focused on destroying French farms, cutting lines of communication and punishing Algerian collaborators — sixty thousand Algerians were fighting in the French army. The attempt to carry the war to the capital in September 1956 ended in a débâcle; the French paratroopers smashed the ALN apparatus in Algiers with great losses to the insurgents. But the political objective was largely achieved — the internationalization of the conflict and the political isolation of the French government. The FLN gained increasing support in the Arab world and it was joined by Algerian political leaders who had initially been hesitant. The French army had at first underrated the extent of the rebellion; but after 1957, strong reinforcements were brought in and systematic measures employed to combat the insurgents. The ALN lost the initiative; the Morice line along the Tunisian border made crossing difficult, and the "regroupment" of villages cut the ALN off from much of its support. By 1961 the number of *fellaghas* inside Algeria was down to five thousand men, scattered in small groups; they could still vex the French but do nothing much of harm otherwise. If FLN morale was low, however, among the French it was at breaking point. They could not keep huge garrisons indefinitely in all the major centers, along with large mobile reserves besides. Twenty thousand Algerian guerrillas were concentrated in Tunis beyond the reach of the French. The European population of Algeria was up in arms against the *défaitistes* in Paris, the military commanders in Algeria paid no attention to the orders emanating from the capital. France, in brief, was on the verge of a civil war as General de Gaulle took over, nor did the danger pass until he had been in power for several years. Meanwhile the FLN had established itself as a government in exile, recognized *de facto* or *de jure* by some fifteen countries (including China and the Soviet Union). De Gaulle had been ready to cut France's losses without at first making his policy public; he had no illusions, was fully sensible that this meant surrender, the exodus of French Algerians and the loss of French property.

Thus, after seven years of struggle, Algeria attained independence. The exodus of the Europeans did not ruin the country as many had expected, just as the influx of *pieds noirs* did not make for the Algerianization of France. Very much in contrast to what Fanon had hoped, Algeria became a dictatorship, first under ben Bella, later under Boumedienne. Ten years after victory, all but one or two of the surviving early leaders of the revolt found themselves in prison or exile. On the first day of the rebellion the FLN had published a proclamation defining its goal as national independence through the restoration of a sovereign democratic state within the framework of the principles of Islam, and the preservation of all fundamental freedoms. The Algerian state that emerged from the war of liberation was not exactly the country of the rebels' dreams; "*Heureux les martyrs qui n'ont rien vu*," one of them wrote.[32]

According to some observers (such as Charles André Julien), the story of Algeria and of the Maghreb in general is one of the lost opportunities insofar as France is concerned. Much play has been made of Algeria's economic maladjustment, and the failure to integrate the Algerians into a modern economic system.[33] But there is no good reason to assume that Algeria would have remained part of France even had there been a much higher standard of living and no unemployment. The Algerians belonged to a different civilization; given the upsurge of nationalism after World War II and the weakening of the European powers, neither economic or social or even political reforms would have made the slightest difference. It might have postponed the struggle for independence by a few years; the FLN did not demand total separation at the start of the rebellion. But whatever the timing and the means, Algeria would eventually, riding the current of the tide in the affairs of the world, have demanded and obtained its independence.

CUBA

Cuba and Algeria, scenes of the two major guerrilla wars of the 1950s, were different in almost every aspect except that the key to victory was political not military in both instances. The Algerian FLN faced a colonial power, Castro and his comrades fought native incumbents. The struggle in Algeria lasted for seven years, it was waged against an efficient regular army of half a million men and was exceedingly costly. The campaign in Cuba took two years and did not involve much fighting; the Cuban army was small (forty thousand men), ill equipped and lacked both experience and above all the will to fight. The Cuban war is very much the story of one man and his "telluric force"; without him the invasion would not have been launched in the first place, after the initial setbacks it would have been dropped.

Castro's force was so small that it is hard to explain its success even in retrospect. Almost up to the end of the war there were no more than three hundred guerrillas, but they made as much noise and received as much publicity as three hundred thousand might have done. The materialization of three hundred guerrillas induced Washington to declare an arms embargo, weakening Batista both politically and militarily. Even professional military journals were quite deluded about the strength of the insurgents; according to an account in the *Marine Corps Gazette* (February 1960), Castro commanded not less than fifteen thousand men and women.[34] From a Marxist point of view, Castro's success is not easy to explain either. It was neither an agrarian rebellion, certainly not a proletarian revolution, nor was it an opposition movement headed by the "national bourgeoisie," or a combination of all these forces, a people's war. If anything, it was Blanquism transferred to the countryside. Cuba was not an underdeveloped country; it was semideveloped, or, to be precise, suffered from arrested development. Its rate of literacy was high, its standard of living about equal to Italy's (before the *miracolo*) and higher than in the Soviet Union. Some of the most respected observers of the Cuban scene have laid Castro's victory variously to the state of the Cuban sugar industry (Hugh Thomas) or the tensions within Cuban social structure such as the disparity between cities and countryside and the sluggish rate of growth.[35] It is of course perfectly true that Cuba was at the mercy of world demand, that the price of sugar was highly volatile and that the industry was in a state of decline (even though 1957 was a bumper year). There was indeed a great gap between the level of income in urban and rural areas, but there was a similar, even greater gap in many other parts of the world. As Theodore Draper stresses, vast tracts of sugar land belonged to American owners, but this was not one of the central issues in Castro's program; Draper emphasizes that there was less antigringoism in Cuba than almost anywhere else in Latin America. All this is so, but it still does not explain why Batista's regime collapsed like the walls of Jericho at the mere sound of trumpets. Hugh Thomas has it that the institutions of Cuba in 1958–1959 were for historical reasons amazingly weak. But were not the historical reasons largely accidental? Batista was a weak and ineffectual dictator, cruel enough to antagonize large sections of the population, yet not sufficiently harsh (or effective) to suppress the revolutionary movement. Cuba had a long record of political violence and (as in Algeria) of guerrilla warfare. The bureaucracy was weak and lazy, the police and army underpaid and demoralized, corrupt, sedentary and internally divided.[36] But all this could with equal justification be said about a great many other countries. Batista had not been unpopular with the masses when he first came to power; in 1940 he had been, as Castro reminded him, the presidential candidate of the Communist party. But the Batista who came to power in 1952 was a changed man; he had become lazy, ate sumptuously and spent much of his time playing canasta or watching horror movies.[37] Batista's coup in 1952 was by no means inevitable; another slightly more intelligent and energetic ruler or even Batista himself, fifteen years younger, would have realized that in the interest of survival he had to strengthen and modernize the police and the military establishment and to make both more efficient. However tyrannical and unpopular, such a ruler would not have been overthrown by Castro and his three hundred. It has been maintained that an unpopular regime cannot possibly be saved by means of repression, however well organized, but the Latin American experience simply does not bear this out. It was not so much Cuba the country, its economy, society and politics that were unique, but the specific political constellation prevailing there in the late 1950s. This is neither to magnify nor to belittle Castro's undoubted courage, personal magnetism and qualities of leadership; it is to point to the fact that the Batista colossus had feet of clay. It was not through farsightedness or by instinct, but through sheer foolhardiness that Castro dared to challenge the dictator, only to discover to his and everyone else's astonishment how brittle the regime was, and how near to collapse. Castro certainly did not lack self-confidence and Havana University, where he had studied in the late 1940s, had been an excellent training ground. It was, as he noted years later, much more dangerous than the Sierra Maestra. There still could have been an accident — a fatal mishap during the landing of his group, or perhaps a quarrel with Crecencio Pérez, the popular bandit who during the critical period after the landing was Castro's main link with the "masses." It is doubtful whether any of Castro's companions had the qualities to lead the rebels from the Sierra to Havana. It was only in March 1957, four months after the landing, that Castro was joined by new recruits from the towns and became less vulnerable. The key questions with regard to the victory of the Cuban revolution concern not Castro, but his enemies and rivals. Why was there no resistance, why did the middle class, the Church, the foreign supporters desert Batista?

The military operations were few and of no outstanding interest. Guevara, in his *Episodes of the Revolutionary War*, recounts various "battles," such as the battle of La Plata or the battle of Arroyo del Infierno. But these were either minor ambushes or attacks against small police or army posts carried out by twenty or thirty men.[38] The decisive "battles" of the war were fought by a hundred men or less; there was only one serious counterinsurgency operation by Batista's force, the "big push" in May 1958. In mid-June the government forces made contact with the rebels, but Castro's combat intelligence was excellent, Batista's forces did not find their way in unfamiliar territory, they were bombed by their own aircraft, and within a few weeks the fighting was over. The "rebels" fought well on the rare occasions they had to fight; there is evidence that in some cases money was offered, and accepted, and that Castro's men did not owe their success entirely to their military prowess. There were certainly more victims during the fighting in the towns (as during and after the naval mutiny at Cienfuegos) than in the Sierra, where police or army posts often surrendered after being exposed to no more than a few minutes of shooting.

Castro's officers and men showed infinitely more fighting spirit, initiative and intelligence than their opponents. Some of the regular army commanders were superannuated, having entered service before Castro and Guevara were even born. A capable and efficient officer was likely to be replaced because his superiors either envied or distrusted him. There was no overall plan and strategy on the part of Castro; neither he nor Guevara had read Mao at the time.[39] If anything, they were guided by the experience of nineteenth-century Cuban guerrilla warfare — every Cuban child was familiar with the exploits of Antonio Maceo and Maximo Gómez; the heroes of the war of independence provided inspiration for the fighters in the Sierra Maestra.

Some American observers insisted from 1957 on that Castro was a Communist, or surrounded by Communists, and Castro in later years himself declared that he had been far more radical in his political views from the very beginning than was generally known. His reasoning went that if he had come out openly in favor of Marxism-Leninism, the rebels would not have been able to get down to the plains, "because there would have been no support for them." But these are rationalizations after the event. The Castro who landed in Cuba was certainly not a Marxist-Leninist, but a radical who could have moved "left" or "right" with equal ease. Many Cubans who supported Castro expected a different revolution from the one they got; it is no less a certitude that Castro and his comrades were primarily men of action, and that while the fighting was going on in the Sierra Maestra they had little inclination to engage in ideological hairsplitting. Gradually they moved toward Communism. This conversion was not altogether surprising, for Fascism was in disrepute and liberalism was out of place in Cuba. On the other hand, there was a strong residue of free-floating radicalism in Cuba, and a growing estrangement from the United States. But all this belongs to a subsequent chapter in Cuba's history. The Cuban revolutionary war was not fought under the banner of Marxism-Leninism, its leaders were not members of the Communist party, and the Cuban Communists established contact with Castro only toward the end of the war. It was fought under the pennon of patriotism, national unity, of freedom from tyranny and corruption.

THE MIDDLE EAST

The Palestinian attacks against Israel have attracted far more attention than other guerrilla wars in the Middle East chiefly because of their international ramifications and the involvement of other Arab states. But there were other wars such as the insurrection in southern Sudan during the 1960s or the fighting in Kurdistan which punctuated most of the postwar period. Guerrilla operations in the Persian Gulf (Oman) have lasted for more than a decade and there was sporadic "urban guerrilla" warfare in both Turkey and Iran. The first armed raids into Israel by Palestinian *fedayeen* occurred in the early 1950s and provoked immediate Israeli reprisals. On a larger scale, attacks began only with the creation of *Fatah in* 1965. Its activities became more widespread after the Six Days' War (June 1967), for though the Arab armies had been routed, Israel had occupied lands with a population of more than a million Arab inhabitants. More important yet, the Palestinians now received very substantial support from Arab governments, whereas before 1967 such aid had been given only grudgingly and selectively. The refugee camps in Israel and outside provided a unique reservoir for the mobilization of new recruits, as well as centers for training and as hiding places. Between 1967 and 1970 Fatah expanded from a few hundred to between fifteen and twenty thousand members. Immediately after the Six Days' War it had attempted to stage "revolutionary guerrilla warfare" both in the cities of Israel and elsewhere about the country. But the terrain was unsuitable, the local Arab inhabitants not too cooperative and the Israeli counter-measures quite effective. (The Gaza region became the classic example of a successful counterinsurgency campaign.) After only a few months Fatah headquarters and most of its members had to be removed from the West bank to the other side of the Jordan; Fatah became a guerrilla movement in exile. Sporadic terrorism continued on a limited basis for many years and there were occasional demonstrations and strikes, but this was certainly not the "general insurrection" Fatah had been waiting for. Located abroad, there were three potential avenues open to the Palestinians for pursuing their war. They could infiltrate guerrillas into Israel either for hit-and-run attacks or in the hope that these would be able to establish *foci*. Alternatively, the Palestinians could shell Israeli settlements from beyond the border; they had missiles in their arsenal which reached fairly deep into Israeli territory. The Israelis would be unable to retaliate without putting themselves in the wrong vis-à-vis international law; Israeli reprisals moreover would aggravate relations between Jerusalem and its Arab neighbors and help prevent a "sellout" by some Arab governments. Lastly, the Palestinians could attack Israelis, Jews and even non-Jews, as well as Israeli installations and institutions outside the country; the "acts of despair" would demonstrate that unless justice were done to the Palestinians, there would never be peace in the Middle East.

Fatah and the other Palestinian organizations tried all three approaches with varying success.[40] Small units were infiltrated into Israel from Jordan, and later from Syria and Lebanon. But despite the covert sympathy for them among some of the domestic Arabs, the terrorists' position was more like that of goldfish in a bowl than fish in an ocean. Only very small units (up to four or five members) could be infiltrated. They were usually intercepted within a few hours, at most within a few days; only one or two groups managed to stay undetected in Israel for as long as two months, and this primarily because they refrained from outright violence. Between 1968 and 1971 there were nonetheless innumerable cases of infiltration, or random shootings, of bombs, resulting not alone in losses to the Israelis, but in sizable ones to the raiders, and gradually this type of tactic was restricted to a very few hit-and-run attacks with clearly defined aims. Because of their dramatic character, they were to attract far more publicity; instances of this were the Lod Airport massacre (carried out by Japanese terrorists), the attack against a school at Ma'alot (1974), and a hotel in Tel Aviv (1975). The shelling from across the borders began early in 1968 in the Jordan valley, spread in October 1968 to southeast Lebanon, and during 1969 to the whole of southern Lebanon (Fatahland). Again there was Israeli retaliation, first against the Jordanians and later against the Lebanese. A certain pattern emerged. The Palestinian terrorists would shell Israeli settlements from across the border. The Israelis would then retaliate, but since the Palestinians would have evaporated, the Jordanian and the Lebanese regular army units would have to bear the brunt of the Israeli attack, which did not improve relations between the Palestinians and their hosts. Following heavy clashes with the Jordanian army in 1970, the Palestinians had to transfer their activities to Lebanon, which became their major springboard for attacks against Israel. In the south of Lebanon the Palestinians established a virtual "state within a state," leading to severe tension and to bloody encounters in turn with the Lebanese.

The shelling of one country from the territory of another is certainly a warlike action; whether it can be defined as guerrilla warfare is a moot point. But the most controversial aspect of the Palestinians' activities were those carried out in third countries — the killing, for instance, of members of the Israeli Olympic team in Munich in 1972, the hijacking of airplanes, most of them not belonging to Israel, the dispatch of letterbombs, and other gambits such as the attacks against foreign ambassadors in Khartoum. It looked — and has so been argued — as though the Palestinians had simply found by trial and error that there were better means than the traditional ones of guerrilla warfare for furthering their cause, that publicity was the vital weapon, that what counted beyond all else in the last resort was to keep the Palestine issue alive. However widely condemned, all these outrages were given enormous notoriety. It is nevertheless unlikely that this strategy would have worked but for the growing dependence of the industrialized countries on Arab oil. There were far more kidnappings in Brazil but it led the urban terrorists nowhere. The Palestinians, however, had powerful allies and benefited from exceptionally auspicious international circumstances. Militarily, they failed, but as the Algerian example had demonstrated, military failure *per se* meant nothing. Politically the Palestinians succeeded; they were recognized by many member states of the United Nations and an assortment of other international organizations besides.

The splits within the Palestinian resistance did not bring any immediate harm to the common cause. More serious were the long-term effects of the terrorist operations. These stiffened Israeli resistance, made a dialogue between Israelis and Palestinians virtually impossible and, in addition, hampered any attempt to work out any unified Palestinian policy for the future. A policy aimed at the destruction of the Jewish state might have conceivably worked in the pre-atomic age, but with the development of the means of mass destruction the rules of the game had changed. If Fatah and the other Palestinian organizations had little to fear directly from the nuclearization of the conflict, this was not so with regard to Israel's Arab neighbors, and without their support the Palestinians could not continue their struggle in the long run. Worse yet, in the case of a nuclear attack against Israel, the Arab residents of that country were as likely to perish as its Jewish citizens. Further, there was a growing discrepancy between Palestinian theory and practice. Much of the fighting against Israel was done by others. If other guerrilla movements throughout history never had enough money, the Palestinians, thanks to the oil windfall, had almost too much of it.[41] The abundance of funds made it possible to engage in various kinds of operations, military and propagandistic both, beyond the reach of other, less affluent guerrilla movements. At the same time a surfeit of money bred corruption; guerrillas must be lean and hungry, a condition which exposure to life in Hilton hotels did nothing to encourage.

While Fatah proclaimed resoundingly that the shame of the defeat was to be washed away by the mass struggle of the Palestinian people, it became only too manifest after 1973 that not the armed assaults, let alone the masses, but the profits of the oil-producing states had brought about the change in Palestinian fortunes. It could be argued that whether the Arab masses did or did not in fact participate in the striving against the Zionist enemy was beside the point, all that mattered, again, was the result. It is unlikely, however, that a Guevara or a Fanon would have approved such a rationalization; they would have held that a people that owed its national liberation to financial manipulations could scarcely be accounted free.

The Israelis tended to belittle the role of the Palestinians, and the fact that there was so much "guerrilla by proxy," that is, terrorist acts committed by Japanese, French or Latin American mercenaries, only strengthened their contempt for the military qualities of their opponents. There is no denying that in contrast to other guerrillas, rural and urban, the Palestinians usually avoided clashing with the Israeli security forces and directed almost all their attacks against the civilian population. But all this does not change the fact that the Palestinian organizations were by no means totally ineffective, and that individual infiltrators did show courage; realizing that they could not conduct guerrilla warfare along conventional lines, they had to look for other means to harass the enemy even if this approach led them beyond guerrillaism and even urban terror, however liberally these terms are interpreted.

If the ups and downs of Palestinian resistance point up the overriding significance of foreign help, the fate of the Kurds only accents that importance the more. The Iraqi Kurds, who constitute about twenty percent of the population of the country, fought for their autonomy from 1960 to 1970. They had taken up arms on many previous occasions but never for so long a period. The Iraqi army was fought to a standstill by the *Pesh Merga* in the hills of Kurdistan despite their numerical superiority and the fact that the Kurds had only light arms; it was only during the last year of the war that the Kurds acquired some anti-aircraft guns. The Kurdish war also proves that a mastery of guerrilla doctrine is not really of decisive import; their leader Mulla Mustafa Barazani had in all probability never read a single manual on the subject, he and his men simply knew all there was to know about it by instinct. There was far more fighting, and many more casualties in the war in Kurdistan than in many, much better-publicised guerrilla wars, including the Palestine-Israel conflict. But it never attracted much attention, perhaps because the Kurds failed to appreciate the great strategic importance of oil and did not attack the Kirkuk oilfields. When the war was renewed in 1974 the Kurds were defeated with relative ease, the international situation having changed in their disfavor. Up to 1970 the attitude of the Soviet Union had been one of friendly neutrality. But with the emergence of a pro-Soviet dictatorship in Baghdad, the Kurdish struggle no longer served any useful purpose from the Soviet point of view, and the Iraqi army, supplied and trained by the Russians, was now able to cope with the problems of mountain warfare. Furthermore, Iran, which had hitherto provided arms and supplies to the Kurds, closed its border, the Shah fearing that an escalation of hostilities with the Baghdad regime, and indirectly with the Soviet Union, might endanger his regime; the stakes in the game had suddenly become higher. Thus Kurdish resistance collapsed, not because *Pesh Merga* was fighting any less bravely but because, to quote an old Kurdish proverb, "Kurds have no friends."[42]

The Eritrean Liberation Front (ELF), on the other hand, did have friends. Founded in Cairo in 1958, this separatist organization launched in 1961 a terrorist campaign which, until 1975, was on a relatively small scale. There was certainly much less fighting in Ethiopia than in Kurdistan, and for years the ELF had no more than a few hundred active members. It had, however, strong backing in the Arab world, particularly in Syria, and it had the political support of the Muslim states in Africa. It was well supplied with arms and money. Thus, in the course of a decade, a minor army came into being, and as Ethiopia faced a major domestic crisis, the ELF could stake its claims with much greater vigor.[43]

A third example of the decisive impact of outside help is PFLOAG (People's Front for the Liberation of Oman and the Arab Gulf). Established as the Dhofar Liberation Front in 1965 by nationalist opponents of the Sultan of Oman and active in Dhofar province, it was taken over by "scientific socialists" three years later. Although the rebels at no time numbered many more than a thousand, the sultan had only twenty-five hundred men at his disposal up to 1970, altogether insufficient on any count to crush the insurgents. Later the sultan's small army was reinforced by British advisers and Iranian and Jordanian troops. The headquarters of PFLOAG were located in South Yemen, which served also as a sanctuary, the main supplier as well as the fountainhead of ideological inspiration. The Chinese and the Soviets competed for a stake in this interesting attempt to apply Marxism-Leninism (or Leninism-Maoism) in conditions varying between those of the stone age and the feudal era; Chinese influence was on the decline after 1970. PFLO (the AG was subsequently dropped) was the most antireligious of all Arab extremist movements, but this did not deter Colonel Ghadafi from providing financial assistance any more than it did the Iraqis from proffering help. Seldom in guerrilla history has such a small war in such a remote country attracted so many foreign powers. Russian artillery operated by Chinese-trained guerrillas in South Yemen territory shelled Iranian forces engaged in counterinsurgency operations on Oman territory. The original initiators of the revolt had invoked their belief in Allah and pan-Arabism, but they were bitterly criticized by the professionals who took over the leadership from them in 1968 for having chosen "the mistaken path of spontaneous action under a leadership incapable of leading armed struggle."[44]

Attempts to launch guerrilla warfare in Turkey and Iran in the 1960s were, on the whole, unsuccessful. The Iranian peasant was too conservative and the Shah's agrarian reform had to a certain extent taken the wind out of the sails of the revolutionaries. "Armed confrontation will start in towns and their suburbs," wrote Farahani, "as the Iranian peasantry with its rustic environment is not conducive to revolutionary preparedness."[45] The Iranian revolutionaries were split into several factions (Maoists, *Siahkal,* and the National Liberation Movement). One of the peculiar features of the Iranian resistance is the collaboration between the extreme left and the (Shi'ite) religious fundamentalist sectarians (NLM) who, led by *ulema*, established a little guerrilla army of their own.[46] This political alliance dates back two decades and is based on opposition, albeit for different reasons, to the Shah's reforms.[47]

Guerrilla warfare in Turkey was similarly impeded by internal dissent. Various small sects that would emerge from time to time engaged in kidnappings or assassination, but there was no coordination between them. Most of them derived from *Dev Genc,* the Federation of the Revolutionary Youth of Turkey, a roof organization for an assortment of radical groups.[48] They all favored armed struggle, but some were Maoists tending toward rural guerrilla warfare and the creation of a Vietnam situation in Turkey, others called for a second Turkish war of independence. Some wanted to infiltrate the Turkish armed forces and to conquer them from within; others, on the contrary, looked for a confrontation with the army. Yet for all that rural Anatolia with its backwardness and abject poverty should have been fertile soil indeed for the recruitment of guerrillas, the students of Ankara and Istanbul failed to gain any substantial foothold outside the universities.

Guerrilla warfare in the Middle East was most successful in the very place in which it seemed most unlikely. At the height of the 1950s Cyprus insurgency, twenty-eight thousand British soldiers were chasing some two hundred and fifty terrorists on an island of half a million inhabitants. The leader of EOKA was an old man by military standards; Grivas, a native of Cyprus, an ex-officer of the Greek army, was fifty-seven when the campaign started. He had some support from this and that influential politician in Greece, but the Athens government was far from enthusiastic about his venture and on several occasions threatened him and demanded that he stop it. Archbishop Makarios, the political leader of the Greek Cypriots, was also initially opposed to Grivas's move; he would much have preferred a campaign of sabotage, and (according to Grivas), the throwing from time to time of a few hand grenades into Turkish mobs, just to teach them a lesson.[49] A year after the outbreak of the rebellion, however, Makarios was to declare that the terrorist operations had been more effective than seventy-five years of "paper war." Grivas had the Cypriot Communists against him, the strongest single party on the island; early on they revealed in their manifestos the real identity of "Dighenis," Grivas's pseudonym. That the Turkish minority saw in EOKA a mortal enemy scarcely needs saying. International public opinion did not support EOKA; the Communist bloc, the Chinese, the Third World countries, all the traditional sympathizers of guerrilla movements showed a lack of interest, and quite often downright hostility. But despite all these handicaps, Grivas succeeded in a three-year campaign (from 1 April 1955 to Christmas 1958) in ousting the British from the island, which led to the declaration of Cypriot independence. The British suffered relatively few losses, but since Britain was reconciled to liquidating the remnants of its empire anyway and since it found the task of policing a rebellious island too burdensome, not much force was necessary to persuade Whitehall to surrender.[50] The long-term results of the Grivas campaign were nonetheless disastrous, for victory in 1958 was followed by tragedy in 1974. The EOKA campaign had sharpened the old conflict between Cypriot Greeks and Turks. Eventually the Turks invaded Cyprus and the country was *de facto* partitioned. Grivas's old partner and antagonist, Makarios, bore an equal measure of responsibility for these tragic developments, for he had shown no greater willingness than Grivas to work for an accommodation with the Turkish minority.

LIBERATING AFRICA

Power was transferred in a more or less orderly fashion by the colonial administration to the new rulers in most African countries. There were exceptions, such as in the Congo, and guerrilla warfare occurred in the Portuguese colonies. In addition, there was a good deal of internal fighting in the postindependence period; some of it was tribal in character, or separatist (Sudan, Nigeria, Chad, Eritrea), elsewhere it was a conflict between various contenders for power. Not all these opposition movements were strikingly effective (the Nigerian Sawaba or the Senegalese AIP), but in other cases, as in Biafra and the Congo, the internal feuds were much more bloody than the anticolonial guerrilla wars themselves. In contrast to Latin America, China and Southeast Asia, the African guerrilla leaders hardly ever lived and fought with their troops; their headquarters were almost invariably in some neighboring country. The anticolonial guerrilla movements were usually split; in almost every country there were two or more such groups battling each other even more fiercely than they fought the common enemy; their "mass basis" was in essence tribal rather than national. The Algerian FLN and the Vietnamese Communists also had to face competition early on in their struggle, but they destroyed their opponents and thus were able to monopolize the field. In Africa, on the other hand, the splits persisted in many instances, and this affected both the character and the course of the guerrilla war.

The Mau Mau revolt in 1952 was the first of the postwar insurrections. Dating back in origin to the 1920s, it was led by educated members of the Kikuyu tribe. They complained, not without justification, that some of their best land had been taken away from them; they resented the fact that there was no secular education and that female circumcision had been banned. In a solemn oath, the members of Mau Mau swore never to sell land to a European or an Asian, not to smoke foreign cigarettes or drink foreign beer, never to sleep with a prostitute and to behave in general as a patriot and a decent citizen.[51] Considering the relatively small number of people involved and that it was geographically restricted to a part of the country, the Mau Mau revolt was quite bloody; more than eleven thousand Kikuyu (but fewer than a hundred Europeans) were killed. The revolt failed, but within a matter of only a few years the British departed and Kenya became independent. The survivors of the Mau Mau land army were given a hero's welcome, but after that a determined effort was made to erase Mau Mau from Kenya's public memory.[52] It was not that anyone doubted that the Mau Mau had made a contribution to Kenya's independence, but many of their practices had been repugnant in the extreme. Above all, the massacres perpetrated by the Mau Mau against Kikuyu loyalists (as in the Lari massacre), and members of other tribes, were divisive and did not augur well for the future of a country in which a variety of tribes would have to live peacefully side by side. The next major rebellions to occur, in Angola in March 1963 and in the Congo in 1964, were also basically tribal in nature. While Europeans were killed, the number of blacks of other tribes, and of *mestizos* and *assimilados* who came to grief was far greater.[53]

After concluding a trip of African capitals in 1963, Chou En-lai observed that the prospects for revolution in Africa were excellent. Events in the suceeding years did not quite bear out his prediction. True, a great number of liberation movements came into being and were duly registered on the payrolls of the African Liberation Committee, the coordinating body of the OAU. They would publish victory communiqués from nonexistent war fronts and celebrate the establishment of liberated and semiliberated zones; the Mozambique Frelimo had a particularly bad record, but others were not far behind.[54] As in so many other cases, they would more often clash with each other than with the declared enemy. To some extent this was the result of old tribal feuds which made the formation of national movements difficult. Thus SWAPO was essentially an Ovambo organization (despite all disclaimers), and the UPA and FLNA had their power base in the Bakongo tribe. Existing dissension was fanned by Sino-Soviet rivalry for influence in the continent. At one time or another almost all African liberation movements split into a pro-Soviet and a pro-Chinese wing, beginning with the first of the "modern" (i.e., quasi-Maoist) guerrillas such as the Camerounian UPC which, founded in 1947, started armed struggle in 1956. Lastly, the conflicting ambitions of leading personalities often collided. The list of leaders of African liberation movements assassinated by political or personal rivals (sometimes with a little help from the colonial powers) makes depressing reading. It includes some of the most gifted leaders, such as Amilcar Cabral (of PAIGC) and Eduardo Mondlane (of Frelimo). Some guerrilla movements practiced almost constant internal "purges" (Frelimo again was one of the worst offenders). The sad events in Angola in 1975 brought into the limelight a state of affairs that had existed, on a smaller scale, for many years previously. Least affected by internal disputes and tribal rivalries was PAIGC in Guiné-Bissau, headed by Amilcar Cabral, a talented leader, about whom more below.[55] But in Guiné as well, after a decade of fighting, a sympathetic observer noted that "a clear-cut military victory that would expel the colonial forces would . . . be a miracle."[56] Another historian of the African liberation movements, writing in 1971, prophesied victory over the Portuguese not before the 1990s.[57]

In retrospect it is easy to understand the reasons for these misjudgments. After the initial upsurge in the early 1960s, the tide turned against the guerrillas almost everywhere in Africa. The ANC-ZAPU units which had infiltrated into South Africa were destroyed, SWAPO activities were largely ineffective, the MPLA campaign in Angola collapsed in 1966, and Roberto Holden's GRAE was largely inactive after its initial operations in the early sixties had petered out. Frelimo failed to prevent the building of the Cabora Bassa Dam. Only in Guiné-Bissau, PAIGC made progress; the number of guerrillas there increased from four thousand in 1964 to (allegedly) ten thousand in 1970. After ten years of strife, the three Angolan independence movements had altogether some ten to fifteen thousand fighters, and the Portuguese armed forces had fewer casualties in a decade than the French in Algeria had in a single year. The leaders of the African guerrilla movements spent far more time attending international congresses than in stepping up guerrilla warfare. The Liberation Committee of the OAU was taken to task in 1967 for incurring "excessive administrative expenses and subsidizing certain individuals."

In view of these and other weaknesses, the eventual successes of the liberation movements against Portugal seem almost inexplicable. But however small the guerrilla forces and infrequent their operations, they enjoyed certain distinct advantages such as secure bases in neighboring African countries, sufficient financial help from the OAU, the Soviet bloc and China, and a steady supply of arms. Frelimo and the MNLA were as well armed toward the end of their war as their Portuguese opponents, save that they had no artillery (which was useless in the bush) and no air force (which would have been of limited assistance only). Above all, they were facing the poorest European nation, which could ill afford to pay for a protracted guerrilla war in colonies which, with the exception of Angola, were of little economic value. Even the suppression of a small insurrection such as the Mau Mau had cost some hundred and thirty million dollars, while the Algerian war cost anything from five to ten billion dollars. The very presence of guerrillas in neighboring African countries made the stationing of considerable forces necessary (a hundred thousand men in the three main colonies), and this the Portuguese simply could not in the long run afford. By 1970 the Portuguese had spent two billion dollars on their colonial wars. Furthermore, there was considerable international pressure which the Portuguese notwithstanding all their defiant gestures could not ignore. The Portuguese commanders knew, in brief, that they were fighting a rear-guard action, and this scarcely made for any great enthusiasm or high morale.

Attention has been drawn to the intense rivalry between the Soviet Union and China in Africa. During the cultural revolution and for several years thereafter, the Chinese were almost totally preoccupied with their domestic affairs, and by the time they returned to the African scene in 1971–1972, the Soviet Union had made considerable strides in winning support in most major African liberation movements. But how deep was Soviet and Communist influence?

Radical African leaders have frequently described themselves as "Marxist-Leninists"; on the other hand, they have claimed that their movements were "authentically African."[58] They have declared at one and the same time that their movements had no official ideology, yet that only "scientific socialism" could serve as their lodestar. Such contradictions are more apparent than real. While individual leaders had acquired the rudiments of Marxism in European universities, this certainly did not apply to their followers, and in any event, the problems facing the guerrilla movements either during their struggle, or after victory, were such as neither Marx nor Lenin, nor even Mao, had ever envisaged. Eventually military leaders came to power both in those African countries in which the transition had been peaceful and the others which had fought for their independence. The poor countries were still poor after independence, the rich remained rich, and the importance of ideological pronouncements should not be overrated.

LATIN AMERICA

Guerrilla operations in Latin America reached their climax in the early 1960s. They were mainly concentrated in the countryside, but with the failure to establish secure rural bases (Argentina and Brazil in 1964, Peru and Venezuela in 1965, Bolivia in 1967), urban terrorism became the fashion, principally in Uruguay (MLN — the Tupamaros), Brazil (ALN), and Argentina (ERP and Monteneros). The political doctrine and overall strategy of these movements are discussed elsewhere,[59] but the causes of success and failure remain to be analyzed.

The chances of guerrilla warfare in Latin America were excellent in many ways. Capitalist development in the continent had few achievements to its credit but all its defects were only too manifest. There was poverty, glaring exploitation and widespread anti-Americanism. The establishment was not usually noted for its social conscience or its reformist fervor. A comparatively large class of intellectuals violently opposed the *status quo*, looking on the urban slum dwellers and landless peasants as their revolutionary reservoir. There was a long history of political violence, and Castro's victory had given fresh hope to all revolutionaries; victory was possible, after all, even in a single country. The ruling strata were weak, disorganized and devitalized, the forces of repression inefficient. In short, there was a revolutionary situation with all its classic ingredients, "objective" and "subjective." There was no lack of discontent nor of idealism; there was mass support on the part of the younger generation in the universities and even in sections of the army. And yet, without exception, the guerrillas failed to reach their goals and the intriguing question is why.[60]

An analysis of the development of guerrilla movements in Latin America indicates above all that they were nearest to victory in the least repressive countries such as Venezuela and Uruguay. Bolivia was a military dictatorship when Guevara tried to establish his *foci* there, but President Barrientos was a populist of sorts with considerable backing among the *campesinos.* The new upsurge of guerrillaism in Colombia in 1975 occurred precisely at a time when the relatively liberal government of the day was engaged in carrying out a policy of reform. The insurrection in Venezuela in the early 1960s, spearheaded by the MIR and the Communists (who together established the FALN — Armed Forces of Liberation) and the urban terrorist operations of the UTC, came closer to success than in any other country in the continent. But even they had no real mass basis, as the results of the 1963 elections proved. The brutal character of many guerrilla operations antagonized the masses and isolated the insurgents. Usually the guerrillas assumed that the regime could be overthrown with one forceful push (*golpe*). The inevitable setbacks caused splits in their ranks; if the Tupamaros' strategy of provocation in Uruguay at least brought about the downfall of liberal democracy and the rise of a military dictatorship, the Venezuelan guerrillas did not carry even that much weight.

It is difficult to generalize about Latin American guerrilla movements because conditions varied so much from country to country, and the movements themselves were so disparate. Some countries were predominantly rural, others primarily urban; the Guatemalan MR-13 was launched by young officers, the Tupamaros by students, the Venezuelan guerrilla groups by political parties. In Peru and Colombia there was a connection with spontaneous peasant uprisings; in Venezuela the Communist party supported the guerrillas, elsewhere it opposed them. Yet for all these differences, certain staple patterns emerge:

1. If the guerrillas were inspired by Castro's victory and had assimilated its lessons, the government forces had also learned from the Cuban example. Initially unprepared for counterinsurgency, they became quite adept at it; sometimes they knew more about it than the guerrillas themselves. The armies were built up

and modernized, the use of helicopters made guerrilla activities in the open country very hazardous indeed. Moreover, as Malcolm Deas has noted, soldiers in Latin America are not as unpopular as policemen — the army has a different relationship with the population. "No Latin American army had the combination of vices to be found in Batista's army."[61] But the establishment had not only learned from the military tactics used by the guerrillas; in Peru, far-reaching agrarian reforms carried out by the army stole the guerrillas' thunder and a sizable part of their forces went over to the government, or at least became a loyal opposition.

2. In most Latin American countries (as in the Arab world, Ulster and Africa), there was not just one guerrilla movement but several; their internal splits and the tortuous relations between nationalist, pro-Moscow Communist, Trotskyite and Maoist parties were an ever-present source of friction. On rare occasions the guerrilla movements would make common cause: in February 1974 the Bolivian ELN, the Tupamaros, the Chilean MIR and the Argentinian ERP set up a Junta of Revolutionary Coordination (JCR). But far more often there would be disunity, internal strife and purges. The Colombian ELN was notorious for the acts of terror committed in its own ranks; the commander José Ayala was shot by his own men, and Fabio Vasquez had many of his rivals liquidated, sometimes by "war tribunal," sometimes without such legal niceties. There were bitter, though less bloody mutual recriminations among the Venezuelan, Peruvian and Bolivian guerrillas. Above all there was the Communist problem. The Cuban revolution had developed in its early phase quite independently of the Communists, but in other Latin American countries the revolutionary potential usually belonged to one political party or another, and no incipient guerrilla movement could afford to ignore this elementary fact of political life. The guerrillas (with the notable exception of Uruguay) became involved in sectarian in-fighting on ideological, organizational or simply personal lines.*

3. Following the Cuban example, the guerrillas at first envisioned the countryside as their main field of action, yet strategic considerations quite apart, they found it unexpectedly hard to rally the support of the peasants. There were exceptions, such as Hugo Blanco's POR in Peru, although this created problems of its own, for while the *campesinos* were willing to defend their homes and

* The climate of dissension in these groups has been vividly described in Régis Debray's novel *L'indésirable* (Paris, 1975).

land, they were reluctant to operate outside their immediate neighborhood. The leaders of the guerrilla movement were, for the greater part, city people of middle- or upper-class origin, young men (and in a few cases, women) who spoke, quite literally, an altogether different language from the peasant population, which belonged to all intents and purposes to another race. Just as the *mestizo* leadership of the Angolan MPLA did not understand the tribal tongues of their own warriors, so the Peruvian and Bolivian city revolutionaries were at a frustrating loss to communicate with their Indian recruits. For all their enthusiasm, they also found it anything but easy to adapt themselves to the hard life of the countryside; they were genuinely shocked by the miserable lot of the peasants, but at the same time they shared the contempt for manual labor deeply rooted in Latin American (and African) society. Since as revolutionaries they had to live, work and fight side by side with peasants and manual workers, this instinctive attitude, which only a few of them could completely overcome, did not exactly make for smooth relations with the toiling masses. Although any number of sound tactical reasons could be cited for the later withdrawal of the guerrillas from the country areas, there is no doubt that on the whole they were only too happy to get back to a city milieu more familiar in every respect and certainly less arduous.

4. The peasants (*campesinos*) had a tendency to adopt a wait-and-see attitude rather than embrace the revolutionary cause on sight. If the *guerrilleros* successfully defied the government forces, they could count on at least the passive support of the rural population. But if they suffered a setback and it appeared that the forces of law and order were after all stronger, there would be no help for the insurgents and they would find themselves betrayed to the authorities. And even if the peasants were prepared to join in their fight, they would do so, as mentioned already, only within the district in which they lived. This has been one of the traditional weaknesses of rural guerrilla warfare; peasants cannot easily be turned into professional revolutionaries willing to give up their ties and roots.

5. The guerrillas were caught on the horns of several dilemmas from which there were no easy disentanglements. Guerrilla ideologists have claimed since time immemorial that rugged and forest country is the most favorable for guerrillas. But if the country was too rugged or the forest too thick, the guerrillas would be hard put to it to get supplies. If they retreated into remote, unpopulated areas that were not easily accessible, they would be secure but ineffectual — as Guevara had noted. If they opposed elections (as

in Venezuela), this would damage their image as staunch fighters for democracy. If, on the contrary, they contested elections, they would be defeated — as in Uruguay — and by extreme reactionary forces at that. If they waged guerrilla warfare in small elitist conspiratorial groups, consisting primarily of students (or recent graduates), they would be reasonably safe from detection, but the moment they tried to broaden their urban base and "mobilize the masses," they would expose themselves to infiltration by enemy agents.

6. A successful guerrilla movement could weaken or bring about the overthrow of a relatively democratic regime, or an ineffectual autocracy. The strategy of provocation predicted that the democratic regime would be replaced by a reactionary and ineffective military dictatorship which would within a short time antagonize the middle class and especially the intelligentsia so that, as in Cuba, the guerrillas would gain the support of the majority of the population. But not all military dictators were as inadequate as Batista; others showed far greater determination and ruthlessness. Through terror and counterterror, they succeeded in paralyzing or altogether destroying the guerrilla movements (Guatemala, Brazil). Antiguerrilla murder squads came into being and torture was practiced with considerable effect.[62] Worldwide public opinion was enlisted against these atrocities, but appeals and manifestos from foreign lands showed diminishing returns and were in any case a poor substitute for guerrilla victory. This raises the general question of the strategy of destroying the stable image of the government and creating a "climate of collapse." Modern governments, as Robert Taber has observed in his *War of the Flea*, are highly conscious of "world opinion," they do not like to be visited by human rights commissions,

> their need of foreign investment, foreign loans, foreign markets, satisfactory trade relations, and so on requires that they be in more or less good standing with a larger community of interests. Often too, they are members of military alliances. Consequently, they must maintain the appearance of stability, in order to assure the other members of the community that contracts will be honored, that treaties will be upheld, that loans will be repaid with interest, that investments will continue to produce profits and be safe.

The weakness of this strategy is that it usually works only up to the point where the terrorist ceases to be a mere nuisance and becomes a real danger to the regime. Once this point is reached, the state —

however concerned it may be about the U.N., a human rights commission and adverse publicity in the press — will react in kind, cruelly, unhampered by laws and conventions or humanitarian considerations. Once the Brazilian ALN and other such groups opted for individual terror, the government responded with all but indiscriminate counterterror. Terrorists, sympathizers, and no doubt some innocent people as well disappeared without a trace. There were no trials and no death sentences, which made it all the more difficult to organize protests at home or abroad. Despite the only too justified outcries about repression and torture, the stable image of the government was not destroyed; the terrorist movement collapsed, not the state.

7. When fortune smiled on them, the guerrillas were on top of the world; victory appeared to be just around the corner. But they were not good losers, even though they knew most of the time, certainly in theory, that their struggle would be long and punishing. They had been spoiled by the Cuban example; by temperament, most Latin American guerrillas were *golpistas*, burning to topple the system with one big shove. Psychologically, they needed quick results, and if these did not come, there would be despondency and mutual recrimination. They were capable of great sacrifice and exertion for a short time, but not of sustained effort or of fortitude in adversity. This applies to both rural and urban guerrillas. The rapid success of the Guatemala army in November 1966, of the Peruvian security forces against Hugo Blanco (1963) and Hector Bejar (1965), the collapse of Douglas Bravo in Venezuela and of the Bolivian guerrillas, are a few of the many examples. The one exception was the Columbian ELN under Fabio Vasquez which continued its struggle for more than a decade with varying success; but most of the time they operated in remote areas and were no real threat to state security.

With the defeat of the rural guerrillas, action was transferred to the cities with great initial effect. From 1968 to 1972 hundreds of banks were attacked in Brazil and Uruguay, stores were robbed, political leaders assassinated, businessmen and foreign leaders were kidnapped, "enemies" were executed or kept in "people's prisons." After the defeat of the urban guerrillas in Brazil and Uruguay, Argentina in 1974 became the chief scene of such action. The objective was not alone to spread fear and confusion, but to establish a "parallel government" comparable to that of the Soviets in Petrograd in 1917.

The immediate successes were astounding. The slums of the big cities — but also the upper-class residential areas — provided far better cover for operations than the Sierra. It was easier to get money and weapons in the city than in the countryside, and to collect information about the targets for attack. All this daring found an echo far beyond Latin America — in the United States, Canada and even in some European countries. If the old-style guerrilla tactics had been applicable only to backward countries, the new urban guerrilla warfare seemed to offer immense possibilities to almost every country in the world, including the most developed ones. Nevertheless, the startling successes of the first years were again followed by grave setbacks and in some cases by total collapse. Unable at first to cope with this new danger, the forces of order were learning quickly. However strictly conspiratorial rules were observed, sooner or later the traces of a single Tupamaro would lead the police to a whole group, to its arsenal, and eventually to its headquarters, and once this happened, escape was difficult, far more so than in the Sierra. Marighela's assertion that the police "systematically fail" was overly optimistic — they certainly did not fail to shoot him, as well as the other urban guerrilla leaders in Brazil and Chile, and to arrest the Tupamaros. But individual failures quite apart, the entire urban guerrilla strategy was found wanting. It is true that urban guerrillas would get more publicity in a day than rural ones in a year; as far as the media were concerned, their exploits were far more newsworthy. But with repetition interest inevitably diminished; once a consul had been kidnapped for the fifth time, the news no longer automatically commanded the headlines. The terrorists had to think of new, sensational and even more bizarre exploits, such as the theft of the remains by the Monteneros of former President Aramburu (whom they had killed in 1970) from his grave. But there were limits to human imagination, and in any case publicity could not in the long run replace an overall policy. Urban guerrillas frequently referred to the Algerian example, but it was precisely in the city of Algiers that the FLN had suffered its greatest defeat. The FLN could not compel the French colonial government to evacuate Algeria; in no circumstances would a French guerrilla movement have been able to take over France even under the weak governments of the Fourth Republic. It was one thing to appear as the spearhead of a national movement against the hated foreigner; it was another, infinitely more difficult task to compete with other native political parties in the struggle for power.

This was the overall lesson learned by the Latin American guerrilla movements in the 1960s and 1970s. But seen from the guerrilla point of view, the picture was still not all dark. The old Latin American social order with its inefficiencies and inequities could not last, and armed struggle was certainly one of the possible ways of changing it. Cuba had been a success, perhaps there would be a victory elsewhere, given some fortunate juxtaposition — a prolonged political and/or economic crisis, able guerrilla leadership, and bungling by the forces of law and order. "Objectively" there still is a revolutionary situation in many Latin American countries.

URBAN TERRORISM

In the late 1960s rural guerrillaism gave way to urban terrorism in many parts of the world. The major exceptions were Vietnam (where the war had, however, proceeded far beyond the guerrilla stage), the Portuguese colonies, and some minor theaters of war such as Burma, Thailand, the Philippines and Eritrea. Elsewhere the hijacking of airplanes, the bank raids and the kidnapping of diplomats and other public figures rather than the ambush in some remote jungle village became the symbol of armed struggle. Skyjacking had taken place since the early 1930s, with an average of about two to three cases annually, most of them not even political in character. But there were thirty-five cases in 1968 and eighty-seven in 1969. They affected twenty-three countries; Israel was the victim in only one case. After 1969, the number dropped, both as a result of diminishing returns and more stringent security measures. The wave of kidnappings and assassinations started with the murder of the U.S. ambassador to Guatemala and two American senior army officers by the FAR in Guatemala in 1968. This was followed in 1969 by the kidnapping of the U.S. ambassador to Brazil (by the ALN), the murder of the West German ambassador to Guatemala (by the FAR) in 1970, the kidnapping of the West German ambassador to Brazil (again by the ALN), the murder of the Quebec minister of labor by the FLQ, the kidnapping of the Swiss ambassador to Brazil by the ALN and of the British ambassador by the Tupamaros, of the Israeli consul general in Turkey in 1971 by the TPLA, the murder of the U.S. ambassador and other diplomats in the Sudan by Black September. After 1972, Argentina became the main site of kidnappings with the ERP concentrating on businessmen; Mr.

Aron Bellinson was released in June 1973 after the ransom of a million dollars had been paid, the release of Mr. Charles Lockwood, the following month, cost two million dollars, and of Mr. John R. Thompson, the same month, three million. The ERP allegedly received fourteen million dollars for an oil executive in June 1974. A record was established in 1975 with the kidnapping of two Bunge and Born heirs for whom some sixty million dollars were reportedly handed over.[63]

Hijacking and kidnapping were also the favorite ploys of the Palestinian organizations and of some of the European terrorist groups. But there was competition in this field by angry or mentally unstable individuals; to launch rural guerrilla war at least a small group of people was needed, whereas any single madman or criminal could put a time bomb on a plane. The clandestine nature of urban terrorist operations made it difficult to establish how many members the terrorist groups numbered, in whose name they were operating or speaking, whether the motive was political, or whether the love of excitement or money was the driving force, and to what extent the whole phenomenon belonged to the realm of psychodrama rather than politics. Some Latin American terrorist groups consisted of no more than a few dozen members, the Japanese URA had at its height three hundred, but the number was reduced by defections and mutual assassinations, the FLQ had fewer than a hundred and fifty members, the TPLA fewer than a hundred, the British "Angry Brigade" eight, the Symbionese Liberation Army about ten, and after its shootout with the police, only three or four. But even if there were only three members left, to the media the "Army" was still an "army," "bulletins" were published, "ideological platforms" hammered out and spectacular exploits, whether an assassination or a bank raid, still caught the headlines. Ideologically there was often utter confusion; leading members of the pro-Fascist Tacuara (Argentina) moved over to neo-Trotskyite groups, the Monteneros (also in Argentina) stressed at one and the same time both their "Christian" (anti-Semitic) and radical-socialist character; the Canadian FLQ and the Official IRA presented a mixture of Marxist-Leninist doctrine and religious-nationalist sectarianism. Urban terrorist interest in political philosophy was strictly limited — the deed was more important than the thought.

The most active of the urban terrorist groups was the Provisional IRA which had split from the "Officials" in 1969.[64] The Provos had some five to six hundred militants, the Officials about four hundred, but they enjoyed considerable support among the Catholics in

Northern Ireland against a background of deep-rooted anger about national and social discrimination. The IRA had a sanctuary and a supply base in the Irish Free State. Officially it was banned there and the Irish government regarded their activities with disfavor, but being doctrinally committed to the idea of a united Ireland, it could not drop the northern activists entirely. Like other urban terrorist movements, the IRA had international connections; money came from well-wishers in the United States, money and arms from Libya's Colonel Ghadafi — a protector of terrorists from Northern Ireland to the Philippines — and from Eastern Europe. The ideological differences between Provos and Officials are discussed elsewhere in this study;[65] cynics had it that the only difference was that the former went to church each week, and the latter only once a year. In practice, as opposed to doctrine, their operations were not so much directed against the British government, but against the Protestant community. As the fighting progressed, it became a straightforward sectarian civil war, with the British army in the thankless role of an arbiter trying to limit the fighting and to isolate the gunmen from the community. The Protestants had paramilitary organizations of their own (the UVF and others) and their terrorist record resembled that of the IRA. Looked at in historical perspective, the armed struggle in Northern Ireland since 1969 was not a novel phenomenon but simply a new stage in the age-old struggle between two neighboring communities. The methods used in this civil war were on the whole old-fashioned; the IRA tried to extend its operations to England, but this, too, had been tried before World War II, and on a small scale even back in the nineteenth century. The IRA was more successful than other urban terrorist groups because it had a fringe, albeit small, of supporters; and it was probably not accidental that this base was sectarian-religious rather than "revolutionary" in makeup. The FLQ lacked both a sanctuary and a clear program and, since Canadian political culture was considerably less murderous than in Ireland, its movement was much more short-lived. The Basque ETA, with its bank robberies, holdups, bombings and kidnappings, was a little more effective because, like the IRA, and unlike the Latin American urban terrorists, it had its base in a national minority. Urban terrorist operations in the United States and West Germany (on which more below) had no major political impact even though they greatly preoccupied journalists, psychologists, lawyers, judges and law enforcement officers.

The international character of urban terrorism has already been remarked: the Palestinian PFLP engaged in combined operations with the Japanese URA, various Latin American urban terrorists would cooperate with Palestinians, who in turn collaborated with Baader-Meinhof and other gangs, as the 1975 "Carlos Affair" (the case of Ilich Ramirez Sanchez) demonstrated. Foreign governments would take an active interest; the Iran urban guerrillas were financed for years by Iraq; Cuba continued to contribute to various Latin American terrorist groups even though, in principle, it favored rural guerrillas. "Carlos" had been trained at the Lumumba University in Moscow. The Soviet attitude to urban terrorism was ambivalent; on the one hand they would welcome and support movements likely to cause disruption in the West, on the other, they could not fail to realize that small ineffectual factions such as Baader-Meinhof would bring a political backlash that would be directed not only against members of that specific group but against Communists in general. Terrorism came to resemble the workings of a multinational corporation. An operation would be planned in West Germany by Palestine Arabs, executed in Israel by terrorists recruited in Japan with weapons acquired in Italy but manufactured in Russia, supplied by an Algerian diplomat, and financed with Libyan money.[66] With the improvement and greater accessibility of modern technology, the potential for destruction for small groups of people became much larger.[67] As technical progress continued, society became more vulnerable to destruction. A single individual could spread alarm and confusion even by means of a telephone call about a bomb that had allegedly been placed in some vital place. This new power acquired by a few has, however, its limits; it could paralyze the state apparatus but it could not take over. Urban terrorism faced its practitioners with an insoluble dilemma — to reduce the risk of discovery they had to be few in number. The political impact of a small anonymous group was bound to be insignificant. Urban terrorists are not, as the Palestinian (1948) and Cypriot experience had shown, serious contenders for power; once the foreign enemy had withdrawn, they dropped out of the picture since they were so few and had no political organization. Prospects are better perhaps where terrorists operate as the military wing of a political movement rather than on their own initiative. But in this case there are always the seeds of conflict between the military and the political leadership of the movement. In urban terrorism it is the action that counts, not consistent strategy or a clear political purpose. It is a Herculean task to disentangle its rational and irrational components, and not always a rewarding one.

An aristocracy and a proletariat emerged among the urban terrorists of the 1970s. The "proletarians" were shunned by the Russians and the Cubans because their activities did not fit into Communist policies; they were rejected by the Libyans and Algerians because they belonged to the wrong religion or nationality, and refused a haven even by South Yemen. Yet the "proletarian" terrorists were unquestioningly sincere whereas there were a great many question marks with regard to the political bona fide of the terrorist aristocracy, those with powerful protectors and rich financial backers. Were they not mere pawns manipulated by outside forces? It was by no means always clear whose interests they tried to serve and their ideological declarations could not be taken at face value.*

* The attack against OPEC headquarters in Vienna in December 1975 and the abduction of oil ministers was a typical illustration of the mysteries of transnational terrorism. The leader of the group was said to be a Venezuelan who had been in close touch with Cuban intelligence, according to the French authorities who had first established his identity. But according to the Egyptian press the operation in Vienna had been paid for by the Libyans. In this maze of great and small power rivalries, of big business and intelligence intrigues, there were no longer any certainties apart perhaps from the fact that the official purpose of the attack (to save the Palestinian revolution) was most probably not the real one.

8

Guerrilla Doctrine Today

The great upsurge of guerrilla warfare in the 1950s and 1960s was reflected in the hundreds of books and thousands of articles devoted to the strategy and tactics of wars of national liberation, *foci*, revolutionary warfare, the advantages of the rural over the urban guerrilla, and vice versa. Most of this new body of doctrine emanated from Latin America and was left wing in inspiration. There were heated polemics about the "correct approach," about "subjective" and "objective" conditions, about the place of the vanguard, and the role of the masses in the struggle. But the number and even the quality of books produced was not necessarily an indicator to the efficacy of the movements sponsoring them. Some of the most protracted and bloody guerrilla wars such as those in Algeria, the Middle East or Ulster, produced few theoretical reflections on the subject. The Kurds fought for twenty years and knew all there was to know about guerrilla warfare even though they probably never read a book on the subject. If they were defeated in the end, it was not because of any doctrinal shortcomings. Conversely some guerrilla movements, which barely functioned, were very strong on doctrine. Broadly speaking the more remote the Marxist-Leninist inspiration, the more limited was the interest in ideological disputations. This is not to say that all the theoretical writings were innovative: most of the literature presented variations on the same theme.

The Marxist-Leninist tradition had a profound influence on the vocabulary of the Latin American apostles of guerrilla warfare, but the real impact of patriotic-nationalist traditions was quite unmistakable. (The slogan of most Latin American guerrilla movements was *Patria o Muerte-Venceremos.*) About the revolutionary character of these movements there is no doubt; on the other hand they did not just "deviate" from Marxism on essential points but did to Marx what Marx had done to Hegel: they stood him on his head. That any political dogma is bound to be adapted and modified in the light of new historical developments goes without saying; Marxists in particular have always stressed the necessity of "creatively employing" their method. But if these changes are fundamental and far-reaching, if some of the basic tenets are given up, the point is bound to be reached sooner or later when the old label no longer conforms to the new content, when, in fact, it misleads and becomes a source of misunderstanding. Other ideologies and creeds have faced a similar fate: Christianity without God is an interesting phenomenon, but is it Christianity any longer?

The new doctrines of guerrilla warfare must be studied within their context of time and place, and be subjected to a critical analysis for they by no means provide a true reflection of guerrilla experience. The writings of Guevara and the speeches of Fidel Castro contain much of interest about the Cuban revolution; they contain even more myths and *post facto* rationalizations that only can be explained in the light of their authors' subsequent political careers. They went into the struggle with one ideology and emerged with another. While the Cuban revolutionaries were fighting in the Sierra Maestra they were a movement in search of an ideology; some of them belittled revolutionary theory altogether and thought that all that mattered was revolutionary struggle. As the Tupamaros put it in later years: "Words divide us, action unites us." Castroist-Guevaraist doctrine is the product of a later period, and offered no guidelines while the struggle lasted. It fails to explain what happened and why, it is not the key to the lessons of the Cuban revolution.

Guevara's views in 1965 differed to a considerable extent from those he had pronounced five years earlier, and Debray's estimation likewise underwent radical changes. The unity of theory and practice is always highly imperfect and ideology frequently serves as a smokescreen. To provide two extreme examples: firstly, from the ideological pronouncements of the leaders of the IRA one would not learn that sectarian elements were prominently involved in their struggle; secondly, the one major tactical innovation of the Palestinian guerrillas was that their operations were mainly conducted from outside Israel. But official pronouncements almost invariably proclaim the opposite. The IRA and the Palestinian Arab spokesmen no doubt have sound political reasons for preferring fiction to fact.

There was a tendency among the guerrilla theorists of the 1950s and 1960s to emphasize the universal applicability of their doctrines. This was as true of Lin Piao's thesis about the struggle of the world villages against the world cities, as of Frantz Fanon's analysis of the anticolonial struggle, or of the propagation of the Cuban model as befitting all Latin America. But conditions varied so much from country to country that these wholesale prescriptions were usually quite unrealistic. They led to major setbacks for the guerrillas and became a source of confusion to those trying to understand the dynamics of guerrilla warfare. What Boris Goldenberg wrote about the Cuban revolution applies *mutatis mutandis* to guerrilla warfare everywhere: in view of its unique character it is a topic for the historian and not the sociologist.[1]

Goldenberg's remark remains true despite the fact that the background of most Latin American guerrilla leaders was remarkably similar. They all hailed from middle- or upper-class families, they were young, had received higher education and, if they were not city-born, they had lived in towns before joining the armed struggle. The biographies of Castro and Guevara are of course common knowledge. Castro was not quite thirty when *Granma* landed in Cuba, Guevara was two years younger. The former came from a wealthy landowning family and had studied law; Guevara was a physician by training and had specialized in allergy. Debray was twenty-six years of age when he wrote *Revolution in the Revolution*, he came from an upper-class French family, was a *normalien* and had taught philosophy for a short time before he arrived in Latin America.

Of the other major guerrilla leaders, Douglas Bravo who came from a Venezuelan landowning family was also a lawyer by training; Camilo Torres, a priest and professor of sociology, hailed from one of Bogota's leading families and was thirty-six when he joined the guerrillas; Hugo Blanco, the son of a Cuzco lawyer, had studied agriculture and was in his late twenties when he became a peasant leader; Yon Sosa, son of a Chinese father, was an army lieutenant and like his comrade and rival, Lieutenant Turcios Lima, had undergone antiguerrilla training in the Guatemalan army; Raul Sendic, leader of the Tupamaros, had almost completed his legal studies when he began his revolutionary career — he too came from a landowning family; the Peruvian Hector Bejar was a poet, painter and engraver, aged about twenty-five at the time of his guerrilla exploits; Javier Heraud, the poet, was only twenty-one when he was killed; Cesar Montes, leader of the guerrillas in Guatemala, was also in his early twenties; the Brazilian Leonel Brizola was an older man, once a professional politician (and *bon vivant*), who served for a period as governor of Rio Grande do Sul; Carlos Lamarca, the Brazilian urban guerrilla leader, was an army captain when he joined the insurgents and he too had been trained in antisubversive warfare. The only two men who were considerably older were the engineer Carlos Marighela, the author of the *Minimanual*, a mulatto, who became a guerrilla fighter at the ripe age of fifty-seven after more than three decades as a Communist party official, and Abraham Guillén, who provided the *Philosophy of an Urban Guerrilla*, a native of Spain, had fought in the Spanish Civil War and was fifty-three when his book was published. But unlike Marighela, he was a noncombatant in the guerrilla movement.

The bourgeois background of most, if not all Latin American guerrilla leaders is not in doubt, but this does not mean, as their orthodox-Communist rivals sometimes argued, that their mentality was "petty bourgeois," or that the revolution they aimed at was bourgeois in character. There are limits to the usefulness of class analysis; throughout history revolutionaries have failed to act in accordance with "class interest." They were *spostati* by choice.

The theory of the Castroist revolution was formulated only in part by the *Maximo Lider* who was primarily a man of action (and speeches). It was given its fullest and most systematic expression by Guevara and subsequently by Regis Debray.[2] During the war Castro was mainly preoccupied with the tactics of fighting, and it is interesting to note that he changed his approach more than once as he gained experience. When he first landed in Cuba, he did not assume that his small band could possibly defeat Batista's army, but he did anticipate that his initiative would trigger off a general strike in the cities that would, in turn, lead to the overthrow of the regime. There was no such strike but, as subsequent events demonstrated, the regime proved far weaker than he imagined. Once the guerrilla war was launched, Castro announced that he would burn Cuba's entire sugar crop ("including my own family's large sugar-cane farm in Oriente") as a previous generation of Cuban guerrillas had done.[3] But he later changed his mind and found more effective means of weakening the regime than a scorched-earth policy. After he came to power he would proclaim the basic lesson: "*El deber de todo revolucionario es hacer la revolución*; the duty of every revolutionary is to make the revolution . . . it is not for revolutionaries to sit in the doorways of their houses waiting for the corpse of imperialism to pass by. The role of Job doesn't suit a revolutionary."[4] There were other such pronouncements but for a comprehensive exposition of Castroism-Guevaraism, we must turn to Guevara's handbook *La Guerra de Guerrillas* published first in April 1959 and to some of his subsequent articles, the most important of which was *Guerra de Guerillas: un Método.*[5]

GUEVARA

The essence, the three fundamental lessons of the Cuban revolution as Guevara saw it, are boldly stated at the very beginning of his handbook:

1. Popular forces can win a war against the army.
2. It is not necessary to wait until all conditions for making revolution exist; the insurrection can create them.
3. In underdeveloped America the countryside is the basic area for armed fighting.

At the time he conceded that not all the conditions for a revolution could be created through the impulse of guerrilla activity:

> Where a government has come into power through some form of popular vote, fraudulent or not, and maintains at least an appearance of constitutional legality, the guerrilla outbreak cannot be promoted, since the possibilities for peaceful struggle have not yet been exhausted.

Three years later in his article he withdrew this reservation: the conditions for armed struggle existed everywhere in Latin America. Imperialism and the bourgeoisie tried to keep in power without using ostensible violence. But the revolutionaries had to compel them to remove their mask, to expose them in their real *Gestalt* as a violent dictatorship of the ruling classes, thereby intensifying the revolutionary struggle. In other words, a democratically elected, constitutional government had to be compelled by provocative guerrilla attacks into using its inherently dictatorial powers.

Guevara's three basic tenets are fundamentally opposed to the teachings of Marxism-Leninism and, to a certain extent, even to Maoism. For he regarded the armed insurrection not as the final, crowning phase of the political struggle but expected, on the contrary, that the armed conflict would trigger off, or at least give decisive impetus to the political campaign. In Marxist-Leninist thought, as in Maoism, the political party is the leading force and there is a heavy emphasis on ideology and indoctrination. In the Castroist-Guevaraist concept the political party does not play the central role, and there is no such emphasis on ideology and political education. True, according to Guevara the guerrilla must be a social reformer (above all an agrarian revolutionary), because this is what distinguishes him from a bandit.[6] But the revolutionary spirit is somehow taken for granted, and so is support by the people.

In later years, Debray was to put it even more succinctly: a successful military operation is the best propaganda. The guerrilla force is the party in embryo. The vanguard party would not create a popular army, but it would be for the popular army to create the political vanguard.[7] He based himself on Castro who declared on one famous occasion: "Who will make the revolution in Latin America? The people, the revolutionaries, with or without a party." Guevara argued that the guerrilla must triumph because of his moral superiority over the enemy and because of the mass support he enjoyed; the fact that he might be inferior to the army in firepower was of no great consequence. At the same time Guevara, like Mao, regarded guerrilla operations as the initial phase of warfare; the guerrilla army would systematically grow and develop until it acquired the characteristics of a regular army. The aim was victory, annihilation of the enemy, and this objective could be reached only by a regular army, even though its origins lay in a guerrilla band.[8]

The assumption that the people could defeat a regular army was not entirely new; it had been borne out, to give but one example, in Bolivia in 1952. What seemed to be new was the concept that thirty to fifty dedicated revolutionaries were sufficient to launch an armed struggle in any Latin American country. Debray was even more optimistic: ten to thirty professional revolutionaries could pave the way, preparing the masses. But was this not exactly what Blanqui had preached one hundred years earlier? Debray argued that there were two essential differences: the revolutionaries did not aim at a lightning victory, nor did they want to seize power for themselves.[9]

The strategic concept of Guevara and Debray differed from that of even the most militant Communists, Trotskyites, and Maoists in that they belittled revolutionary spontaneity (such as advocated by the Trotskyites) and discounted the self-defense units of the workers and peasants (such as existed in Colombia and Peru). Debray thought that the peasant *syndicats'* struggle was essentially defensive in character, and did not aim at seizing political power. Even if such defense associations did manage to survive for several years (as they had in the south of Colombia), they would be defeated in the long run because they constituted a fixed target for the government forces. Furthermore, the peasants merely wanted to defend their families and their possessions whereas only total partisan warfare stood any chance of success. The Chinese Communist bases, according to Debray, could not serve as a model for Latin America: China was a far bigger country, and the enemy forces there had been relatively weak. The Latin American *foci*, the centers of insurrection, on the other hand, had to be military in character rather than territorial; he excepted only the universities, but these were regarded as of secondary importance, mere centers of recruiting and propaganda. By itself a *foco* could not overthrow the system, it was merely a detonator planted in the most exposed enemy position, timed to produce an explosion at the moment of choice. The Latin American guerrillas would not survive the early stages of the armed struggle if they were to engage in static defense; they would (as Fidel put it) have to carry their *foci* with them like supplies in their knapsacks. It might be possible and even desirable to establish territorial *foci* at the very beginning of the struggle but this ought not be a strategic aim; the Cubans only set up their first *foci* after seventeen months of fighting in the Sierra Maestra and they would have ceded them if this had been necessary from an operational point of view.

Castro and Guevara firmly believed in the absolute primacy of the armed struggle, and their conviction grew, if anything, in the years after the Cuban revolution. They maintained that the struggle would have to incorporate many Latin American countries, creating two, three, many Vietnams. Castro stated quite specifically that the Andean region would be the Sierra Maestra of Latin America. His call was based on the assumption that the chances of success existed almost everywhere if only there were enough revolutionary enthusiasm. On the other hand, there was the realization that, in view of the political situation, that is the growing pressure of imperialism, it would become increasingly difficult to defeat the enemy and stay in power in any single country (Guevara). The regular armies had after all learned the Cuban lessons and were psychologically and militarily much better prepared for irregular warfare than they had been at the time of the Cuban insurrection (Debray).

These theses of revolutionary strategy were based on a genuine belief that an almost unlimited revolutionary potential existed over the entire continent, though the desire to reduce the pressure on Cuba must have also been a factor. Fidel, in his second Havana declaration, provided the keynote: the conditions of each country could either hasten or impede the revolution, but sooner or later it must occur everywhere. Guevara added that it was criminal not to make use of the opportunities that offered themselves. The weakest link in the eyes of the Cubans was Venezuela and, after the struggle there had failed, Guevara chose, with disastrous results, Bolivia as the most promising area for installing his "detonator."

This strategy was bound to lead to bitter controversies with the Marxist-Leninist parties. The two main bones of contention which led to open schism were the relationship between the military guerrillas and the political party, and the Cuban insistence that, as a matter of principle, the rural areas would have to be the main battlefield. The Cubans argued that the countryside had much more to recommend itself than the cities from a military point of view, if only because access was more difficult. Furthermore, the revolutionary potential of the peasantry had hitherto been virtually untapped. In the early days of the Cuban revolution Fidel's slogan had been "All guns, all bullets, all reserves to the Sierra," despite the fact that resistance to Batista in the towns had all along been far more intense and better organized. Ten years later he was, if possible, even more emphatic about the subject: it was absurd, even criminal, to try to lead a guerrilla movement from the city. Given the Cuban example, an urban guerrilla movement could not develop into a revolutionary force, capable of seizing power. The urban guerrilla was at best an instrument for agitation, a tool for political maneuvers, a means for political negotiation.[10] The city, as Fidel put it, was the "grave of the guerrilla." In the towns, where there could be no single command and centralized leadership, the guerrillas were forced to disperse, particularly in the early phase of the struggle, weakening the insurgents far more than it hampered the government forces. This is not to say that the Cubans regarded the struggle in the mountains as a peasant war; essentially it would be a revolutionary partisan war which the peasants would support and which some would gradually join.[11]

There were other, equally weighty reasons in favor of waging war in the countryside; a war which would be expanded to the small cities and, in the end, be carried to the metropolitan centers. "As we know," Debray wrote, "the mountain proletarianizes the bourgeois and peasant elements, and the city can bourgeoisify the proletarians."[12] The spokesmen of the Cuban revolution regarded the urban working class on the whole as a conservative element and they did not except the Communist parties. Living conditions in the towns were fundamentally different from those prevailing in the countryside; even the best comrades were corrupted in the cities, infected by alien patterns of thought. Life in town was tantamount to an "objective betrayal." The guerrilla movement was the real proletariat, with nothing to lose; the guerrilla leadership in the towns or aboard was the "guerrilla bourgeoisie." The city was the place where politics were made, the countryside the scene of revolutionary action. And the guerrillas, needless to say, wanted to get away from urban politics. Unitl about 1965 Debray believed that it might be possible to win over most Communist parties to the idea of the armed struggle and that this would lead, of necessity, to the old-guard Communist leadership of mere politicians being replaced by a younger, more dynamic leadership. Only after the setbacks in Venezuela and elsewhere was he inclined to write off the Communists as altogether hopeless. He regarded the guerrillas as representing the interests of the proletariat even if their social background was anything but proletarian.

Guevara's handbook of guerrilla warfare contains much practical combat advice: how terrain should affect an attack against an enemy convoy or position; the order of fire in battle; the establishment of a good supply system and medical service; the planning of acts of sabotage; the setting up of a war industry in the liberated zones. Special sections deal with the role of women in the struggle, with the conduct of propaganda, the establishment of an intelligence network, the civil organization of the insurrectional movement. But there is not much that is novel in these suggestions. What Castro and Guevara knew about guerrilla combat, they had learnt from Alberto Bayo, a native Cuban, who had emigrated to Spain and acquired his expertise in the war against Abdel Krim. Bayo served as an air force officer in the Spanish Civil War, later moved to Mexico where, in 1955, he gave a crash training course to the insurgents about to embark on their expedition to Cuba. His instruction manual, *150 Questions to a Guerrilla*, was widely read (among the Weathermen, for instance) and contained a wealth of practical information.[13] Bayo died in 1967, having attained the rank of "commandante," the highest in the Cuban army.

Guevara noted three stages in guerrilla warfare; first, the tactical defense, when the small guerrilla force would be hunted by superior enemy forces; gradually, the point of equilibrium, when the possibilities of action for both the guerrillas and the enemy become equalized. At this point, large columns would be employed by the guerrillas in a war of movement; this would not replace the guerrilla war but would merely be guerrilla warfare on a larger scale — something akin to Mao's "mobile warfare." Finally, a popular army would crystallize, overrun the government forces and seize the big cities. The critical period, as Guevara saw it, is the very early one, and he posited three preconditions for the guerrilla's survival — constant mobility, constant vigilance and constant distrust.[14] These are sensible observations but guerrilla fighters throughout the ages have instinctively known these home truths, and they were noted by many writers before Guevara.

Of greater interest are his remarks about the human, psychological factor, even though his thoughts on this subject emerge only incidentally from his writings. Emphatic stress is given to the "political will" and "decision": "Generally, guerrilla warfare starts from a well-considered act of will; some chief with prestige starts an uprising for the salvation of his people, beginning his work in difficult conditions in a foreign country." But what if there should be no great leader such as Fidel? Guevara's answer is simple but not altogether convincing. The leaders would learn the art of warfare in the practice of war itself — struggle is the greatest teacher. In short, Napoleon's *on s'engage, puis on voit*. Religious terminology is frequently invoked, reference is made to a "special kind of Jesuitism"; the revolutionary, clandestinely preparing for war "should be a complete ascetic." There is much preoccupation with honor, courage, vengeance, hatred and death; hatred for the enemy drives a man beyond his physical limits and transforms him into an effective, selective and cold machine for killing — "our soldiers have to be that way." Or elsewhere, "Death will be welcome to us wherever it will surprise us if only our call will be taken up by the others. . . ." Or, on another occasion, the reference is to "the people willing to sacrifice itself in a nuclear war so that its ashes might serve as the cement for a new society. . . ."[15]

These are appeals by a prophet or a mystic, reminiscent of D'Annunzio or Codreanu rather than Lenin or Trotsky. They lead us a little closer to the metapolitics of Cuban guerrilla doctrine. It is not of course that Fidel was a paradigm of asceticism or that the guerrillas all wanted to commit suicide. But there certainly is a revolutionary romanticism, with its pessimistic, even tragic undertones, to which Raoul Castro and others have referred.[16] Purity is the ideal, revolution is a mystical adventure, a struggle to the death for something enigmatical. Although these observations refer more to Guevara (and to Camilo Torres, about whom more below) than to Fidel Castro, almost all Latin American revolutionaries have certain features in common: their character and mental makeup present a curious mixture of contradictory qualities, admirable and not so admirable. In many ways they were the most attractive revolutionary heroes to have emerged in Latin America; enthusiastic, courageous, idealistic, willing to make sacrifices, genuinely concerned about the fate of the poor, impatient with bureaucracy. They had a sense of mission and were free of the corruption and the egotism of the society surrounding them. They became the heroes of many of the best young people in Latin America and beyond. But there was another side to it. Their heroism was by no means free of *machismo* or showing off, their concern for the poor was paternalistic, they were anything but democrats and the idea that the masses could or should be trusted in free elections was quite alien to them. The dividing line between selfless heroic action, *caudillismo*, terrorism, and gangsterism has never been quite distinct in Latin America and the same lack of delineation was true of the Cuban revolutionaries. Castro, who began his career as a university gun fighter, wrote about the companions of his youth:

> . . . the young men who, moved by natural yearning and the legend of a heroic era, longed for a revolution that had not taken place and at the time could not be started. Many of those victims of deceit who died as gangsters could very well be heroes today.[17]

Conversely, it would appear that many, perhaps most, of the apostles of Communism, Latin American style, in the late 1950s and 1960s could with equal ease have become fighters for Fascism, Latin American style, twenty years earlier. This applies to elements in their ideology and traits in their character alike. They were children of their time; in the last resort action was more important than ideologies which had always been imported to Latin America anyway. One element remained constant — nationalism, populist and elitist at the same time; the rest was given to continual change, according to the prevailing intellectual fashion. It is quite true, as Theodore Draper has noted, that ideologically Castroism never lived a life of its own:

> historically, it is a leader in search of a movement, a movement in search of power, and power in search of an ideology. From its origins

> to today, it has had the same leader and the same "road to power" but it has changed its ideology.[18] . . . A caudillist movement of a new type, it uses power not for power's sake and needs an ideology to justify power ideologically, a mixture of the Latin American revolutionary tradition and European communist elements.[19]

The apparent break in the ideological continuity of Castroism can be understood only against this background. The same goes for the constant changes in doctrine, the debates whether objective conditions were of any importance for a revolutionary, whether armed struggle was the inevitable road to power everywhere, whether Cuba was an example for the whole of Latin America. There is, as already noted, a great and growing discrepancy between the doctrine of Castroism as it developed after 1960 and the realities of the Cuban revolution. It is true that similar discrepancies exist between the doctrines of all revolutions and the real course of events, but it is particularly striking in the case of Cuba. The Castroist doctrine is a myth, which is not to say that it is irrelevant. Taking the Castroist doctrine of revolution at face value, one would never glean the facts that Castro and his comrades had not originally intended to launch guerrilla warfare, that the "masses" played an insignificant role in the fighting, that, generally speaking, there was little fighting at all, that the Batista regime collapsed, in the final resort, because it was rotten to the core and not because the insurgents were so strong. There is no recognition of the fact that there was relative prosperity in Cuba at the time and even relative freedom; Castro's Sierra Maestra appeal for a rising was reported in the Havana press.[20] It is never acknowledged that Castro's intelligent manipulation of the mass media (including the foreign press and television) was of the greatest importance, that he received decisive financial help from bourgeois political leaders inside and outside Cuba, that America turned against Batista and imposed an arms embargo. The importance of the revolutionary struggle in the towns, which involved more fighting and cost more lives than in the countryside, is systematically played down in Castroist literature. A Marxist critic later wrote that Fidel's victory over Batista's army was not achieved by force of arms; corruption, the great Cuban vested interests, the Church, and in the final analysis, Yanqui imperialism, all assisted in Batista's defeat.[21] In moments of candor, Cuban official sources have provided explanations that come close to this analysis: that the revolution was won largely because of Castro; that U.S. imperialism was disorientated; that support was given by a large segment of the bourgeoisie and some big landowners; and lastly, that most sections of Cuba's peasants were proletarianized. Only the last part of this official formula is quasi-Marxist in character and it is also the one which is incorrect. In short, the revolution prevailed in Cuba because of a unique set of circumstances, and thus attempts to reproduce it elsewhere in Latin America were bound to fail — not for lack of "objective revolutionary situations" or courageous guerrilla leaders, but because it was unlikely that the United States or the non-Communist circles would support a movement of this kind in the post-Cuban situation. By 1965 one observer noted that the counterinsurrectionists knew more about guerrilla warfare than the guerrillas themselves.

DEBRAY

The orthodox Communists rejected Castro's "adventurism," partly because they opposed his policy for tactical reasons, but mainly because they could not accept the subordination of the party apparatus to the military leaders. The polemics usually proceeded by proxy: Castro did not want to attack Moscow openly so instead accused the Latin American Communists of cowardice — for not supporting Douglas Bravo's guerrilla operations in Venezuela, for instance. The Latin American Communists on the other hand were reluctant to engage in a dispute with the Cubans because of their tremendous prestige in radical circles all over the continent. In the circumstances, Debray became the main butt for their attacks; he was merely an unofficial spokesman of the regime and could be criticized with greater impunity. *Inter alia* Debray was charged with not presenting a detailed Marxist class analysis in his writings, with trying to prescribe for the whole continent, with not taking into account decisive local peculiarities, with "liquidating" theory, with having a wrong model of revolution. It was pointed out to Debray that the armed struggle was no panacea; it did not necessarily unite the revolutionaries as events in Venezuela and elsewhere had shown. It was explained to him that urban revolutionaries did not enjoy a *dolce vita*, that he was an "ultra-voluntarist," an elitist, distrustful of the masses.[22]

Some of this criticism was quite to the point. Castro, Guevara and Debray had admitted the existence of "national peculiarities" in principle but had paid them scant attention in practice, assuming, apparently, that the Cuban model was equally applicable to Honduras and Brazil. Some criticism was correct but irrelevant: all revolutionaries are elitists and voluntarists even though some admit this more openly than others. Other charges were quite unfounded: a Marxist class analysis of Latin American society, however interesting *per se*, was not, as experience proved, the answer to the feasibility of a revolution. The Cuban experience demonstrated that the "subjective" factor was of decisive importance, that revolutionary war was a contest of will: if Batista lost his nerve while Fidel maintained supreme self-confidence, this had little to do with the social tensions in Cuban society. Tensions exist in every society and it would be impossible to prove that there were more tensions in Cuba than elsewhere in Latin America. Orthodox Marxists could argue that the Cuban experience was unique, an exception, but the exception had been successful, whereas guerrilla movements operating in objectively favorable conditions had failed.

The Castroist-Guevaraist doctrine was most fashionable throughout Latin America until about 1968, the year after Guevara's failure in Bolivia. In the years that followed, Cuba began to toe the Soviet line more closely (a fact for which Soviet economic pressure might account in part). The Cuban leaders also lost some of their illusions about an imminent victory of the revolution in Latin America in view of the guerrillas' minimal progress on the continent; the revolutionary spirit of 1960–1968 gave way to internal splits, interminable polemics, and mutual recriminations. Ironically, Castro, who had bitterly attacked the Communist parties because most of them had not opted for the armed struggle, found himself by 1970 at the receiving end of similar charges. Douglas Bravo contended that Cuba had retreated, declared a truce, choosing to sacrifice the revolutionary cause in favor of economic development. Cuba, he argued, had refused to unleash a war on the grand scale. True, Cuba would be lost in such a war but the Cuban revolutionaries would be able to carry the revolutionary struggle to other Latin American countries.[23] The Cuban leaders were in no mood to accept such advice; they preferred the establishment of closer relations with nationalist regimes, such as in Peru, which had carried out "progressive changes." Finally, the vacilations of Cuba's overall political strategy quite apart, Cuba's prestige declined among those who had enthusiastically welcomed the revolution a decade earlier. As they saw it, the great promise of 1960 had not been quite fulfilled, the regime had become bureaucratized despite all the good intentions, the revolutionary spirit was slowly petering out. It was still regarded as a progressive regime, but the image of Cuba no longer quickened the heart.

Debray's approach also changed markedly over the years. After his release from a Bolivian prison, he went to Chile, eventually returning to France where he joined the socialists. In *La Critique des Armes*, published in 1974, he stated that the hypothesis he had previously advanced had been belied by experience. The theory of *foci* had been wrong insofar as it had dissociated the military from the political, the clandestine from the legal, struggle. Debray now considered the political training of cadres and contact with the masses as of paramount importance. One had to return to the ABC of the great revolutionary teachers, beginning with Marx. For the zealous revolutionaries who continued to believe in theories he had advocated only a few years earlier, he felt nothing but contempt: "Schizophrenia as a norm of organization is the last stage of individual megalomania." He saw nothing but irresponsibility and revolutionary delirium in their pronouncements and noted sarcastically that the revolutionary phrase begins its flight, like the owl of Minerva, when night falls and when revolution has reached the stage of agony.[24]

Thus ended a chapter in the history of Latin American revolutionary doctrine. The same Debray, who had rejected the very idea of the urban guerrilla in 1965, now came to regard the Tupamaros as the most intelligent and politically sophisticated of all Latin American guerrillas — but even they were defeated. Venezuela, as he saw it, had acquired a hypertrophied and omnipresent repressive apparatus, but despite this, it still functioned as a liberal republic with elections and a normal political life; in these conditions, too, revolutionary violence failed.[25] He had come back full circle in 1974, to the position adopted all along by Communist leaders like Prestes or by Guevara in 1960 before he announced the inevitability of the armed struggle.

REVOLUTIONARY STRATEGY IN LATIN AMERICA

Although the Cuban leaders were at the very center of the debate on revolutionary strategy, they had no monopoly in this field. The political splits of the Latin American radicals into dozens of factions and hundreds of *groupuscules* were reflected in a bewildering multitude of theses and platforms; even an encyclopedia would not do them justice. They ranged from those who advocated cooperation with "progressive" bourgeois-nationalist groups within a parliamentary framework, to the Posadistas, a Trotskyite faction, who took the Soviet Union to task for not waging nuclear war. Of the guerrilla ideologists only the advocates of urban terrorism deserve more than cursory mention; the views of the others can be summarized briefly, since they coincide, broadly speaking, with the concepts developed by Fidel, Guevara and Debray.

Douglas Bravo, the Venezuelan guerrilla commander, was in the forefront of the armed struggle on the continent for several years. A former leading Communist, who broke with the party, he established his first rural *foco* in 1962. The conditions facing him and his men differed in some important respects from those in Cuba: there was a greater degree of urbanization in Venezuela, where more than sixty percent of the population lived in cities; the enemy confronting the insurgents was not a Batista but a democratically elected mass party, the *Acción Democratica*. Guerrilla strategy manifested itself on the one hand in collaboration with criminal elements and the indiscriminate use of terror (such as attacks against trains carrying urban holiday-makers). The slogan was to kill at least one policeman a day. Since the policemen were usually of lower class background than the guerrillas, this topsy-turvy class struggle did not always endear them to the very groups they wanted to win over. Bravo's manifestos on the other hand were quite moderate; one would look in vain for any radical program of social change. He demanded agrarian reform and criticized the government for not conducting a friendlier policy vis-à-vis Cuba and North Vietnam. He referred at length to the glorious struggle for national liberation of 1810 and the fight against Yanqui imperialism, but these same motifs could be found in the programs of most Latin American parties.[26]

In later years the guerrilla movement split and some of Bravo's erstwhile lieutenants sought explanations for the mistakes committed: the guerrillas could have won in 1962, said Teodoro Petkoff, if they had combined the armed struggle in the towns with an insurrection in the army. The decisive battles were fought in the cities, not in the countryside, where the rural guerrillas could not survive without help from the towns. Hence the conclusion drawn by another Venezuelan guerrilla leader: in future it would be necessary to concentrate on the urban centers, building a powerful civilian and military base on the strength of clearly formulated, concrete political goals. Foreign models were only of limited assistance and a specific Venezuelan road to socialism would have to be developed.[27]

Despite the fact that Peru was far more rural in character than Venezuela the Peruvian guerrilla leaders, Hector Bejar and Hugo Blanco, reached very similar conclusions when they too looked back on the reasons for their failure. Bejar wrote that the guerrillas had fought in the forests, while the peasant population was concentrated in valleys and high zones; unless they mastered the tactics of operating on high open plateaus, they would have to stay in the forests, militarily secure but politically ineffective. More importantly, "our attitude was based on an underestimation of the cities," therewith blocking the road to successful revolutionary agitation among the urban masses. He also critically reappraised the apolitical attitude of the guerrillas: "Our groups were oriented towards action and found in it their only reason for existence."[28] Bejar mentioned in passing yet another important handicap: the cultural gap between the city-born, educated guerrillas and the *campesinos* was enormous. Quite literally, they did not speak the same language; the former, with a very few exceptions, did not understand Quechua (the Indian language) and the latter hardly spoke any Spanish.

Hugo Blanco attained fame as a successful organizer of peasant associations. The struggle for land control was carried out under his leadership, with the slogan *Tierra o Muerte* (Land or Death). These associations were to be the nucleus of a new society, they would run their own schools, courts of justice, health services. To all intents and purposes they would constitute a "dual power" on the lines of the Soviets in Petrograd in 1917. But again, in retrospect, Blanco admitted that the basic weakness of the guerrilla struggle had been its lack of support by a mass party: "We did not attach sufficient importance to the fundamental role of the party." The peasantry in Peru, Blanco said, was the major revolutionary force, but in the long run, once they were given land, they would become bourgeois. In the final analysis, therefore, the working class was the only guarantor of a socialist revolution.[29]

Camilo Torres, the revolutionary priest, who joined the Colombian guerrillas, made no major theoretical contribution to the actual conduct of guerrilla warfare; he was killed in the very first engagement in which he took part. His writings were devoted to the necessity of implementing social change and land reform, and to attacks against the establishment of the Catholic Church, which identified itself with the propertied classes. He tried to prove that the revolution was not just compatible with Christian ethics but was a Christian imperative.[30] He realized that the precondition for the success of the revolutionary cause lay in a united front of the various opposition forces who were engaged in intense internal feuds. But his great prestige as the most eloquent spokesman of the left was quite insufficient to achieve this aim; the Colombian insurgents, more than any other Latin American guerrilla movement, remained deeply split and frequently engaged in bloody purges of their own ranks. They probably lost more of their cadres in killing each other than by enemy action.[31]

PHILOSOPHY OF THE "URBAN GUERRILLA"

During the late 1960s the center of gravity in Latin American guerrilla fighting shifted from the countryside to the cities and this soon gave rise to the new doctrine of the "urban guerrilla." The basic idea was of course as old as the hills, and in Latin America in particular there was a hallowed tradition of urban insurrection, assassination and kidnappings. A radical critique of rural guerrilla fighting was provided by the Bolivian Trotskyite leader, Guillermo Lora: the guerrillas in Bolivia (and elsewhere) were an alien body in the countryside. As Guevara wrote in his Bolivian diary: the *campesinos* were "impenetrable like stones." When one talked to them one could not be sure whether or not they were ridiculing the guerrillas. Revolutionary impatience had been the reason for neglecting political work among the peasants; but, even if the guerrillas had succeeded in raising the banner of truth in a whole district, they would still have failed because they were cut off from the workers in the cities. Guerrilla "impertinence and adventurism," therefore, made a decisive contribution to the defeat of the whole left. As for the Cuban model with Castro as an undisputed leader of the Latin American revolution, this was according to Lora as much an imposition as the Russian Communists' pretensions after 1917 to dominate the international working-class movement.[32]

Michael Collins and Menahem Begin were "urban guerrillas" as, in a way, were many nineteenth-century revolutionaries. But the nineteenth-century insurgents believed in *golpismo*, the seizure of power following one short, violent battle. This belief had been shared by the old IRA in 1916. The Irgun believed that the struggle would be a protracted one but it did not develop any specific new guerrilla doctrine. In Venezuela there had been urban terror on a large scale during the early 1960s; in Caracas in 1963 the insurgents came very near to victory. Defeated, they transferred their operations to the countryside. It was only after the spectacular failure of Guevara in Bolivia, and the earlier setbacks in Peru, Venezuela and elsewhere that urban terror in Latin America came into its own. Cuban revolutionary doctrine was inconsistent on the subject; although the towns were rejected as the grave of the revolutionaries, there are occasional references in the writings of Guevara and others to the unjustified neglect of urban operations. Mention has been made of the fact that the Cubans must have been aware of the fact that they owed far more of their eventual success to assistance from sympathizers in the towns than they were willing to admit.

If the Cubans were opposed to "urban guerrilla" warfare but did not altogether exclude it, few of the advocates of urban terrorism rejected rural guerrilla operations in principle. It was simply a matter of different conditions and priorities. They considered their city-based operations either as the first stage of a general insurrectionary movement or as part of an insurrectionary pincer movement based on the cities and the countryside. The starting point for an "urban guerrilla" doctrine was the undisputed fact that the rural guerrilla movements had been unsuccessful on the whole. They pointed to basic social and demographic facts which their predecessors ignored to their detriment. Latin America has not only the highest rate of population growth in the world but also the fastest rate of urbanization and there is an enormous, constant inflow of poor, unskilled, jobless people to the towns.[33] Of the population of Argentina, forty-five percent live in Greater Buenos Aires; forty-six percent of all Uruguayans reside in Greater Montevideo. The population of Mexico City and São Paulo is nearly ten million, and Rio de Janeiro will have twelve million inhabitants by 1980. In view of these facts, the idea of the countryside "encircling" the cities seemed outdated, however propitious the "objective" revolutionary situation in the villages. "Urban guerrilla" strategy is based on the recognition of the fact that the political-military-economic center of power is in the great conurbations, that it could and should be attacked there, not from the periphery.[34]

The first, but also the least known of the advocates of this strategy was Abraham Guillén, an "anarcho-Marxist" (Hodges) of Spanish origin, who settled in Uruguay, after working for many years in

Argentina. He did not exclude cooperation with the rural *milizias* but did maintain that in highly urbanized countries revolutionary battles ought to be waged in the urban areas "for the revolutionary potential is where the population is."[35] These observations referred primarily to Uruguay and Argentina; in Brazil, on the other hand, revolutionary warfare should preferably be rural (Guillén later revised his views and criticized Marighela for assigning a merely tactical character to the urban guerrillas and strategic significance to the rural guerrillas). Guillén argued that Carlos Lamarca (the other important Brazilian urban terrorist leader) would not have been killed had he stayed among the nine million inhabitants of São Paulo instead of venturing into the countryside, where he was betrayed by a hostile population.[36] To endure the struggle the small armed minority would have to lead a consistently clandestine existence with the support of the population. Guillén does not clarify how this contradiction, clandestine existence and mass support, might be overcome in practice. Their basic principle should be to live separately and fight together. Urban guerrillas should use light arms, but machine guns and bazookas would have to be employed as well to give them the advantages enjoyed by a highly mobile infantry. They should not try to seize large objectives and engage in "Homeric battles" but concentrate on small, successive actions. As a result of facing a hundred guerrilla cells of five persons, the police would have to cede terrain, especially at night: "If at night the city belongs to the guerrilla and, in part, to the police by day, then in the end the war will be won by whoever endures longest."[37]

Writing shortly after the Paris events of May 1968, Guillén attributed a leading role in the revolutionary process to the students; he was one of the few Latin American guerrilla strategists to give them first place in the list of revolutionary forces. (Students were, of course, the strongest element in most guerrilla movements but guerrilla strategists usually felt self-conscious about this fact and preferred not to mention it.) The support of eighty percent of the population was needed according to Guillén. If they received such support, the guerrillas could win the war even though imperialism held an overall superiority of a thousand to one; for at a given place and time the guerrillas could still be superior to the enemy in numbers and firepower by five to one. Guillén agreed with the Guevara-Debray thesis about the role of the vanguard. In Brazil there was no working-class vanguard, but there was a Marxist vanguard of professional revolutionaries and, in the final analysis, it was of no importance from which class the cadres hailed. Guillén's impact on the Tupamaros was considerable in the early years of their struggle but he was by no means their uncritical admirer. He repeatedly dwelled on certain errors, tactical and fundamental, which he thought they committed.[38] They had rented houses in the cement jungles of Montevideo, thus establishing a "heavy rearguard" and "fixed fronts" for billeting, food, medical supplies and armaments. This exposed them to mass detentions and the seizure of their arms. They had excelled during the first hit-and-run phase of the struggle but then failed to escalate their operations by using larger units. Such advice smacks of armchair strategy. The main problem for all "urban guerrillas" was one of broadening their ranks. While their cadres were few and the scale of their operations small, they were relatively secure. The more numerous they became, the more difficult were the problems of housing and supply, and the easier they could be identified and captured.

Guillén opposed unnecessary violence: in a country in which the death penalty had been abolished, it was self-defeating to condemn to death even the most hated enemies of the people. A popular army that was not a symbol of justice, equality, liberty and security could not win popular support in the struggle against a dehumanized tyranny. Hence his opposition to the Tupamaros' "prisons of the people," to indiscriminate execution of hostages, to the use of violence against subordinates: surely there was little point in defeating one despotism only to replace it by another. Guillén opposed the cult of leadership (of which, however, the Tupamaros were much freer than other Latin American guerrilla movements), and he complained about their ideological shortcomings. In many ways they had become overly professionalized and militarized, and did not really know what kind of revolution they wanted. On the one hand they forbade their members to criticize the pro-Moscow Communists, on the other they gave publicity to conservative nationalists. The kidnapping of the alleged CIA agent Dan Mitrione was a success, his execution a mistake. When the Brazilian consul in Uruguay was kidnapped, his wife appeared as a heroine of love and marital fidelity: "Every *cruzeiro* she collected" in an appeal for his release "was a vote against the Tupamaros and indirectly against the Brazilian guerrillas." By demanding large sums of money for political hostages, the Tupamaros came perilously close to resembling a political Mafia. There was a historical irony about would-be liberators who indirectly lived off the surplus of the very people they wanted to liberate.

In later years, Guillén became more appreciative of the operations of the Chilean MIR and the Argentinian ERP, who demonstrated their ability to mobilize large masses, and who, unlike the Tupamaros, were more critical of both right-wing nationalism and Communist opportunism. Towards the end of 1972 he was inclined to write off the Tupamaros altogether; they had served as the best revolutionary academy in the world with regard to "urban guerrilla" warfare but their tactical brilliance was unmatched by their strategy and politics. Their supreme command had become centralized, it knew all, said all, did all. Such excessive centralization proved fatal in the end.

Carlos Marighela is more famous than Guillen, partly in view of his active participation in the armed struggle, partly because his *Minimanual* was banned in so many countries — though it contained very little that was not known to any experienced urban terrorist. After he left the Communist party in 1966 he was primarily concerned with tactical questions. His views on strategy were quite inconsistent: in 1966 he regarded guerrilla warfare merely as one form of mass resistance and did not expect that it would be the signal for a popular rising; he thought that the center of gravity should be in the countryside. Yet he proceeded to act quite differently.[39] Three years later he still argued that the decisive battles should be fought in the rural districts (the "strategic area") and that the fighting in the city was tactical and complementary only.[40] His own activities were however concentrated on the "complementary" front and he would frequently and incongruously refer to the strategic importance of the great conurbations.

Marighela's basic approach was as radical as that of Fidel and Guevara. Not only was it the duty of every revolutionary to make the revolution, not only was his single commitment to the revolution, not only did the guerrilla constitute both the political and military command of the revolution, "the urban guerrilla's reason for existence, the basic action in which he acts and survives, is to shoot" (*Minimanual*). Towards the end of his life, he no longer revealed any interest in political goals, let alone political agitation. Robert Moss rightly noted that his later writings read like manuals of military drill, not political manifestos.[41] A left-wing Brazilian critic later wrote that the "fetishist attachment" to, and the overestimation of, unlimited terrorism led to confusion, profound mistakes and, ultimately, to defeat.[42] Marighela's approach, very briefly, was one of provocation, compelling the enemy to "transform the political situation into a military one." He assumed that, in the process, the government would alienate large sections of the population, particularly the intelligentsia and the clergy. North American imperialism would have to be called in for help, and this would add to the popularity of the insurgents' struggle. The fundamental objective was to shake the basis upon which the system rested — the Rio–São Paulo–Belo Horizonte triangle (whose baseline runs from Rio to São Paulo) — for it was there that the economic, political and military power was concentrated. In Marighela's scheme, a great deal of freedom of action was left to the small units; they were to decide whether to launch an attack without reference to the high command. They were perfectly entitled to assassinate not just the commanders of the security forces but also low-ranking "agents." The struggle should proceed on three fronts — the guerrilla front, the mass front (meaning a combat front, not agitation among the masses) and the support network. Ideally, all these fronts ought to be equally effective, but Marighela realized that the revolutionary movement was bound to develop unevenly. He insisted that the constantly expanding guerrilla front carry out a scorched-earth policy to create alarm among the dictators. His aversion to any bureaucratic hierarchy dominated by *apparatchiki* came to the fore time and again: in the revolutionary organization only missions and operations were to be prized, not rank and position; only those prepared to participate actively in the struggle and bear the sacrifices had the right to be leaders. No complex chain of command, no political commissars or supervisors should be set up; a strategic command and regional coordination groups would direct the military organization. The regional command, in Marighela's scheme, would not be allowed permanent contact with the mobile units: no one should know all about everything and everybody. Like many other guerrilla leaders before him, Marighela stressed the importance of training ("everything depends on marksmanship"). The personal qualities needed by an urban guerrilla were, above all, initiative, unlimited patience, and fortitude in adversity.

The basic unit in the urban guerrilla army "is the firing group," consisting of no more than four or five people; a "firing team" constituted two such groups operating separately. Motorization was absolutely essential in the logistics of urban terrorism. The great advantages for the urban guerrilla were surprise in attack, a better knowledge of the terrain, greater mobility and speed, a better information network. Basic tactics always employed the hit-and-run principle, to attack and to get away. Attacks should be launched from all directions, in an endless series of unforeseeable operations, thereby preventing the enemy from concentrating his apparatus of repression; combat and decisive battle should always be avoided. Bank raids, Marighela noted, were the most popular form of action: "we have almost made them a kind of entrance exam for apprenticeship in the technique of revolutionary war."[43] In addition there were to be ambushes, occupation of schools, factories and radio stations, provided that the withdrawal from fixed targets was well planned. The list of the urban guerrilla's revolutionary assignments was long and varied; he should defend popular demonstrations, liberate prisoners, seize weapons from army barracks, execute agents of the government, kidnap policemen and Americans. Public figures such as artists or sportsmen should be kidnapped only in special circumstances when one could be reasonably sure that popular opinion would favor such action. Transport should be sabotaged, oil pipelines cut, and fuel stocks systematically depleted. Bomb attacks should be undertaken only by those technically proficient, "but they may include destroying human lives." Spreading baseless rumors was part of the war of nerves waged by the guerrilla, and "information" should be supplied to foreign embassies, the U.N., human rights committees and other such bodies. Marighela ended his treatise with some reflections about the political results of urban guerrilla war: the people would blame the government not the terrorists for the various calamities that befell them. He apparently regarded democratic reforms as the great danger on the road to revolution, and he hoped that in the chaos brought about by the "urban guerrilla" war, elections would appear a mere farce and the political parties thoroughly discredited. The future society, as he saw it, would be built not by long-winded speakers and signers of resolutions but by those steeled in the struggle, an armed alliance of workers, peasants and students. The participation of intellectuals and artists in urban guerrilla warfare would be of the greatest advantage in the Brazilian context. Of great importance, too, was the support of the clergy, with regard to communication with the mass of the people. "This is especially true of workers, peasants and the women of the country [*sic!*]."

It is only too easy to detect major ideological inconsistencies in Marighela's writing. But there is no doubt whatsoever about his fanatical dedication to the cause, the burning fervor pervading all his writings, and it was this single-minded advocacy of the revolutionary deed which attracted so many young followers, willing to engage in suicidal operations. In theory, urban terror was only one element in a broader revolutionary strategy, but Marighela was not prepared to wait in vain for the rural guerrilla *foci* to emerge. So great was his preoccupation with his spectacular exploits in the towns that there was no time and energy left to promote insurgency in the countryside.

Marighela was killed in a gun battle in São Paulo in November 1969; his successor met a similar fate, and Captain Lamarca, the head of the even more militant and action-orientated VPR, was shot two years later. By 1971 the Brazilian security forces had defeated the terrorist challenge even though sporadic operations continued. Economic prosperity may have played a part in this, but the growing modernization of counterinsurgency and the use of torture by the police were more decisive. Debray mentions American financial assistance to the Brazilian police; according to the figures he provides, these grants amounted to a million dollars per year. Latin American guerrillas are known to have seized bigger sums in a single bank raid or as ransom for their hostages. Under torture, even some of the most trusted and steadfast guerrillas betrayed their comrades and, since urban terrorists were far more vulnerable than rural guerrillas once one link of the organizational chain was broken, they suffered irreparable losses. The Brazilian "urban guerrillas" could effect the downfall of a quasi-democratic regime and promote the emergence of a dictatorship which would not hesitate to apply torture and other means of counterterror, but they were not strong enough to survive the backlash. Torture evoked much protest but did not provide new recruits; the terrorists became more and more isolated, their heroism was admired but not emulated; even the left came to consider their action as, at best, pointless.

Urban terror in its most sophisticated and, for a while, most effective form, made its appearance in Uruguay, the very country which Guevara thought the least likely scene of armed struggle because it was the most democratic. The Tupamaros (MNL) came into being in 1963; the heyday of their movement was between 1968 and 1972.[44] The movement did not, however, produce a great deal of literature; its programatic writings were few and far between. Their doctrine was first outlined in the form of a catechism, "Thirty questions to a Tupamaro," circulated in late 1967.[45] The basic difference between the Tupamaros and other left-wing organizations was, as they put it, the emphasis on revolutionary action rather than theoretical statements; revolution is not made by the elegant phrase. Once accepted that the basic principles of socialist revolutions are given and tested in countries such as Cuba, "there is nothing more to discuss." An armed movement can and should start operating at any time. The Tupamaros shared the conviction of the early Fidel that action should proceed "with or without a party." Strategies adopted elsewhere in Latin America could not be applied to Uruguay if only because the prospects for rural *foci* were almost nonexistent. This was compensated for, however, by the fact that the enemy was exceedingly weak. The Uruguayan security forces counted only some twelve thousand men, badly equipped and trained, "one of the weakest organizations of repression in America." Montevideo was a city sufficiently large and polarized by social conflict to make it an arena more suited to the struggle than many other centers on the continent. Overcoming their distaste for political programs, the Tupamaros produced a platform in 1971 but it contains little of interest since it closely resembled the programs of other left-wing movements on the continent, referring to agrarian reform, the nationalization of big factories, the expulsion of "imperialism," etc. More revealing were their occasional pronouncements on practical issues. The Tupamaros specialized in kidnappings, and for three years the government failed to retrieve a single hostage. A Tupamaro leader stated that the kidnappings were part of an overall strategy designed both to obtain the release of captured comrades and to undermine the foundations of the regime.[46] In contrast to other Latin American guerrilla movements, the Tupamaros endeavored to keep out of the ideological quarrels of the left; this was probably easier for them to do successfully than for guerrillas elsewhere, for the Tupamaros had a virtual monopoly as far as guerrilla operations in Uruguay were concerned. They were the most internationally minded of all Latin American guerrillas and took up Camilo Torres's idea of a continental organization of armed forces. This internationalism was probably not unrelated to the exposed situation of Uruguay; they very much feared an invasion from Brazil or Argentina. Their social composition was even more middle and upper class than that of other guerrilla movements: these were the sons and daughters of the establishment. (According to an apocryphal account, a Ph.D degree was a *conditio sine qua non* for membership.) In 1971 they decided to join the other forces of the left in a broad popular front (*frente amplio*) to contest the election; their candidate, however, did rather badly, receiving less than twenty percent of the vote.

Organizationally, the experience of the Tupamaros is interesting; they managed to combine "strategic concentration" with tactical decentralization and *compartamentación* of the basic units. This made it difficult for the security forces to paralyze their organization despite frequent arrests. The Tupamaros were the only Latin American guerrillas to establish something akin to a countergovernment with "prisons of the people" and hospitals. They greatly undermined the authority of the government, disrupting the civil administration and the economic life of the country. Eventually, the democratic regime was replaced by a dictatorship, the army was brought in and liquidated Tupamaro activities with surprising ease. Some observers have explained the defeat of the Tupamaros in 1972 with reference to their mistaken decision of the previous year to open their ranks to many new members. But while this may have contributed to their downfall, the main reason was apparently the loss of revolutionary *élan*. Although the decline of the Tupamaros took place (unlike the downfall of the ALN in Brazil) against a background of severe economic crisis, there appears to be no close connection between guerrilla success (Cuba) or failure with the economic situation. Debray correctly noted in retrospect that, in digging the grave of liberal Uruguay, the Tupamaros also dug their own.[47]

THE USES OF TERROR

The shift from rural guerrilla warfare to operations in the cities was by no means limited to Latin America. There had been urban terrorism in Palestine during the last years of the British Mandate, in Cyprus and Aden, and, of course, in Ireland. In some instances only sporadic actions by very small groups were involved, elsewhere the struggle lasted for years and was well organized. "Urban guerrilla" factions were plagued by internal division not less than their precursors in the countryside. Ulster and Spain provide typical examples. The IRA split into two factions in 1969: the more militant "Provisionals" advocated the establishment of a thirty-two county Democratic Socialist Republic based "on the 1916 Proclamation, justice and Christianity," and attacked the "Officials" of the IRA for their Leninist ideological bias and their failure to launch massive terrorist action in Northern Ireland.[48] The Basque nationalist movement split into several factions. ETA VI propagated a Trotskyism of sorts, dissociating itself from Basque "bourgeois nationalism";[49] while supporting terrorist operations in principle, it concentrated on political activities. Meanwhile ETA V, another faction of young militants (who derived their inspiration from the Algerian insurgents) engaged in spectacular terrorist acts, decrying the Marxist "vegetarianism" of their opponents. Unlike Guillen and Marighela few of the leaders of these urban terrorist groups attempted to provide a new doctrine, but they were very ready to furnish personal accounts of impressions and explanations. Pierre Vallières, one of the leaders of the Quebec separatists (FLQ) combined Paris Left Bank anarcho-Communism with a Fanonian belief in the cathartic effects of revolutionary violence.[50]

For a somewhat more systematic ideological exposition and an attempt to present a coherent strategic concept one has to turn to the advocates of urban guerrilla warfare in West Germany and the United States. The groups involved were small in numbers, the effect of their operations insignificant, but they attracted a great deal of publicity and, at least to that extent, succeeded in achieving their aims. The Baader-Meinhof group (RAF — *Rote Armee Fraktion*), whose origins were in the student movement of the 1960s, stated in their first manifesto that the formation of armed resistance groups for purposes of "urban guerrilla" operations was both possible and justifiable.[51] The overall aim was the seizure of power; the main obstacle facing the RAF was, according to the leaders of the group, the unfortunate fact that the public had been immunized by counterrevolutionary propaganda.[52] With attacks against state oppression however, the masses could be revolutionized — bombs would help to awaken their consciousness. The state apparatus was to be demoralized, and partly paralyzed, thereby destroying the myth of its invulnerability and ubiquity.[53] In the first phase of the struggle the main task would be to disseminate the idea of an armed struggle, to collect arms, and to organize small units of three, five or ten members. During the second phase the actions of a minority would turn into a mass struggle; with the support of the masses, militias were to be formed in those areas where the enemy was so much weakened as to be no longer able to concentrate his forces. Critics noted that the concept up to this stage closely followed Maoist strategy — simply transferring it to the cities.[54] During the third and final phase, mass action (street demonstrations, strikes and barricades) would support the terror of the guerrilla units. Following the Latin American example, the RAF aimed at provoking the authorities into using increasingly brutal retaliation, massacres, and "Fascist concentration camps." The RAF regarded their struggle as part of a worldwide campaign against American imperialism, but this emphasis on the connections with national liberation movements in Asia, Africa, the Middle East and Latin America evoked criticism from within. A former leading member of the group, the lawyer Horst Mahler, noted that operations in such a context were no longer urban guerrilla warfare but the establishment of a fifth column of the Third World inside West Germany — a concept unlikely to appeal to the masses.[55] He also criticized the elitist character of the RAF, "the total isolation from the masses." In their very first declarations the RAF announced the intention to combine legal political activity (*Basisarbeit*) with the armed struggle. Such a combination, however, soon came to be regarded as impractical be it only because it would not escape the attention of the police. Mahler argued that even though it was exceedingly difficult for individuals to engage simultaneously in legal and illegal activities, the RAF as such should have found a way to overcome this obstacle. The movement needed a (civilian) base; some of the operations initially envisaged against absentee landlords and speculators were highly appropriate and it had been a mistake for the RAF to discontinue them.

According to their original concept, the Baader-Meinhof group acknowledged that the revolutionary proletariat was the only force capable of guaranteeing victory over capitalism. In time the RAF would become the mailed fist of a (new) Communist party. This concept too fell by the wayside and was replaced by a new strategy which (Mahler claimed) was rooted in the anti-authoritarian phase of the student movement, with its counterculture, its "moralistic" attitude towards politics and other "petty bourgeois" ideological remnants.

The leaders of the RAF group were neither "instinctive" guerrillas nor well-educated theoreticians, but activists whose imagination had been caught by the armed struggle in other parts of the world. They wanted to apply the "lessons" of the Far East and Latin America to conditions that were utterly different. It would be unrewarding to submit their military concepts and political ideas to rigorous analysis. They never made it clear what kind of revolution they had in mind, or whether a political party was needed to carry it out. They argued that the urban insurrections of the past such as the Paris Commune or the Russian revolution of November 1917 were of no use as a model for Germany in the 1970s. But they never indicated a more appropriate historical model, nor did they develop one of their own. On the other hand, they were heirs to a German tradition in which at least lip service ought to be paid to theory. They could not possibly reject ideology as did segments of the American New Left. ("Fuck programs! The goal of the revolution is to abolish programs and turn spectators into actors. It's a do-it-yourself revolution. . . .[56]) What mattered in the last resort, however, was the thirst for action not a conviction based on "scientific theory."

But for the Germans' belief that all self-respecting terrorists needed a theory there were obvious similarities between the RAF and the small groups of white American "urban guerrillas" who developed out of the civil rights and anti-Vietnam demonstrations, and the university sit-ins. They were rooted in the age of radical chic when instructions for the fabrication of Molotov cocktails were featured on the cover of a journal dedicated to the critical study of English letters. Ideologically, these groups were inchoate and their manifestos were illiterate — the illiteracy of liberal, middle-class schools, not of the ghetto. Their common denominator was the destruction of the present political and social order, but beyond this it was a case of everyman-his-own-urban-guerrilla.

The main white "urban guerrilla' faction, the Weathermen, emerged when the SDS (Students for a Democratic Society) split into three groups in late 1969. Many of its members had visited Cuba and Vietnam and wanted to bring the war back home: "We are adapting the classic guerrilla strategy of the Viet Cong and the urban guerrilla strategy of the Tupamaros to our own situation here in the most technically advanced country in the world."[57]

While guerrilla movements elsewhere fought for national liberation, the Weathermen maintained that in America the urban guerrilla has to be antinational. For the revolution to be defined in national terms within so extreme an oppressor nation as the U.S. would be tantamount to "imperialist national chauvinism." But there was at least one section of the American people to which the concept of national liberation was applicable. Hence the appeal to build a movement which would support the blacks who, in the past, had fought almost alone. The Weathermen contacted the Black Panthers but these rejected the call for a joint urban guerrilla war, instead suggesting a mere alliance.

The Weathermen were not sanguinely optimistic about the prospects of revolution in America; they saw it occurring, if at all, as a belated reaction to a successful world revolution. They were aware that it was pointless to appeal to workers or peasants, and they regarded the university campus as their main base. If Guevara (*sic*) had taught them that revolutionaries moved like fish in the sea, the alienation and contempt of the young people for America had created the ocean they needed. Guns and grass were united in the youth underground, freaks were revolutionaries and revolutionaries were freaks, as one of their communiqués put it. The manifestos of the Weathermen and of the SLA (Symbionese Liberation Army) betrayed the influence of a hedonistic counterculture; whatever they had learned from Guevara, it was not his appeal for an ascetic life. Making love, smoking dope and loading guns were all part of the revolutionary process, as they envisaged it. The "most potentially explosive conflict" brewing in America was between men and women and the program of the SLA stated *expressis verbis* that a system had to be created whereby people would not be forced to stay in personal relationships when they preferred to be free of them.[58] The Weathermen's interest in politics was strictly limited; they were not concerned with training revolutionary cadres, let alone the education of the masses. The main aim was to scare and shock "honky America" and to this end all violent means, however barbaric, were appropriate. They approved of the murder by the Manson gang of the actress Sharon Tate, eight months' pregnant; to shoot a "genocidal robot policeman" was regarded as a sacred act.[59] Subsequently, the Weathermen copied, without marked success, the strategy of certain black groups who saw the *Lumpenproletariat* and criminal elements as their natural recruits. Even later the SLA, with its emblem of a seven-headed cobra ("a 170,000-year-old sign signifying god and life"), aimed specifically at enlisting nonpolitical convicts into its ranks; they, after all, were only the victims of the system. What attracted a few young men and women (more women apparently than men) was not ideology but a life style, above all, "togetherness." A member later recalled that what had struck him was "that they were a family, a big very tight family. I wanted to be part of that. People were touching each other. . . ."[60]

The aims of the black urban extremist groups by comparison were far more tangible. They did not complain about "our colossal alienation" but did demand full employment, decent housing, education, and the power to determine the destiny of the black community. Some of them, such as the Cleaver faction of the Black Panthers and the Black Liberation Army, were advocates of urban terror. The policeman was the representative of the occupation army in the black ghetto; the weapon was needed to educate the masses. Negro youth were called upon to show their mettle by brandishing guns. America was to be burned and looted, to be cleansed with fire, blood and death. Founded by Huey Newton and Bobby Seale, the Black Panthers were subsequently plagued by as much internal division as their white counterparts. Stokely Carmichael, their erstwhile prime minister, a "cultural nationalist," was denounced for "not maturing to embrace" the ideology of the party, i.e., "the historical experience of Black people in America translated through Marxism-Leninism."[61]

The fullest exposition of their doctrine was provided by Eldridge Cleaver, minister of propaganda. Marxism-Leninism was an outgrowth of European problems and a new ideological synthesis was called for to suit American conditions. There was no all-American proletariat just as there was no all-American *Lumpenproletariat*. The working class, as he saw it, was the right wing of the proletariat, the *Lumpen* constituting the left wing. It was not the *Lumpen* who were the parasites, but the working class. The streets belonged to the *Lumpen*, and it was in the streets that the *Lumpen* would rebel. They could not strike because they had no secure relationship with the means of production: they had been locked outside the economy. Their immediate oppressor was the pig police who confronted them daily. Thus the *Lumpen*, who had been analyzed out of the revolution by the (white) Marxists-Leninists, would hit out at all the structures around them.[62]

The Cleaver faction was eventually ousted from the Black Panther party which, under Huey Newton, moved more and more towards community action. The Cleaverists on the other hand advocated the combination of above-ground political action with antipolice terror, bank robberies, the execution of businessmen and the kidnapping of diplomats. George Jackson, who was killed during an attempted jailbreak in August 1971, envisaged in his book *Blood in My Eye* (1972) resistance to the Fascist American government as a "fluid, mobile, self-impelled attrition of people's urban guerrilla activities lying in wait inside the black colony." Other spokesmen stressed that they regarded themselves as the "Babylonian equivalent" of the Tupamaros, Frelimo and the NLF.[63] If there was only a thin line between bandits and revolutionaries this (they assured the flock) should not cause undue apprehension, for many famous revolutionaries had started their careers as bandits before becoming politicized. Furthermore, there were good tactical reasons for letting revolutionary acts seem like acts of banditry.[64]

The bark of the American "urban guerrilla" was considerably worse than his bite. By late 1970 the Weathermen had failed; the operations of the SLA in 1974 were no more than the actions of a few unstable individuals, the like of whom have always existed on the margins of a violent society. Eldridge Cleaver, after a prolonged stay in Algiers, discovered that the "Babylon" from which he had fled in anger and disgust had much to recommend itself. Thus the appeals to conduct "urban guerrilla warfare" petered out.

During its heyday urban terrorism had presented a problem for the American and West German police but politically it failed to become a force to be reckoned with. The white American revolutionaries, in the words of a friendly critic, were the children of middle-class families who knew no oppression. They substituted their own personal hang-ups and moods for the demands of "the people," claiming to speak for a people they had never met. The black militants were acting in a milieu far more congenial to violent action and their revendications were much less far-fetched. But they too found little sympathy for their cause within their community. Their protest was nationalist in inspiration, or, to be precise, racialist — a response to the racial oppression American Negroes had encountered throughout their history. The occasional invocations of Marxism should not be given too much weight, and the same goes, *a fortiori*, for the Weathermen and the SLA. Their problem was not the lack of freedom but a surfeit of freedom; their godfather was neither Lenin nor Mao but liberalism running wild.

FANON

Except when they were made in conjunction with separatist national movements, all attempts in the 1960s to conduct guerrilla warfare in the economically developed, industrialized societies of Western Europe and the United States generally failed. Guerrilla warfare was more successful in Africa against the remaining European colonial outposts. Most African countries attained national independence without an armed struggle but some did not, and it was in these parts that fighting occurred and that attempts were made to formulate a specific African guerrilla doctrine. Of the ideologists of armed struggle, Frantz Fanon, a native of Martinique and a psychiatrist by profession, was the most important by far — not so much as regards his actual impact on guerrilla warfare, but certainly with regard to the repercussions of his writings outside Africa. Fanon did comment on occasion on military issues in the narrow sense — such as the question of arms supply to the Algerian FLN (which had become difficult to secure following French counterinsurgency measures).[65] But it was not for such technical advice that his fame spread; rather he provided a new ideology on the cathartic role of violence in the African revolution. Violence, as he saw it, was a cleansing force, liberating the African from his inferiority complex, his despair and inaction; it made him fearless and restored self-respect; it bound the Africans together as a whole.[66] "Violence alone, violence committed by the people, violence organized and educated by its leaders makes it possible for the masses to understand social truths and gives the key to them." The Mau Mau insistence that each member of the group strike a blow at the victim was a step in the right direction, since it required each guerrilla to assume responsibility for the death of a settler. Marxism-Leninism was acceptable as far as it went, but it did not encompass the colonial situation, ignoring its racist aspects.

The idea of violence is central to Fanon's thought but he was a stranger to Africa, and it is interesting to note that he has been far more widely read and admired among American blacks than among Africans. He also had many white admirers among the European and American left even though certain aspects of his views were sometimes regarded as embarrassing or disturbing: in part because such murderous humanism was difficult to digest, and partly because the origins of the cult of violence were not, to put it cautiously, altogether respectable. It is not really material whether Fanon had been familiar with Sorel's writings[67] — the uncomfortable fact still remained that the ideological precursors of Italian Fascism, such as Corradini, had already argued that the proletarian nation has a moral obligation to resort to noble violence. Mussolini himself had declared that there was a violence that liberated and a violence that enslaved, a moral violence on the one hand and a stupid, immoral violence on the other. Whereas Sorel had created his mystique of violence in the context of the struggle of a working-class vanguard, Fanon saw the peasantry and the *Lumpenproletariat* as *the* spontaneously revolutionary forces. The peasants, with nothing to lose and everything to gain, he argued, were the first to discover that violence paid off. The working class, however, pampered by the colonial regime, was in a comparatively privileged position. The Marxist idea about the history-making role of the urban proletariat had been disputed before by Wright Mills and others: but Fanon went even beyond Bakunin in his enthusiasm for the role of "the hopeless dregs of humanity," the pimps, petty criminals, hooligans and prostitutes. He thought that their revolutionary potential was enormous and, if the insurgents did not give it full attention, colonialism would make use of it.[68]

This concept was not likely to endear him to the Communists and neither was his argument that there was "no true bourgeoisie" in Africa.

The critics argued that Fanon had got his facts wrong, that he brought confusion and division to the revolutionary movement.[69] In fact, Fanon was more acute than his critics; the *Lumpenproletariat* (given a broader definition than his own) was about as patriotic as the rest of the population, and did play a part in the armed struggle. There were few workers in the guerrilla units simply because there were not many of them in Africa in the first place; the rank and file consisted of peasants. In other respects Fanon revealed a naïveté uncommon among students of the human psyche. He argued that the people who played a violent role in the national liberation would allow no one to set themselves up as "liberators"; one could hardly think of a more mistaken prediction of the political future of Africa. It is true that on various occasions he expressed grave misgivings about the political regimes likely to emerge after decolonization which, he feared, would only constitute a "minimal readaptation." He suspected that the new leaders would not heed his appeal to turn their backs on European civilization, to destroy European institutions, to make an end, not only to colonial rule, but also to the corruption of the settlers, to the brightly lit towns with their asphalt and garbage cans.

Fanon died at the early age of thirty-six. His influence on the African political elites was not lasting; their main interest was not in the cultural aspects of the African revolution; they were preoccupied with economic and administrative problems, or with simply bolstering their own positions. What Fanon had written on the evils of bureaucracy and one-party, one-leader dictatorships was not of the slightest use to them.[70]

CABRAL

Passing on to the writings of Amilcar Cabral, one descends from the rarified heights of existentialism to the well-trodden paths of Marxism-Leninism, from a Dostoyevskian novel to sober and unexciting political-socioeconomic analysis. Cabral, who hailed from the Cap Verde islands, was assassinated in Conakry in 1972 by political rivals. Like many other guerrilla leaders in the Portuguese colonies, he was of mixed mulatto rather than of pure Negro origin and, like Mondlane, the head of Frelimo, he married a white woman. He studied agriculture in Lisbon, became a Leninist, and for several years worked as an adviser to the Portuguese government in Africa. Cabral agreed with Fanon about the necessity of an armed struggle. Although he wrote from time to time about the liberating role of violence without which there could be no national liberation, he never engaged in the fetishization of violence. From time to time, he even submitted offers to the Portuguese authorities to negotiate a settlement. Like Fanon, he stressed the importance of the participation of African women in the struggle for liberation, and agreed with Guevara's thesis that there was no need to wait for a revolutionary situation, one could create it.[71] In contrast to Fanon's emphasis on psychological and cultural issues, however, he was far more interested in economic development. While Fanon hardly ever mentioned the role of the political party in the struggle, this was a central issue for Cabral, whose assessment of the forces likely to support the armed struggle also differed greatly from that of Fanon.

To begin with Cabral and his comrades of PAIGC faced great difficulties in winning over the peasants. The slogan "the land to him who works it" could not be applied to Guinea-Bissau, as he admitted, because there the land did belong to the peasants; neither were there big land holdings — the land was village property. Without concentrations of foreign settlers (Cabral wrote), it was not at all easy to prove to the peasant that he was being exploited, as Fanon had argued.[72] Extreme suffering alone did not produce the *prise de conscience* needed for the national liberation struggle.

Cabral's view on the role of the petty bourgeoisie in the armed struggle is of considerable interest; he admitted that in Guinea-Bissau, as in other parts of Africa, it played a leading role in the struggle. Yet, economically, it was without a power base and hence could not seize political power. The historical dilemma facing it was in Cabral's words either to betray the revolution or to commit suicide as a class.[73] This assumption was not however shared by a close and very sympathetic observer of the Guinean scene, who, on the contrary, reached the conclusion that the lower middle class was the natural holder of power in tropical Africa because it was the only class possessing knowledge, know-how and organization.[74]

The political party which Cabral founded was organized according to the Leninist principles of "democratic centralism," and though its statutes provided for "collective leadership," all important decisions were taken by Cabral himself. He acted not only as head of the Central Committee, and as commander in chief, but also as secretary for political and foreign affairs. While expressing strong support for the Soviet Union, he tried not to become implicated in the Soviet conflicts with Cuba and China. The Chinese initially supported PAIGC but later grew markedly cooler in view of Cabral's "opportunism" vis-à-vis Moscow.

Cabral had no military experience when he launched the guerrilla war but he devoted more time and energy to providing guidelines for the armed struggle than most other African guerrilla leaders. PAIGC enjoyed several important advantages in its fight. The army facing it represented the weakest of all European colonial powers; when the insurrection broke out there were altogether one thousand Portuguese soldiers in Guinea-Bissau. (The total number of inhabitants was about half a million.) Furthermore, PAIGC had a permanent base in neighboring Conakry for the training of its cadres and for supplies. (There was no such base to support guerrilla warfare in the Cap Verde islands and no armed struggle took place there.) Early on in the campaign, almost one-half of Guinea-Bissau passed into the hands of the PAIGC. Admittedly, this was the less important part; the Portuguese had never really controlled the whole country, and Cabral noted that "we had established guerrilla bases even before the guerrilla struggle began." Within a year or two after the start of the war, Cabral's forces organized semi-regular units and popular militias, and soon a Northern and Southern front came into being.[75] The Portuguese forces were too few in number to engage in systematic on-the-ground attacks against rebel bases, and their air force insufficiently strong to inflict decisive damage on the rebels. The greatest obstacle with which PAIGC had to contend in the early phase of the struggle was the tribal structure of the population. There was no lack of enthusiasm to join the struggle, but guerrilla chieftains tended to act independently; there were, in Cabral's words, "isolationist tendencies." The command decided to carry out a major purge and at a later stage it became official policy to appoint a member of one tribal group as commander of another.[76] Furthermore, PAIGC had to contend with rival groups supported by Senegal. PAIGC carefully avoided criticizing religious beliefs and superstitions and did not ban the use of fetishes and amulets (*Mezinhas*), on the assumption that the guerrillas would soon learn that a "trench was the best amulet" (Cabral). It should be noted in passing that the religious issue was of considerable importance to most African liberation movements. Replying to European left-wing criticism, an Algerian writer stressed that in a colonial country, where the dominant religion was persecuted, rejection of Islam was a sign of snobbery on the part of a Western-assimilated, intellectual elite, who were not only detached from the people but neutralized and corrupted by the ideology of the oppressor.[77] The journal of the FLN frequently noted that, of all the Islamic peoples, the Algerians were perhaps the most attached to their faith.[78] But a distinction must be drawn between the Algerian rebels, most of whom were devout Muslims quite irrespective of whether Islam was "persecuted" or not, and the Marxist-Leninists, who tolerated religion for tactical reasons, hoping that sooner or later it would disappear.

Cabral's realistic approach can also be gauged from his attitude towards nationalism and Panafricanism. He was on close terms with other African revolutionary leaders, but unlike Fanon, or Nkrumah in his last years, he did not think that African unity was an aim attainable in the forseeable future. He was perhaps the most intelligent, certainly the most sober, of the African guerrilla leaders; comparisons with Guevara, therefore, seem a little far-fetched. Marxism-Leninism was the great formative influence in his youth, but when in later years he referred to the overriding importance of the "historical reality" of each people, the fact that social and national liberation were not for export, the necessity of conducting policy and warfare according to widely varying local and national conditions, these assessments reflected the maturing of a mind not given to slavish imitation of foreign models. Cabral paid his respects both to the Kremlin and the pope; but as the years passed, the specifically African elements in his thought reasserted themselves to a certain extent. Like all intellectuals in backward countries, he was a socialist, but not of the democratic-socialist variety. His socialism was largely synonymous with nationalism, anti-imperialism, and national liberation, an ideology very different from the European socialist tradition and impossible to define within the language of that tradition.

FATAH DOCTRINE

Many guerrilla wars took place in the Middle East and North Africa in the postwar period — from Algeria to Kurdistan, from southern Sudan to Dhofar in southern Arabia. But only the Palestinians in their struggle against the state of Israel developed a more or less coherent strategic doctrine. Its military models were highly eclectic in character; there were innumerable references to the lessons of China, Vietnam, Cuba, Algeria and other wars, even when these were of little relevance to Arab-Israeli conflict.[79] In their choice of tactics, the Palestinians were more innovative: the Cubans apart, they were the first to hijack planes on a large scale; they dispatched letterbombs; they attacked Israeli nationals abroad and non-Israeli Jews. Their operations were launched from bases outside Israel and some were undertaken by foreigners rather than Arab nationals (the Lod airport massacre for instance). But these innovations would not be discussed on the theoretical level. On the contrary, the leading Palestinian organizations dissociated themselves from operations which made for bad publicity. They would be attributed to some new *ad hoc* organization, of which no one had heard before, and which would evanesce as suddenly as it appeared. Fatah doctrine adopted guerrilla warfare as the most suitable approach for the destruction of Israel; it was skeptical of the Arab governments' adherence to conventional warfare. The Palestinians did not trust the Arab governments on the assumption that even if they were to decide to go to war and succeed in defeating the Israelis, this would not result in the destruction of Israeli society — the ultimate aim of the Palestinians. Furthermore, the conflict had to be Palestinianized for psychological reasons: having tasted the bitterness of defeat, the shame ought to be wiped out by the Palestinians themselves. Echoing Fanon, the spokesmen of Fatah argued that violence has a therapeutic effect, inculcating courage, purifying the individual, and forging a nation.[80] For a variety of reasons the Palestinians wanted to emulate the Cuban example. The wars in China and Vietnam, unlike the Palestinian struggle, had been sponsored by Communist parties; furthermore, Mao had uttered doubts as to whether a protracted war was at all possible in small countries. Conditions in Algeria had been different, where a great majority of the Arab inhabitants supported the FLN against a minority of French settlers. Moreover, unlike Algeria, most of Israel was a plain and thus unfavorable guerrilla territory. Again unlike Algeria, the insurrection had to be prepared for from outside the borders of the state.[81]

Palestinian Arab doctrine frequently referred to the formation of a "revolutionary vanguard," to a "revolutionary explosion," to various stages in the struggle for liberation, but these phrases were simply copied from other guerrilla movements. Less vague were the explanations about the aims: there would be a long series of small battles, the Israeli enemy would be worn down, the army would constantly have to deploy strong forces against the *fedayeen*. The financial burden would become intolerable, foreign investment would cease, immigration would be discouraged, and there would be growing political polarization within Israel. The rise in casualties would create a climate of confusion and fear, the "grievance community" would widen, and eventually the Israelis would realize that unless they successfully resisted Zionism, they would be crushed by it.[82]

The critics of the Palestinians have argued that theirs was neither a war of national liberation nor a guerrilla war. It was not a war of national liberation because the Palestinians did not want to liberate the inhabitants of Israel, but to replace them: it was, in other words, a conflict between two peoples for the same territory. The Palestinians counterclaimed that the Israelis are neither a nation nor a people. The Jews were to be thrown into the sea according to an earlier Palestinian guerrilla doctrine; after 1968 this slogan was no longer used. The aim was not physical destruction but merely the return of the Arab refugees and the establishment of a democratic, secular state. The slogan of the "democratic state," however, created further ideological difficulties: the guerrilla organizations had to insist at the same time on the Arab character of the future democratic state, intending it to be an inalienable part of the wider Arab homeland. Because the Palestinian movements failed to establish bases inside Israel, the guerrilla nature of their operations was open to dispute. If agents who were not Arabs were enlisted to hijack airplances that were not Israeli, this was certainly not a "people's war" in any meaningful way. The spokesmen of the Palestinians would answer that all that counted was the political effect, namely to publicize the Palestinian cause all over the world.

That the political content of Fatah doctrine remained vague was no mere accident. The ideologists of Al-Fatah declared that the aim of the movement was to bring together all the revolutionary forces engaged in the struggle for liberation, and that "Byzantine discussions" about the social structures to emerge after liberation were to be eschewed. Unless there was ideological neutrality, the patriotic effort would be dissipated and the Arab masses alienated: in a battle for survival ideological differences had to be put aside.[83] Other Palestinian organizations, such as the two factions of the "Popular Front of the Liberation of Palestine" (PFLP and PDFLP) refused to accept such ideological neutralism. This group, originally known by the name ANM (Arab National Movement, *Kau-*

miyun el arab) was founded in Beirut in the 1950s. Among its slogans, the concept of "vengeance," with its connotations of Arab tribal vengeance, featured prominently, and it was also known as the "fire and iron" group. In the words of its historian, the ANM gradually gave up these slogans under the pressure of charges of Fascism and fanaticism.[84] They supported Nasserism at first but later many members announced that they had embraced "scientific socialism." In 1969 the group split into two factions — one led by Dr. Habash (PFLP), the other headed by Naif Hawatme. The doctrinal differences between the two factions were insubstantial, but Hawatme's PDFLP did place greater emphasis on political rather than terrorist activity; it also regarded itself as the more Marxist of the two groups, stressing its affinity with Cuban and Vietnamese socialism. In actual fact the cause of the split was not a clash of political views so much as a clash of temperaments between the leading figures. Both factions agreed on certain theoretical formulations, such as the necessity to conduct the war under the leadership of a revolutionary Marxist-Leninist party, and to transform the guerrilla war into a "people's war of liberation." In contrast to Fatah, they inveighed against Arab "reactionary circles," threatening to blow up installations in the oil-producing countries. On several occasions they even threatened that a revolution in the Arab world was the prerequisite for the liberation of Palestine. The leadership of the Palestinian resistance, as they saw it, had to be taken out of the hands of "petty bourgeois elements"; only the working class, in coalition with poor peasants, would safeguard the revolution.[85]

As so often, however, guerrilla practice was by no means coincident with guerrilla doctrine: the PFLP did not carry out its threats against Saudi Arabia or Kuwait, and no attempt was made to put working-class cadres at the helm of the guerrilla movement. If the PFLP spokesmen proclaimed that they did not fear the prospect of a third world war for they had nothing to lose, this assertion also has to be taken with a pinch of salt. Of all guerrilla movements in history, the Palestinian resistance groups appear the richest by far; the annual contributions by Arab governments were estimated in 1973 as being in the range of fifty to a hundred million dollars.[86] The Arab guerrillas have more to lose than their chains. Neither PFLP nor PDFLP showed any intention of forming a Marxist-Leninist party, or of cooperating with the existing Communist parties.

There were many unique elements in the history of the Palestinian organizations. They became involved in large-scale fighting in their host countries (Jordan and Lebanon) which felt threatened by the emergence of a "state within a state." The Palestinians showed great aptitude in the conduct of propaganda warfare abroad; it was a major political victory that by the mid-1970s there was growing acceptance of the fact that the Palestine issue was not just a refugee problem but involved the restoration of a people's legitimate national rights. This achievement was not however the result of a successful guerrilla war but of the oil weapon and the increasing weight of Arab governments in international politics.

NKRUMAH, NASUTION, GRIVAS

Throughout the 1950s and 1960s guerrilla warfare was a topic of the greatest interest; one U.S. Marine Corps officer called the sixties "the decade of the guerrilla."[87] Those recently engaged in guerrilla warfare were asked to make their thoughts and their experiences more widely known. Others like Kwame Nkrumah who had no obvious expertise in the field, practical or theoretical, nevertheless volunteered *obiter dicta* on the subject. Nkrumah wrote his book during his Conakry exile, shortly before his death.[88] On the basis of various diagrams, he tried to show that colonialism was "primitive imperialism," Fascism was "extreme capitalism," that revolutionary warfare was the key to African freedom, and that a new African nation ought to be established within the continental framework. His blueprint envisaged the establishment of an All African People's Revolutionary Army (AAPRA) under the command of an All African People's Revolutionary Party. The main enemy was neo-colonialism, and though Nkrumah called for its unmasking, he finally restricted himself to innuendoes about certain, unnamed regimes. Nor did he elaborate in his *Handbook* on the fact that the guerrillas, in all probability, would have to fight Africans rather than foreign neo-colonialists. Since Nkrumah had no firsthand knowledge of guerrilla warfare, the military sections of his book are not original; he borrowed quite indiscriminately from Mao, Castro, the Algerian and other guerrilla leaders. The resulting admixture was so vague that it could be applicable to every country and none. The Communists were ambivalent towards Nkrumah: they welcomed his attempt to apply Marxist-Leninist ideas to Africa, but were dismayed by his ideological pretensions, his claims to have established an original system ("Conscientism," "Nkrumahism"). In addition, there were differences on matters of substance; to give but one example — neo-colonialism for Nkrumah was "collective imperialism," whereas the Communists always emphasized the contradictions between the various imperialist powers.

Abdul Harris Nasution was not among the trailblazers of guerrilla warfare either, but he commanded more respect as a military authority, in view of his personal involvement. He was twenty-three years of age when the Japanese invaded his native Indonesia and had just been commissioned in the Netherlands Indies armies. He then served in the Civil Defense Forces established by the Japanese. In the war against the Dutch he first commanded a division and was later made chief of the operational staff of the Indonesian armed forces. Subsequently, as Indonesian minister of defense, he had considerable counterinsurgency experience. His guerrilla handbook, written in 1953, reminds one of Mao with the politics left out. As a precondition of success in guerrilla warfare, according to Nasution, the guerrilla's roots must lie in the people. The counterguerrilla had to try to sever the guerrilla from this base, not only by military operations but by political, psychological and economic action.[89] He is not an uncritical admirer of guerrilla warfare, and time and again stressed its limitations: "How great were the setbacks and how great the amount of confusion and difficulty that befell us because we played the role of the guerrilla too long." In his view, guerrilla-mania (the lack of discipline, planning, the belief that everyone could fight as he wished) was the most dangerous enemy of the guerrilla movement, having the effect of a counterguerrilla movement. Like Mao, Nasution accepted the general fact that guerrilla warfare alone could not ensure victory; hopefully, it weakened the enemy by draining his resources. Final victory, as he saw it, could only be achieved by a regular army in a conventional war.[90]

An axiomatic statement of this kind might have been true with regard to China. But it certainly did not apply to Indonesia where resistance against the Dutch never proceeded beyond sporadic acts of violence. General Grivas's experience in Cyprus is further proof that generalizations about guerrilla warfare are of doubtful value; according to the classics of guerrilla warfare, it should never have happened because the territory was too small. Yet a handful of combatants, variously estimated between sixty and two hundred, who never had more than a hundred automatic weapons and five hundred to six hundred shotguns between them, sustained a fight against several divisions of British soldiers for four years and eventually ousted the British.[91] Conditions were not propitious for a variety of reasons: EOKA was right wing whereas many Cypriots gravitated towards Communism; the Turkish minority needless to mention saw in the EOKA fighter the enemy *par excellence*. Metaphorically, EOKA was anything but a fish in a friendly ocean. One unique feature of guerrilla warfare in Cyprus was the smallness of the units involved, which only rarely exceeded eight to ten men. Grivas's original plan had been to concentrate his units in the Olympus and Pentadactylos mountains where the terrain seemed most suitable. But he soon changed his plan; most of the fighting proceeded in the lowlands, and eventually in the towns and the suburbs. On the basis of his experience, Grivas wrote that leadership was more important than terrain. Some of his best results were achieved in flat and nearly treeless terrain: "It remains axiomatic that in guerrilla warfare, with able and courageous leadership, one can take on any undertaking, whatever the nature of the terrain."[92] Most of his observations are in the mainstream of guerrilla doctrine: there is not set, textbook approach and no universal tactics, each case is special. Attacks should be sudden blows of short duration, boldly executed, and followed by instant and rapid withdrawal. The entire territory should be a single field of battle, without distinction between front and rear. The enemy should never know where one might strike. The overall strategy should be to wear the opponent down by prolonged attrition. All this was unexceptional but it is most unlikely that Grivas would have made any progress had he faced an enemy more resolute than the British forces in Cyprus. Britain was about to liquidate the last outposts of its empire in any event and, given these circumstances, even very slight armed resistance was bound to precipitate the process. General Grivas's experience shows that guerrilla operations can be launched in the most unlikely conditions, but they are likely to succeed only if the enemy is either weak or refuses to take drastic action. There are no universal lessons to be drawn from the Cyprus experience. The political constellation was auspicious and this more than compensated EOKA for the adverse topographical conditions. Guerrillas trying their luck in similar conditions against a different enemy would have been destroyed.

GUERRILLA WARFARE AND COMMUNIST DOCTRINE

Contrary to widespread belief there is no specific Communist guerrilla doctrine. Communists have of course been concerned with the seizure of power, be it as the result of armed insurrection, civil war, or political process. Soviet and European Communists assumed that more probably than not this would involve the use of military force, but they have never argued that it was the only possible way to power. Guerrilla warfare for the Russians was just one manifestation of the revolutionary process, which ought to be utilized in the fight for the worldwide victory of Communism. For the Chinese it was one specific stage in an armed struggle which would inevitably lead to national and social liberation in the Third World and thus prelude the triumph of Communism in the industrially developed countries.

Such general observations do not, however, suffice for an interpretation of Soviet and Chinese policies vis-à-vis guerrilla movements. For the approach towards individual groups depended upon a great many factors, doctrinal considerations being only one of them. The Soviet policy of détente did not in principle preclude support for movements of national liberation in Asia and Africa. On the contrary, the Soviet leaders had a genuine interest in their success for they assumed that as a result there would be a shift in the overall global balance of power in their favor. They knew, furthermore, that unless they supported these liberation movements the Chinese would appear as the champions of national liberation. On the other hand détente, or to be precise, Soviet reluctance to become involved in a world war in the nuclear age, did inhibit to a certain extent the amount of help that could be given to the proponents of armed struggle in various parts of the world. For the Soviet leaders assessed, quite correctly, that it would be difficult in the long run to prevent a major war unless some control was exercised over the conduct of small wars outside their own immediate sphere of interest. This assessment gave rise in Latin America, and to a certain extent in the Middle East, to complaints about "Soviet betrayal." Most guerrilla movements would have instinctively turned to China but for the unfortunate fact that China could be of much less help as a supplier of arms and money, and could not lend them much political support either. They also did not like certain aspects of Chinese theory and practice which were thought too primitive for Latin America or simply inapplicable to other parts of the world.

On the whole, Soviet leaders took a dim view of guerrilla warfare in developed countries. In their book it was the task of the Communist parties in these countries to make the revolution; meddling petty bourgeois elements only caused trouble. More likely than not their endeavors would fail, bringing harm upon the Communists too. On the other hand Communist spokesmen justified an armed struggle in Third World countries and, on certain conditions, elsewhere;* Communist parties have on occasion engaged in guerrilla warfare not only in Asia but also in Greece, Venezuela and the Philippines. But more often than not the initiative seems to have come from the local leadership. During the last decade Moscow appears to have counselled prudence and caution: more than a spark was needed to kindle the flame in countries in which the working class was weak and disorganized. The main problem facing the newly independent nations was, in the Soviet view, to overcome backwardness and poverty and to consolidate their political and economic independence. The leaders of the main Communist parties in Latin America, such as Prestes, Corvalan and Arismendi, have echoed these warnings against "adventurism." Only mass movements led by the experienced vanguard party of the working class, armed with Marxist-Leninist theory, could guarantee the victory of the revolution. Or to quote the leader of the Peruvian Communists, Jorge del Prado: international experience has shown that "revolutions are made by the masses" and though "the majority of our people feel the need for radical changes . . . the masses have not yet come to see the need to fight for political power."[93] Venezuela in 1962 was the one major exception to the rule in Latin America but the leaders of the party soon had second thoughts there too and discovered that an armed struggle was after all only an "auxilary form" of the general fight. Orthodox Communists quite justifiably suspected the loyalty of pro-guerrilla elements for the very same Venezuelan leaders who had been most enthusiastic about guerrilla warfare (Bravo, Petkoff, Marquez) were later to display a disturbing lack of loyalty, criticizing Soviet policy in Czechoslovakia, for instance. Eventually they all left, or were excluded from, the party.

The Soviet leaders faced problems whenever they had to deal with radical elements who had come to power after a successful guerrilla war. In both Cuba and Algeria the Communist parties had stood aside initially or had even opposed the struggle. In Cuba the problem was solved, after some minor altercations, by the merger of Fidel's supporters and the local Communists. The Algerian FLN, on the other hand, did not wish to transform itself into a Marxist-Leninist party, and the Soviets had to desist from giving open support to the Algerian Communist party. In Angola, Mozambique and

* On some occasions Soviet support was apparently given to urban terrorist groups in Western countries. Such assistance was not, however, given openly but through various intermediaries such as the Cubans.

Guinea-Bissau there were no Communist parties of standing and, for that reason, no complications arose in the relationship between the Russians and the local guerrillas. In the Arab world the Communist parties supported the Palestinian guerrillas in principle without however actively participating in the fighting. There was political rivalry but it did not reach a critical stage.

Ideological reservations quite apart, their own experience with the Yugoslavs, the Albanians and the Chinese has taught the Soviet leaders to view the guerrillas with concern. One only had to scratch these self-styled Communists to discover that they were fiercely nationalistic underneath. Worse still, they disputed Soviet hegemony over the Communist camp and even pursued policies contrary to Soviet interests. There was a real danger that victorious guerrillas elsewhere would prove no more amenable. On balance, Soviet policymakers found it much easier to cooperate with non-Communist military dictators than with radical revolutionaries unwilling to accept guidance, let alone control.

Soviet spokesmen did stress that they favored the armed struggle provided conditions were right. But this raised the question as to what "ripeness" really meant. According to the orthodox interpretation guerrilla war in Malaya and the Philippines, in Burma and in Greece had failed because the mass base of the insurgents was too narrow. It was admitted, in retrospect, that in Malaya and the Philippines a revolutionary situation had not existed, and the hope that it would come about under the impact of a guerrilla war had been misplaced: "The maturity of the national liberation struggle had been overestimated."[94]

The Chinese were not plagued by such doubts and reservations; if a guerrilla struggle failed it simply meant that the guerrillas had not tried hard enough. Violent revolution, as they saw it, was the universal law. This approach found its extreme formulation in Lin Piao's famous article of 1966 on the international significance of Mao's theory of People's War. Lin Piao wrote that to despise the enemy strategically was an elementary requirement for a revolutionary, for without the courage to despise him and without daring to win, no revolution could be made. All over Asia, Africa and Latin America, where the basic political and economic conditions resembled those of old China, the people were being subjected to aggression and enslavement. Only the countryside provided the broad areas in which revolutionaries could maneuver freely and proceed to final victory: "Taking the entire globe, if North America and Western Europe can be called 'the cities of the world' then Asia,

Africa and Latin America constitute the 'rural areas'. . . . The contemporary world revolution presents a picture of the encirclement of the cities by the rural areas."[95]

Pronouncements of this kind are closely studied in the West without, however, sufficient attention being paid to the fact that there is no greater congruence of theory and practice in China than in the Soviet Union, and that, furthermore, theories change as do the fortunes of those who enunciate them: Lin Piao did not survive his famous article for long. Though the Chinese are committed to support the forces of revolution all over the world, they also have a solemn commitment to coexist peacefully with the countries of the Third World. Chinese experience has shown that guerrillas can create a mass base while fighting an enemy, but this happened in very specific circumstances, during a full-scale war against a foreign invader. With all their belief in voluntarism the Chinese leaders never stated that everyone could start a revolution. Only a truly popular revolutionary movement, with a mass base, would stand a chance of prevailing over its enemies, and therefore a secretly organized coup could be successful in the Third World only in exceptional circumstances.[96] Chinese willingness to support revolutionary movements was not unlimited; it did not, for instance, extend to "urban guerrilla" groups.

In practice Chinese policy was conducted on pragmatic lines: the competition with the Soviet Union for influence in the Third World no doubt played a greater role than did ideological considerations. The Chinese have supported certain guerrilla movements because, in the main, their rivals were assisted by the Russians; the political orientation of the factions was by no means the deciding factor. Thus the Chinese supported ZANU against ZAPU, UNITA and GRAE against MPLA, SWANU against ANC. In the ideological exchanges between Peking and Moscow, the question of revolution and the armed struggle in Africa, Asia and Latin America has been one of the central issues. The Chinese argued that these were the most vulnerable areas under imperialist rule and "the storm centers of the revolution."[97] They accused the Soviets of revisionism, pusillanimity, defeatism and capitulationism, of trying to demoralize the revolutionary movements. But in the final analysis, the policies pursued by the Soviet Union and China vis-à-vis national liberation struggles and guerrilla wars were not that dissimilar, whatever the doctrinal differences.

COUNTERINSURGENCY AND THE INTERPRETATION OF GUERRILLA WARFARE

The spread of guerrilla warfare after 1945, and the many setbacks suffered by Western armies and local government forces against the insurgents, caused much heartsearching among political leaders and military commanders; it also precipitated the emergence of new doctrines of counterguerrilla warfare. In the present context these doctrines are of interest only insofar as they attempt to explain the essence of revolutionary warfare. Since some proponents of these new theories were apparently unfamiliar with previous guerrilla wars and of the history of revolutionary movements, they tended to assume that the phenomenon was of recent date. It was designated "subversive warfare" or "revolutionary warfare," whereas counterinsurgency was termed "modern warfare."[98] In the words of one author, "Mao Tse-tung was the first to treat guerrilla battlecraft as a proper subject of military science and nobody has made a greater contribution to the guerrilla strategy than he. . . ."[99] At the very least Mao was said to have been the first to pull together into a single operational theory the disparate ideas and data previously available, and to have abstracted a set of principles.[100] Such false assumptions about historical origins may well appear of little practical consequence, but this is not so; for Mao, after all, was a "Marxist" and hence the conclusion that the study of Marxism would provide the key to modern, revolutionary, subversive, guerrilla warfare. The works of Marx, Engels and Lenin were subjected to minute examination, all their sayings about war and civil war were collected and analyzed, as though their writings have relevance to what happened in Indochina, Algeria, Cyprus, Kenya, Latin America and the Middle East in the 1950s and 1960s. French generals and colonels were particularly attracted by Mao's thoughts on total war and its political function and they became almost as enthusiastic as the young Chinese Communists waving the Little Red Book at the time of the cultural revolution.[101] A strong philosophical-theological element in French thought developed in the 1950s; this was reflected, for instance, in General Nemo's aside on guerrilla warfare: "There is no true war but religious war." The new doctrine became a new orthodoxy and was accepted by the *Ecole de Guerre* in 1956; the Algerian war was waged according to the lessons learnt in Southeast Asia.[102] The proponents of the doctrine of "modern warfare" were among the first to realize that in the age of the bomb, nuclear war was out of the question, conventional war unlikely, but there were good prospects for revolutionary wars. They noted, again correctly, that the French army (and Western armies in general) were quite unprepared to counter such wars from a military-technical point of view and, even more so, in view of their lack of psychological preparation and political sophistication. As they saw it, the third world war already had begun. There was one enemy from Hanoi to Algiers — international Communism.[103] True, in some instances the enemy was merely the unconscious tool of Communism but this hardly mattered in practice. The *guerre révolutionnaire* could be effectively combatted only if, at an early stage, strong measures were taken against subversion, for otherwise it would lead inevitably to guerrilla warfare. This meant, in practical terms, turning the rebels' organizational weapons and propaganda against them — to combat fire with fire.[104] Some theorists of *guerre révolutionnaire* saw the main problem as the indoctrination of the masses, the conquest of hearts and minds; whereas others, such as Trinquier and Godard, thought that the FLN had succeeded through terror and coercion, and that (Western) propaganda could only succeed once the physical threat was removed. Yet other proponents of this school chose to emphasize the very real grievances underlying the Algerian revolution.[105]

Policies such as those proposed by these theorists could not possibly be carried out within the framework of a democratic society — hence the great frustration they felt. They were sure that they could win the war, but only on condition that they were given a free hand. Politically some of them tended towards right-wing Catholicism, others to "national Communism." While Communism was the enemy from which they wanted to save France, they held the highest admiration for their foes and the greatest contempt for liberal democracy with its self-deception about the nature of the danger facing it, its irresolution and cowardice. One of the leading advocates of revolutionary warfare told an American correspondent that he was a "Communist without doctrine," and Trinquier stated that if he was forced to choose between Communism and international capitalism, he would opt for the former.[106] Ten years later France partly opted out of NATO, ideological-military fashions in Paris changed, and the defense of Christian (or Western) civilization against international Communism in North Africa gave way to close military and political collaboration with yesterday's enemies.

The impact of Maoism on the French generals was by no means unique; perhaps even more dramatic was the influence on the young Portuguese officers fighting in Africa of Frelimo and Cabral's theories. From Frelimo they learned the principles of conspiracy, from Cabral the theory about the "progressive role" of the petty bourgeoisie in the social struggle. Since they too were of petty bourgeois origin this doctrine suited well their social position and their political aspirations. Thus the counterinsurgents of 1973 turned into the "revolutionaries" of 1974, a modern edition of the story of Saulus-Paulus.

American thinking on guerrilla warfare was influenced by the approach of modern political science, mostly behaviorist in character; there was, as one critic noted, a tendency to concentrate on techniques of manipulation and control and administrative measures.[107] There was also the same trend as among the French to see in guerrilla warfare more than meets the eye; the military philosophy of Mao Tse-tung "is much more than it at first seems to be" wrote one author, and another noted that "modern guerrilla movements are armed with elaborate psycho-political weapons."[108] Compared with Pancho Villa or with Zapata these observations were no doubt of a certain validity; in China, Vietnam and elsewhere, guerrilla warfare was not just a localized insurrection or old-fashioned banditry but part of an overall political strategy. But this had been the case, *mutatis mutandis,* in other guerrilla wars in ages past. The key to Mao's success and to that of the Vietminh lay not in the elaborate character of their psycho-political weapons but on the contrary in its simplicity. The failure of Communist guerrilla movements in some countries and the success of non-Communist insurgents in others ought to have been sufficient proof that "Marxism" was not the solution to the riddle. The real explanation is, of course, that the former colonial powers no longer had the strength to hold on to their possessions and, at the same time, classless intellectuals had managed to establish themselves as the vanguard of the masses in the underdeveloped countries. "Since in a backward country all classes of the population with the exception of a thin oligarchic stratum and a few merchants, feel cheated and exploited by foreigners, it is fatally easy to work up a head of steam behind any nationalist movement that promises to end this state of affairs."[109] These national revolutionaries may turn to any radical ideology combining socialist and nationalist elements; they cannot possibly embrace liberalism and democratic rule. Democracy has flowered only in the presence of certain historical conditions which in most Third World countries do not exist and which elsewhere too have progressively weakened.

Some of the writings of American counterinsurgency experts contain much that is of interest and deals with the technique of guerrilla warfare, conspiracy, the preparation of armed insurgency, the motives of guerrilla fighters, the role of propaganda, and other aspects. But for a realistic explanation of the wider political context one looks in vain — not because the military or bureaucratic mind is incapable of understanding the ideological subtleties of highly sophisticated guerrilla movements; there is nothing subtle about them. It is not that these writings are necessarily too prejudiced: on the contrary, there is quite often a tendency to lean over backwards and give the enemy the benefit of all possible doubt. Thus the Field Manual of the U.S. Army on the motives of guerrillas: "Resistance begins to form when dissatisfaction occurs among strongly motivated individuals who cannot further their cause by peaceful and legal means."[110] As if there was no dissatisfaction in every known society, and people unable or unwilling to further their cause by peaceful means. A strong modern dictatorship, whether Communist, Fascist or any other variety, has nothing to fear from these dissatisfied, highly motivated individuals, however deep and justified their grievances. Dissatisfaction there is always, but resistance only has any chance of success against a liberal-democratic regime, or an old-fashioned, ineffectual authoritarian system.

British authors on the subject have been inclined to take an empirical attitude towards guerrilla warfare; while stressing the need for social and political reform, they have reservations about the French and American concepts of psychological warfare. As Julian Paget noted, the cause of the guerrilla has to be simple, inspiring and convincing.[111] If French writers such as Trinquier thought that the support of the population was the *conditio sine qua non* of victory, was it not true that EOKA killed more Greeks than British, and that the same applied, as regards their local populations, to the Malayan guerrillas, the Vietcong, the Fatah, the Mau Mau and others? The great importance of an effective intelligence system and of territorial defense was noted by these authors; the French authorities had not been aware that the Algerian nationalists had established a combat organization in Cairo in 1954 and, as a result, were taken by surprise by the armed insurrection. There was no territorial defense in either Malaya or Algeria; North Africa, an area four times bigger than France, was soldiered by fifty thousand French soldiers, and as a result the French lost Aurès and Kabylia at the very beginning of the rising.[112] McCuen detected three main phases in an urban insurgency: organization, civil disorder, and terrorism. The French author Hogard discerned five stages in guerrilla warfare, with agitators fomenting resentment in the first stage to a general offensive in the last. Brigadier Kitson noted that in most guerrilla wars there was an incubation period of several years; it lasted four years in Cyprus, the Philippines and Kenya, three years in Malaya. It was during this period (of the mobilization of the masses) that the movement was most vulnerable.[113] But in Algeria the incubation period lasted only for eight months, and in Cuba there was no incubation period at all. Robert Thompson, who acquired much experience in revolutionary warfare in Malaya and Vietnam, also pointed to three stages: subversion leading to insurgency, guerrilla warfare, and lastly the takeover. He saw the most vital feature of guerrilla war in organization; he held little sympathy for the view that it was a spontaneous uprising of the people, directed against a repressive, inefficient and corrupt government. The main weakness of the West, Thompson wrote, lay in the attitude of the intellectual community which never gave its own government the benefit of the doubt, even though a Communist regime might prove far more repressive. A similar point was made by the French author Jean Baechler: in a pluralistic political system there was inevitably a party in favor of a negotiated peace. The prime strategy of the insurgents was to try to turn this party into the majority. If the insurgents held out militarily long enough for war-weariness to set in, they would win the war.[114] According to Thompson, the aim of revolutionary war, in contrast to guerrilla war, is political. It might perhaps be more correct to say that in a Communist-led guerrilla movement political and military strategy are more closely connected. For the aim of any war is political even if this is not clearly stated or perceived. Revolutionary war, as Thompson defined it, provides a technique for a small ruthless minority, with neither a good cause nor genuine popular support, to overthrow a government.[115] When the organization was good and the cause weak, the strategy of a protracted war was called for. Thus it was essential to assess at the outbreak of any guerrilla war (i.e., during the second phase of revolutionary war) whether its organization or its cause was the vital factor. If organization were the vital factor, the revolutionary movement could not be defeated by political or social reforms but only by superior organization. Thompson's formula (given here in the briefest detail) is one of the more interesting contributions to the understanding of guerrilla warfare. But it still left a great many questions open, inevitably, perhaps, because reality is always richer and more complicated than any formula, however ingenious. Every movement, revolutionary or not, has a cause, and whether this is "good" or "bad," "strong" or "weak" depends upon a great many factors that defy measurement. It depends above all on the correlation of force — not just military force — in a given country. A gifted leader (or a demagogue) can work up enthusiasm for an almost nonexistent cause. It depends, needless to say, on the political culture of the country. In the early phase of most guerrilla wars, accident is perhaps more important than any other single factor. It is doubtful whether the Chinese Communists would have won if Mao had indeed been killed in the late 1930s, as the Soviet press announced at the time. It is almost certain that the Yugoslav partisans would not have lasted beyond winter 1941 but for Tito, and the Cubans were the first to admit that without Castro the invasion of Cuba would have failed.

Counterinsurgency theorists all agree that guerrilla warfare is cheap, and the fight against it very costly indeed. The budget of the Algerian FLN was about thirty to forty million dollars a year, whereas the French spent a sum of this magnitude in less than two weeks. The cost of killing a single rebel in Malaya was more than two hundred thousand dollars. Writers on counterinsurgency have pointed to the great importance of outside support in guerrilla war, of supply lines and sanctuaries. They noted that guerrilla wars have started almost unnoticed in some countries (Vietnam), and with a big bang in others (Cuba, Algeria). They drew attention to the fact that the size of basic units varied from country to country according to geographical and other conditions; in Cyprus the basic unit consisted of five to eight men, elsewhere it was much larger. The investigations into the origins of guerrilla movements usually raised more questions than were answered. Many writers on the subject have stressed the close connection between guerrilla war and agrarian unrest. J. L. S. Girling maintained that Chinese Communism was based on peasant support, but he also noted that the Chinese (and the Vietnamese) leaders later turned against peasants; in any case, he attributed rural poverty more to overpopulation than feudal abuse.[116] David Galula observed that the slogan "land to the tiller" was unlikely to be very effective in northern China where seventy-six percent of the land was in the hands of the owners, and twenty-two percent more in the hands of part-owners; the redistribution of land, on the other hand, could have been a factor of greater importance.[117] The land grievances on which the propaganda of the Mau Mau leaders focused had been relatively minor. Land tenure was not an issue at all in Guinea-

Bissau. These and other illustrations show that the connection between agrarian unrest and guerrilla warfare is more tenuous than some observers have claimed.

It has been asserted that nationalism alone could not explain the fact that Algerian farmers were ready to risk their lives: they summoned up such resolution only when they felt morally alienated from their rulers.[118] But the great-grandfathers of the Algerian guerrillas had fought for many years under Abd el-Kader without any sense of moral alienation. Liberal observers usually pointed to the link between guerrilla war, social change, and the satisfaction of popular aspirations. This theory has been formulated most succinctly by Eqbal Ahmad:

> Organized violence of the type used in revolutionary warfare is discouraged, rarely breaks out, and so far has not succeeded in a single country where the government made a genuine and timely effort to satisfy the grievances of the people. . . . A regime unwilling to satisfy popular aspirations begins to lose legitimacy. This results in the moral isolation of the incumbents, the desertion of intellectuals and moderates. . . . Popular support for the guerrilla is predicated upon the moral alienation of the masses from the existing government. Conditions of guerrilla warfare are inherent in a situation of rapid social change. The outbreak normally results mainly from the failure of a ruling elite to respond to the challenge of modernization.[119]

This takes us back to the grievance theory. Unfortunately, grievances are part of the human condition: they always exist, however perfect the society. Furthermore, there is no way to measure the intensity with which grievances are felt. Even if a "grievance scale" did exist, it is by no means certain that a revolutionary war is more likely to break out in a country replete with grievances. Guerrilla war succeeded in Cuba but failed in other Latin American countries despite the fact that Cubans had less objective reason to feel aggrieved than many other Latin Americans. The guerrilla victories in Yugoslavia and Albania (and in many other countries) had nothing whatsoever to do with modernization; rapid social change was not the issue in China in 1940, or in Vietnam in 1950. A democratic, or semidemocratic regime unwilling to satisfy popular aspirations, indeed gradually loses legitimacy. A totalitarian regime, on the other hand, can afford to disregard popular aspirations without fearing that it will be "morally isolated," or that the moderates and intellectuals will desert it. If revolutionary war has usually failed in democratic societies this does not mean that the grievances are insubstantial. It simply shows that some societies are less violent than others.[120] History has demonstrated that guerrilla war stands a better chance of success against foreign domination than against one's own kind — nationalism is, by and large, the single most potent motive force. But nationalism *per se,* pure and unalloyed, is an abstraction; in the real world it appears only in combination with other political and social concepts, and programs. It is in this context that the infusion of radical — not Marxist — ideas takes place.

9

A Summing Up

Guerrilla wars have been fought throughout history by small peoples against invading or occupying armies, by regular soldiers operating in the enemy's rear, by peasants rising against big landowners, by bandits both "social" and asocial. They were infrequent in the eighteenth century, when strict rules for the conduct of warfare were generally observed. Guerrilla methods were used in the southern theater in the American War of Independence and in the Napoleonic age by partisans in countries occupied by the French (Spain, southern Italy, Tyrol, Russia). With the emergence of mass armies in the nineteenth century, guerrilla warfare again declined but it lingered on in the wake of major wars (the American Civil War, the Franco-Prussian War of 1870–1871, the Boer War) and in the campaigns of national liberation movements (Italy, Poland, Ireland, Macedonia). Furthermore, guerrilla tactics played an important role in nineteenth-century colonial wars of which the campaigns of the French against Abd el-Kader and the Russians against Shamyl were the most noteworthy. In all these instances the guerrillas failed to achieve their aims except when acting in cooperation with regular armies. The imperial powers, as yet unfettered by moral scruples about the inadmissibility of imposing their rule on lesser breeds, were not deflected from their policies by pinpricks: the Russians did not withdraw from Poland, the Caucasus or Central Asia, the French did not give up North Africa, the British did not surrender India, and if the Italians attained their independence, it was not as the result of a protracted guerrilla campaign. There was not one case of outright guerrilla victory, but in some instances guerrilla campaigns indirectly contributed to eventual political success. Thus, the military outcome of the Cuban insurrection in the late nineteenth century was inconclusive, but by fighting a protracted war the rebels helped to trigger off U.S. intervention which led to the expulsion of the Spanish. The tough struggle of the Boers after their regular armies had collapsed hastened the British decision to grant South Africa a large measure of independence. In Latin America guerrilla war continued to be the prevailing form of military conflict in the absence of strong regular armies.

The First World War saw mass armies pitted against each other; the few instances of guerrilla war (Arabia, East Africa) occurred in minor theaters of war and were certainly not ideologically motivated. The Mexican, the Russian and the Chinese civil wars of the twentieth century saw a good deal of partisan warfare but mainly because neither side was strong enough to mobilize, train and equip a big regular army. Guerrilla war in these circumstances was not so much the war of the weak against the strong, but of the weak against the weak. Revolutionary movements had not yet opted for the guerrilla approach; before the Second World War the prospects for the anticolonial struggle were as yet unpromising. The Soviet Communists established a large regular army as quickly as they could after the revolution; twenty years later the Chinese Communists tried to do the same, though in their case the guerrilla phase was to last much longer.

With the Second World War there came the great upsurge in the fortunes of guerrilla warfare. Hitler's predicament resembled Napoleon's insofar as his forces were dispersed all over Europe and his lines of communication and routes of supply overextended and vulnerable. Like Napoleon before, the Germans had insufficient forces to impose full control on all the occupied territories or even to destroy partisan concentrations. On the other hand, the military importance of the Second World War partisan forces was not very great and did not decisively influence the course of the war. Their main impact was political inasmuch as they resulted in the emergence of Communist governments (Yugoslavia, Albania) or caused protracted civil wars (Greece, Philippines). The European colonial powers, gravely weakened as a result of the war, lacked the financial and military resources and the political will to retain their overseas possessions against the rising tide of independence movements. Public opinion in the metropolitan countries which had once regarded the possession of colonies as a source of pride was no longer willing to shoulder the military and financial burden; imperialism became morally reprehensible. This turn in Western public opinion was of decisive importance for the success of Asian and African national liberation movements. In the Far East and some African countries the leadership of the independence movements was taken over by Communist or pro-Communist forces. Their superior organization and an ideology which corresponded to the cultural level and the emotional needs of the population made them better equipped to act as agents of modernization than their political rivals. Nevertheless, the wars of liberation in Asia and Africa were fought without exception under the nationalist rather than the Communist banner; even in the countries of Latin America, which had been independent for almost a hundred and fifty years, the guerrilla campaigns of the 1960s had strong patriotic undertones.

Guerrilla warfare has not only been practiced since time immemorial, its doctrine too is by no means of recent date. The many illustrations provided in these pages show that the notion that the theory and practice of guerrilla warfare was invented in China in the 1930s is altogether erroneous.[1] Guerrilla techniques were exhaustively described by de Jeney, Decker and other eighteenth- and nineteenth-century authors. The experience gained during the Napoleonic wars provided more systematic and more detailed analyses and prescriptions. Le Miere de Corvey and in particular the Italian and Polish military writers of the 1830s and 1840s were fully aware of the political aspects of guerrilla warfare. Their writings cover almost all the problems that were to preoccupy twentieth-century guerrilla authors — the importance of bases and sanctuaries, the questions as to whether the war would be short or protracted, whether it should be "pure" guerrilla war or be conducted in coordination with regular forces, whether guerrilla units should be gradually transformed into a regular army. Even the relationship between the guerrilla forces and the political movement supporting it was discussed in the writings of Carlo Bianco and Mazzini. These precursors fell into oblivion; Mao and Ho Chi Minh, Castro, Guevara and Debray were not in the least aware of the fact that their ideas had been expounded before and even tried, albeit not very successfully. The twentieth-century guerrilla theorist discovered his strategy quite independently, based on his own experience, instinct, and, of course, native traditions of guerrilla war of which there were more than enough in both Asia and Latin America. As Eric Hobsbawm has noted, there is nothing in the purely military pages of Mao, Giap or Che Guevara which a traditional *guerrillero* or band leader would regard as other than simple common sense.[2] If so, the novelty of twentieth-century guerrilla warfare would seem to be not so much military as political. The author of a valuable recent study has maintained that revolutionary guerrilla war evolved out of Marxist-Leninist modes of political behavior and organizational principles on one hand, and out of the exigencies of anti-Western revolt in predominantly agrarian societies on the other.[3] In the light of the historical evidence this thesis is tenable only subject to far-reaching reservations. The character of guerrilla warfare, needless to say, has changed greatly over the ages, partly through technological developments, partly as a result of changing social and political conditions. But it cannot be maintained that before the 1930s guerrilla wars were apolitical and parochial.[4] Too much importance has been attributed to Leninist doctrine in the guerrilla context, too little to the nationalist-populist component in the motivation and the ideology of these movements. (Populism is used in this context not in the nineteenth-century meaning of the term but as rural and urban opposition to class differentiations and the capitalist form of modernization.) Many twentieth-century guerrilla wars from Pancho Villa to the Mau Mau and IRA, from IZL, Fatah to EOKA owe little to Marxist-Leninism. Neither the Algerians in 1954 nor the Cubans in 1958 were influenced by this doctrine and even Chinese and Vietnam guerrilla warfare evolved more in opposition to classical Marxism than in accordance with its basic tenets. The impact of Marxism-Leninism among contemporary guerrilla movements has been strongest with regard to the role of the political party in mobilizing the masses, the function of propaganda in the struggle, and the emphasis placed, generally speaking, on organization. But political propaganda and organization were not altogether unknown in previous ages. More women have participated in modern guerrilla movements than in the past, but this again is by no means an unprecedented development.[5]

These new developments in the character of guerrilla movements should not be belittled but nor should their ideology be regarded as the master key to their understanding. Communist guerrilla movements have failed, non-Communist groups have succeeded. The importance of guerrilla movements was underrated for a long time; more recently the pendulum has swung to the other extreme, and the general tendency has been to exaggerate their historical role.

Attempts to explain the causes of guerrilla warfare and of guerrilla success have certainly enriched the political language, but they have not greatly contributed to a clarification of the issues involved. "Revolutionary warfare," defined as "partisan warfare plus political propaganda" is an unfortunate formulation which has nevertheless gained wide currency. "Subversive warfare," "internal war," "low intensity warfare," "modern unconventional warfare," "people's war," "subversive insurgency," "guerrilla insurgency," "irregular warfare" — these and many other terms have been used without adding precision or helping our understanding of the phenomenon.[6] There is a wide range of theories to choose from; some take as their starting point the questionable assumption that insurgency is "deviant social behavior" — as if acceptance of foreign occupation or domestic tyrants is the norm, and the decision to oppose them a deviation. Nor is it permissible without qualification to regard political and social harmony as the norm and conflict and violence the unfortunate exception to the rule, to be explained by excessive aggression or ambition, or by deprivation, absolute or relative. The very asking of the question "Why do men rebel?" implies a great many assumptions about both human nature and the perfectibility of society. Quantitative techniques have been of little help in understanding the guerrilla phenomenon, partly because some of the essential ingredients involved cannot be measured, partly because the differences between guerrilla movements are such that even measurable factors become meaningless when added or multiplied. A formula encompassing both Mao and Castro (let alone Pancho Villa, the IRA and Mau Mau) will be of no help as an analytical tool. A comparison between China and Vietnam, or between Angola and Mozambique, or even between the IRA and the Basques may be of value and interest. Moving further afield in time or space, generalizations can be made only with the greatest of caution; for the guerrilla phenomenon presents endless variety. Some were Communist inspired, others were not; some were led by young men, some by old; some of the leaders had military experience, others lacked it entirely; in some movements the personality of the leader was of decisive importance, in others there was a collective leadership; some wars lasted for a long time, others were short; some bands were small, others big; some guerrilla movements transformed themselves into regular armies, others degenerated into banditry; some benefitted from external circumstances (such as difficulty of terrain, or the presence of dissatisfied national minorities), others on the contrary derived no advantage at all from such conditions. Some won and some lost. The possibilities are endless; so are the theories, hypotheses and concepts, monocausal and multicausal, to explain guerrilla warfare, ranging from those stressing socioeconomic conditions to others putting the emphasis on political-psychological factors. The theory that has certainly gained widest currency is the grievance-frustration concept, which has been accepted, in various forms, by liberals and Marxists alike. Men and women will not rebel, risking their lives and property, without good reason — the occupation of their country by foreign armies, economic crisis, a tyrannical political regime, great poverty, or great social discrepancy between rich and poor. The concept seems plausible enough; it can indeed be taken for granted that if people had no grievances and felt no frustration, they would engage in the pursuit of happiness. But the nature of grievances and of frustration is not at all an easy problem.* It has been argued that, if governments only fulfilled popular aspirations, they would not lose legitimacy and there would be no violent opposition and even the intellectuals would happily sing their praises. Unfortunately, the principle of virtue rewarded applies no more to politics than to private life. The state, however well-meaning, may face difficulties through no fault of its own and, as a result of this, its inhabitants will have to suffer. The resources of a government may be limited, it may have to establish priorities, thus discriminating against some people. Nor is there any reason to assume that a state or a social system can be more perfect than the individuals constituting it. It has been argued that the most traditional and the most modern societies are relatively immune to upheaval whereas those in between suffer from instability. As far as rich countries are concerned this is merely stating the obvious; the rural population in these countries is usually small and would-be guerrillas would not be welcomed with enthusiasm in the American corn belt, or among British, French, German or Danish farmers. It would be equally difficult to launch a guerrilla movement in a country in which the bulk of the population suffered from acute starvation and endemic disease, since there might not be enough men and women able to march and to fight. In actual fact, there has been a great deal of guerrilla warfare in very poor countries such as the Congo, Sudan, Oman and Eritrea.

* The inhabitants of Calabria and Basilicata have every reason to feel aggrieved and frustrated for these are among the very poorest provinces of Italy. But they are at the bottom of the scale inasmuch as the rate of suicide and the crime rate are concerned whereas prosperous Piemont, Lombardy and Liguria are on top. Social statistics in other countries show a similar picture.

An extremely unequal distribution of wealth and an acute economic crisis have frequently aggravated political unrest, but it so happens that neither factor was decisive in the major guerrilla wars of the nineteenth and twentieth centuries. Agrarian grievances have played a paramount role in Mexico and the Philippines but elsewhere guerrilla war took place in countries with enough land for all (Africa) or in predominantly urban societies (Venezuela, Uruguay). Guerrilla warfare has occurred in countries with low population density and high density alike, in societies with high social mobility and low mobility with a high rate of literacy and a low rate.

Frequently the break up of traditional societies and the process of modernization has been considered the main agent: social change results in insecurity and the loss of identity, attempts at reform weaken the government's political position. This has led observers to opposite conclusions. On the one hand governments have been advised to go full steam ahead with social reform programs, on the other they have been counseled to slow down so as to reduce the impact of the negative political effects of social change. There is general agreement that socioeconomic improvement does not immediately result in increased popular support; these programs usually give tangible political results only after many years, even decades — except perhaps in the case of a radical redistribution of land. Demographic pressure, growing ecological disequilibrium and the weakening of social ties connecting hinterland and center have been mentioned as important factors conducive to insurgency and revolution. These processes are part of the general crisis in the Third World, but again there is no obvious connection between them and the spread of guerrilla warfare.

So after a great many detours and false scents, the student of guerrilla wars finds himself back at his starting point. If a government has the support of the people it will not be challenged and overthrown. Or, to put it more obscurely, social change will be peaceful if the ruling elites respond to the needs of repudiating the old institutions and relationships and creating new ones. If they fail to do so, it is argued, political violence becomes inevitable. Effective and virtuous governments have nothing to fear, corrupt and ineffective ones are doomed. But "corruption" and "ineffectiveness" are by no means synonyms and in any case the relevance of this thesis to insurgency and guerrilla warfare is not at all obvious. The new institutions and relationships if established may be rejected by part of the population. There may be real but unsoluble grievances, such as the separatist demands of minorities, that would result in the crippling of society and the emergence of a nonviable state.

Other theories have put the emphasis on politics (the presence of conflicting social myths), on cultural-political-moral factors (such as the alienation of intellectuals), or on psychological moments ("terrorist personality").[7]

These theories help to shed some light, at best, on one or a few insurgencies and are quite inapplicable to others. In short, they are quite useless. This sad state of affairs has not escaped the attention of social scientists. One of them has concluded that the actual instances of insurgency observed in our time fail to reveal a correlation with the gravity of socioeconomic, cultural and related ills.[8] Another has said that a Western doctrine on guerrilla war comparable to the principles of war between nations has not developed because the character of insurrections is largely determined by the peculiar social structure of the society in which it takes place.[9] Guerrilla movements, in other words, are an awkward topic for generalization. Yet another observer decrying the "chaotic and inadequate" state of existing etiologies of internal war has pointed to more promising venues to be explored.[10] At the bottom of every protest movement there is a feeling of grievance. But how to measure these grievances, how to account for the fact that at one time a major grievance may be fatalistically accepted, whereas at another time (or place) a minor grievance may produce the most violent reaction? Is it not the perception of the grievance that matters rather than the grievance *per se*? How to explain that conditions perceived as tolerable at one time (or in one country) become intolerable at another time (or in another country)? Such a change in attitude could be produced by a variety of circumstances — the accumulation of reasons for discontent, or a successful revolution in another country (the "echo effect") or the emergence of a new leader, or a new generation of leaders who are driven by a greater sense of justice or ambition or fanaticism than their predecessors. But if such analysis is difficult enough with regard to the spread of political violence in one specific country, it is impossible on a worldwide scale. The great differences in the prevailing conditions are too deep and too numerous to be digested in cross-national surveys.

More fruitful perhaps is the suggestion that the obstacles to internal war should be examined. Students of guerrilla war have almost invariably concentrated on insurgents rather than on incumbents, on the forces which propel societies towards violence, rather than those which inhibit it. This omission explains to a certain extent the misunderstandings that have prevailed among some Western writers. It has been argued both on the left and the right that guerrillas are "invincible" ("Regular armies have almost never succeeded in gaining the ascendancy over guerrilla operations of any importance" — Colonel Nemo).[11] It is easy to understand this pessimism in the light of recent French history but it does not at all correspond with the experience gained elsewhere. Liberal democracies and in particular ineffective authoritarian regimes indeed found it impossible to cope with colonial insurrections in the 1950s and 1960s. Other political regimes have suppressed guerrilla insurgencies with great ease. Guerrilla movements are, as Mao said, the fish that needs water — the water being a minimum of freedom. Such freedom exists if the government is relatively liberal or relatively inefficient. If government control and coercion is really effective, a guerrilla movement cannot possibly develop as the Communist and Fascist experience has shown. Some governments are inhibited in their action by public opinion, others are not. The Iraqi government liquidated the Kurdish rebellion in 1974 with great ease yet the British government failed to suppress the civil war in Ulster. This did not mean that the Iraqi government had greater legitimacy or that it was more virtuous.

Dictatorships, needless to say, are not free of grievances, for all one knows they may be even more acutely felt there. Yet there is no outlet for them; the rebel will be arrested, sent to prison and perhaps shot. His arrest will not be reported by the mass media, it will have no political consequences, his sacrifice will be in vain. If intellectuals are alienated, they keep the fact strictly to themselves for fear of losing their jobs — or worse. Much of Western guerrilla literature is curiously parochial in its stress on the importance of public opinion; in a great and growing number of countries there is no public opinion — or it has no way of expressing itself. It is hardly necessary to describe in detail the means of control and coercion which make resistance in an effective dictatorship very nearly hopeless as long as the leader (or the leaders) have not lost their self-confidence and feel no compunction in using their means of repression to the full. The argument that repression is a two-edged sword, that guerrillas always benefit from government repression, that power is weakest when it uses violence (Merriam) applied perhaps to pre-modern dictatorships, and to liberal regimes whose powers of repression are strictly limited. By and large it is no longer true.

Not only have sociologists and political scientists found it impossible to come to terms with the guerrilla phenomenon, lawyers have encountered the very same difficulties. Since the Second World War many attempts have been made to establish a new basis of legality, a more humanitarian status, for guerrilla forces under the laws of war. For many decades the status of the partisan was based on the Brussels Declaration of 1874 and the Hague Convention. Guerrilla tactics, meaning hostile activities committed by small bodies of soldiers in the enemy's rear during a real war were considered legal, whereas guerrilla war was not. According to this argument organized resistance had ceased and the individuals who engaged in guerrilla war were not bound by the laws of war. Thus private individuals were entitled to commit hostilities against the enemy — international law being a law between states could not issue prohibitions to private individuals. But these individuals did not enjoy the privileges of members of armed forces and the enemy had the right to consider them as war criminals.[12] The Geneva Conventions of 1949 tried to legalize the status of the partisans in internal conflicts, but the lawyers could not agree on what constitutes a state of war and the question whether insurgents could possibly be bound by a convention which they had not themselves signed.[13] For a predominantly terrorist movement the acceptance of the enemy's rules of war would be a negation of their whole strategy. But even guerrilla movements with a lesser emphasis on terror might not be in a position to adhere to the rules of war, for instance with regard to prisoners. A dictatorship may want to give a reward to captured guerrillas rather than execute them, but it will be guided in its actions by expediency rather than law. Guerrilla wars are conducted brutally and Western officers and soldiers have been guilty of excesses and even torture. But in democracies such practices quickly become known, they are denounced and have to be discontinued; in the struggle between a democratic government and a guerrilla movement at least one side is bound by law; in a dictatorship neither is, hence the failure to apply the principles of international law.

There is no theory which can predict the course of guerrilla war, and there is no reason to assume that there ever will be. Concepts and definitions have been postulated but usually this was simply "retranslating from one language of definition to another without hypothesizing anything."[14] One such recent conceptual scheme differentiates between "truly qualitative insurgency" and "unsophisticated armed uprisings," stressing that the basic thrust of a "qualitative insurgency" is psychological, that the guerrilla force component by its very design is not geared to win a military victory, and that there were altogether five insurgencies that correspond with this "sophisticated" pattern (China, Cuba, the two Vietnam wars and Thailand). Such schemes tend to restate the obvious in somewhat arcane language; they are quite harmless but for the tall claims made for these exercises ("A qualitative insurgency becomes an activity lending itself to systematic and reliable analysis"[15]).

Even the broadest formula cannot possibly cover all guerrilla wars. It may be difficult to improve on the definition provided by Professor Huntington:

> Guerrilla warfare is a form of warfare by which the strategically weaker side assumes the tactical offensive in selected forms, times and places. Guerrilla warfare is the weapon of the weak. . . . Guerrilla warfare is decisive only where the anti-guerrilla side puts a low value on defeating the guerrillas and does not commit its full resources to the struggle.[16]

This definition certainly does not apply to Castro's campaign nor can it be maintained that the French in Algeria or the Portuguese in their African colonies put a "low value" on defeating the guerrillas. Various guerrilla movements have succeeded without taking any offensive at all, simply by outlasting the enemy. Guerrilla war is decisive only when the antiguerrilla side is prevented for military or political reasons from committing its full resources to the struggle.

The concept of stages in the preparation of an insurrection figures highly in the writings of counterinsurgency specialists. It implies that the outbreak of guerrilla war is preceded by an incubation period in which the emphasis is on organization, propaganda and conspiracy. That an insurgency in modern times cannot be launched without some form of organization goes without saying: there have been few, if any "spontaneous uprisings of the people." It is also true that the guerrilla movement is particularly vulnerable in the period of early mobilization. But it is equally true that such an organization, be it Communist or nationalist, usually exists already and that not that much preparatory work is needed for launching an insurgency. There was no incubation period in Yugoslavia during the war, or in Cuba, or in many African and Latin American countries. The concept of stages is the result of the Southeast Asian experience but what may be true in Vietnam, Malaya and the Philippines does not necessarily apply to other parts of the world.

Such criticism of existing definitions and models is not based on the assumption that more refined techniques would have resulted in superior insights. The multiple "objective" and "subjective" factors involved in guerrilla warfare and their complicated interaction rule out all-embracing formulas and explanations that are scientific, in the sense that they have predictive value. To recognize these limitations is not to deny that certain patterns are common to many guerrilla movements and that a study of these patterns could be of help in understanding why guerrilla wars have occurred in some conditions but not in others, and why some have succeeded and others have failed. The following attempt to summarize experience is concerned with probabilities not certainties.

1. The geographical milieu has always been of importance. Guerrilla movements have usually preferred regions that are not easily accessible (such as mountain ranges, forests, jungles, swamps) in which they are difficult to locate, and in which the enemy cannot deploy his full strength. Such areas are ideal in the early period of guerrilla warfare, during the period of consolidation, and they retain their uses later on as hideouts in a period of danger. In such areas the guerrillas will be relatively unmolested, but at the same time there are obvious drawbacks. If the enemy has to undergo the hardships of a mountain climate, the guerrillas, too, will have to suffer. It is difficult to obtain food and other supplies in distant, sparsely populated areas. Restricting their operations to these regions the guerrillas will be safe but they will be ineffective, for they will be able to harass only isolated enemy outposts, they will not be in a position to hit at the main lines of communication and they will lose contact with the "masses." Thus the ideal guerrilla territory while relatively inaccessible should be located not too far from cities and villages. Of late, topographical conditions have lost some of their erstwhile importance. On the whole it has become easier for the antiguerrilla forces to locate the rebels. Furthermore, with the rapid progress of urbanization, the countryside has lost much of its original political importance. The village cannot encircle the city if the majority of the population resides in urban areas. For this reason, and for some others, the main scene of guerrilla operations has shifted from the countryside to the city in predominantly urban societies, with a simultaneous shift in strategy from hiding in nature to finding cover in town.

Guerrilla movements need bases and they cannot operate without a steady flow of supplies. Ideally a sanctuary should be on foreign territory outside the reach of the antiguerrilla forces. Bases are needed for guerrilla units to recover from their battles, to reorganize for new campaigns and for a great many other purposes. While movement is one of the cardinal principles of guerrilla tactics a guerrilla unit is not a *perpetuum mobile*. The main drawback of a base is that it offers a fixed target for enemy attack. Thus guerrillas may be compelled to change their bases from time to time, unless they have established "liberated zones" which the enemy, with his resources overextended, can no longer destroy. The question of supply was not of decisive importance before the nineteenth century, when guerrillas (as regular armies) lived off the land, when weapons were unsophisticated and could be locally manufactured. The more complicated the arms, the greater the guerrillas' dependence on supply routes, frequently from abroad. There are but two cases in recent history in which major guerrilla armies survived and expanded without outside supply of arms — China and Yugoslavia. But this was exceptional in that these guerrilla armies came into being during a general war that offered many opportunities of acquiring arms. The decisive victories of Mao's army and of Tito's partisans came only after they had the opportunity of rearming themselves from outside sources.

2. The etiology of guerrilla wars shows that it very often occurs in areas in which such wars have occurred before. The Spanish war against Napoleon took place in the same regions in which Viriathus and Sertorius had fought the Romans. Guerrilla wars had occurred before (or had even been endemic) in the Tyrol, Greece, Serbia, Montenegro, Algeria, the Philippines, north China, North Vietnam, Cuba, Mexico and other countries. This may be partly due to geographical factors, for these are all regions that favor guerrilla warfare. It is also true that the hold of the central government over peripheral areas with a long guerrilla tradition such as Oriente in Cuba, Kabylia in Algeria or Nghe An province in Vietnam has never been as strong as on more centrally located districts. Furthermore, there are cultural traditions favoring or militating against large-scale political violence. Beyond a certain stage of cultural development it is difficult for a guerrilla movement to gain mass support. Neither the middle class nor workers and peasants in civilized countries feel sufficient enthusiasm to "go to the mountains" even at a time of grave crisis. What Engels wrote in 1870 — that our tradition gives only barbarians the right of real self-defense and that civilized nations fight according to established etiquette — is *a fortiori* true

now. Even in the case of foreign invasion and occupation the great majority of the population in a civilized country will not engage in a war risking total destruction.

3. To this extent there is a (negative) correlation between guerrilla warfare and the degree of economic development. There have been few peasant guerrilla wars in modern times in which acute agrarian demands constituted the central issue (Mexico, the Philippines). On the other hand, in many more countries the peasantry has been the main reservoir of manpower for guerrilla armies led by nonpeasant elites. The breakdown of traditional peasant society under the pressure of capitalist development, absentee landlordism, demographic pressure, falling prices for agricultural produce, natural catastrophes and other misfortunes have created in many Asian countries (and to a lesser extent in Africa and Latin America) conditions in which there has been great sympathy among poor peasants, landless laborers, but also middle peasants for popular movements promising land to the landless, even if this promise was not the immediate issue in the war. The difficulty facing the guerrilla leaders has always been to harness this revolutionary potential on a nationwide basis in view of the traditional reluctance of peasants to fight outside their neighborhood. This could mostly be achieved only in the framework of a national struggle transcending the parochial framework such as a war against a foreign enemy (China, Algeria).

4. Throughout the nineteenth and twentieth centuries there have been three main species of guerrilla wars. They have been directed against foreign occupants, either in the framework of a general war or after the defeat of the regular army and against colonial rule. Secondly, guerrilla warfare has been the favorite tactic of separatist, minority movements fighting the central government (the Vendée, IMRO, IRA, ELF, the Basques, the Kurds, the FLQ, etc.). And thirdly, guerrilla warfare against native incumbents has been the rule in Latin America and in a few other countries (Burma, Thailand, etc.). But the national, patriotic element has always been heavily emphasized even if domestic rulers were the target; they were attacked as foreign hirelings by the true patriots fighting for national unity and independence. In China, Vietnam, Yugoslavia, Albania, Greece, the Philippines and Malaya partisan units were established to fight foreign occupants but they became civil war forces with the end of the general war. Throughout the nineteenth century the achievement of national independence has been the traditional goal of guerrilla movements; more recently

social and economic programs have featured prominently. But the patriotic appeal has always played a more important role than social-revolutionary propaganda. Castro's war was fought for the overthrow of Batista's tyranny; most Latin American guerrilla movements have stressed general reform programs rather than clearly defined socialist-Communist slogans in their fight against domestic contenders. As the outcome of these wars show, guerrillas succeed with much greater ease against foreign domination than against native incumbents.

5. The character of guerrilla war has undergone profound changes during the last two centuries, but so has regular war on the one hand, and the technique of revolution on the other. However, there is no justification for regarding modern guerrilla warfare (or "people's war," or revolutionary insurgency) as an entirely new phenomenon which has little connection with the guerrilla wars of former periods. Organization (the role of the political party) and propaganda play an infinitely more important role in present day guerrilla war than in the past, and it is of course true that in some Third World countries guerrilla war is merely one stage in the struggle for power. Guerrilla war was never "apolitical," it was always nationalist in character and became national-revolutionary in an age of revolution. Too much importance has been attributed to the use of Marxist-Leninist verbiage on the part of Third World liberation movements. This has led Western observers to interpret their progress either in terms of a worldwide Communist conspiracy or as a great new liberating promise. While the common denominator of most of these Third World movements is anti-imperialism and the rejection of the capitalist form of modernization, the ideology guiding them is a mixture of agrarian populism and radical nationalism (with "nationalism" and "socialism" often interchangeable). Such political movements have certain similarities with European Communism (dictatorship, the role of the monolithic party) but on a deeper level of analysis they are as distant from socialism as from liberal capitalism. Elsewhere the basic inspiration for guerrilla warfare has been sectarian-separatist (religious-tribal) with revolutionary ideology as a concession to prevailing intellectual fashions and modes of expression.

6. The leadership of nineteenth- and early twentieth-century guerrilla movements was usually in the hands of men of the people (Mina, the Empecinado, Andreas Hofer, Zapata, the Boer leaders, the IMRO). In backward countries they were traditionally led by tribal chiefs or religious dignitaries. More recently they have

become, by and large, the preserve of young intellectuals or semi-intellectuals; this refers particularly to Latin America and Africa with only a very few exceptions (Fabio Vasquez, Samora Machel).

The social origin of the twentieth-century guerrilla elite in Latin America and also in Asia and the more backward European countries is usually middle class, especially the administrative stratum (the "lower mandarins") which has no independent means of its own. Equally frustrated by their own limited prospects and the real or imaginary plight of their country, they have opted for revolutionary violence, the transformation of an old-fashioned, ineffective autocracy into a more modern, more effective and by necessity also more despotic regime. To seize power, the civilian intelligentsia transforms itself into the military leadership. A formula of this kind does not apply to every single guerrilla movement, even less to all of its leaders; nor does it do justice to the idealistic motivation of leading guerrilla cadres. But in historical perspective this has been the political function of radical guerrilla movements. Students were hardly represented in the classical guerrilla movements, more recently their share has been very prominent indeed, and the greater their role, the more radical the character of the guerrilla movement; this is shown for instance by a comparison between Fatah and PFLP, between the Angolan FLNA and the MPLA. Military men have occasionally appeared as prominent guerrilla leaders: Denis Davydov, Yon Sosa and Turcio Lima in Guatemala, the young Prestes and later on Carlos Lamarca in Brazil, Grivas, Kaukji Mihailovic, the Wamphoa graduates among the Chinese Communists. Some guerrilla leaders had limited military experience, for example the Vendéans and Spanish guerrillas who fought Napoleon, the Yugoslav Communists who participated in the Spanish Civil War, Nasution, the Algerians who fought in the French army. But the most important guerrilla leaders of our time, including Mao, Tito, Giap, Castro, Guevara, as well as the foremost theoreticians among them, were self-made men in the military field. Most guerrilla leaders were in their late twenties or early thirties when they launched their campaigns, old enough to impose their authority, uniting the experience of age and the activity of youth, and capable of withstanding the exertions of guerrilla warfare. Some, however, were already in their forties (Tito, Mao), and some even older (Grivas, a few of the Boer generals, Chu Teh, Marighela). Few manual workers have joined guerrilla movements (Korea and Malaya were significant exceptions) and even fewer emerged as guerrilla leaders. Guerrilla leaders, certainly the more successful among them, have always been strict disciplinarians. What Gibbon wrote about Skanderbeg applies to most of them: "His manners were popular but his discipline was severe and every superfluous vice was banished from his camp."

7. Social composition: Attention has been drawn to the fact that peasants traditionally constituted the most important mass basis of guerrilla movements, but conditions varied considerably from country to country even in the nineteenth century and there have been further changes since. The Chouans and the Spanish guerrilla units fighting the French came almost exclusively from rural areas, and the same applied, of course, to the Boer commandos and the Zapatistas. On the other hand there was not a single peasant among Garibaldi's "Thousand." IMRO was initially overwhelmingly a rural movement, whereas the IRA derived most of its support from the cities; IZL and the Stern Gang (in contrast to Hagana) were almost exclusively city-based. Smugglers, poachers, bandits and various déclassé elements played a significant part in certain nineteenth-century guerrilla movements (southern Italy) and also in Latin America and the Far East. Usually the smaller the guerrilla army, the larger the middle-class element. This applies above all to the Cuban revolution and the various urban guerrilla groups such as the Tupamaros. Women have participated in almost all guerrilla movements. They have been most prominent in the small urban guerrilla groups (West Germany, the U.S.) and in Korea (more than a quarter of their total force). Available data are insufficient to establish whether the occupation of insurgents reflects the occupational pattern of the population as a whole. This may have been the case in some countries (Philippines, Algeria) but not in others (Latin America). A poll taken by the French during the first Vietnam war showed that almost fifty percent of their prisoners were classified as "petty bourgeois," and in African guerrilla movements, too, the urban petty bourgeoisie was apparently represented far above their share in the population. The small urban guerrilla movements are preponderantly constituted of students, or recent students, the IRA being the one major exception.

8. The motives that have induced men and women to join guerrilla bands are manifold. Historically, patriotism has been the single most important factor — the occupation of the homeland by foreigners, the resentment directed against the colonial power — often accompanied by personal grievances (humiliation, material deprivation, brutalities committed by the occupying forces). Secessionist guerrilla movements have based their appeal on the discrim-

ination against and the persecution of ethnic or religious minorities. Guerrilla movements fighting domestic contenders stress obvious political or social grievances such as the struggle against tyranny, unequal distribution of income, government inefficiency, corruption and "betrayal," and, generally speaking, the "antipopular character of the ruling clique." Land hunger, high interest rates (Philippines), the encroachment by the *haciendados* on Indian land (Mexico) have been important factors in predominantly agrarian societies. On top of these causes there has been a multiplicity of personal reasons ranging from a developed social conscience to boredom, the thirst for adventure and the romanticism of guerrilla life to personal ambition — the expectation of bettering oneself socially or of reaching a position of power and influence. The dynamic character of guerrilla movements has always exerted a powerful attraction of young idealists — the prospect of activity, of responsibility for one's fellows, of fighting with equally enthusiastic comrades for the national and social liberation of the homeland. As Maguire wrote seventy years ago and Denis Davydov well before him, a partisan must be a kind of military Byron, his enterprise requires a romantic imagination. What induces guerrillas to stay on is above all *ésprit de corps*, loyalty to his commander and fellow soldiers. The feeling of togetherness and team spirit seems to be more important than ideological indoctrination. Guerrilla warfare usually opens larger vistas to personal initiative and daring than regular warfare; it has been said that slavish imitation produces good military tailors but not guerrilla leaders. But the motives are by no means all idealistic; guerrilla war is an excellent outlet for personal aggression, it provides opportunities for settling accounts with one's enemies, and conveys a great sense of power to those hitherto powerless. While sadism has never been official guerrilla policy, there has always been more deliberate cruelty inflicted in guerrilla wars than in the fighting of regular army units, subject to stricter discipline. This is true for the partisan wars of the Napoleonic period and also for many subsequent guerrilla wars. The gentlemanly guerrilla war has been a rare exception (the Boer War); on the other hand there were many guerrilla wars in which sadism was established practice (IMRO, IRA, Arab and African guerrillas). The cause legitimizes both the fulfillment of personal ambition and the infliction of cruelty which in other circumstances would be considered inhuman. As Le Mière de Corvey noted more than a hundred and fifty years ago, there can be no guerrilla warfare without hate and fanaticism. There is a tendency not just to employ violence but to glorify it; in this respect there are parallels between modern guerrilla movements and Fascism. Guerrilla warfare and, *a fortiori*, urban terror implant a pattern of dictatorial practices and brutality that perpetuates itself. Graduates of the school of violent action do not turn into practitioners of democracy and apostles of humanism after victory.

9. Organization, propaganda and terror have always been essential parts of guerrilla warfare, but their importance has greatly increased over the years and the techniques have been refined. Organization implies the existence of a political party or movement or at least a noncombatant fringe, semilegal or underground, providing assistance to the guerrillas — money, intelligence and special services. In some instances the guerrilla movement has been more or less identical with the party (Cuba, Uruguay, Algeria); elsewhere it has acted as the armed instrument of the party. Wherever guerrillas had no such connection with a political party (EOKA, the Stern Gang, many African and Latin American guerrilla movements) they could at least rely on a periphery of sympathizers, which, albeit unorganized, provided support. In most recent guerrilla wars political propaganda has been of equal or greater importance than military operations (Cuba was the most striking example). Elsewhere propaganda has played a subordinate role; this is especially true for guerrilla wars waged by secessionist movements. These had the support of their own people anyway; but no amount of propaganda would have persuaded the Turks of the justice of the Macedonian or the EOKA cause, nor would have made Ulstermen join the IRA, or persuaded the Iraqis to make common cause with the Kurds. But even secessionist guerrillas want to influence world public opinion. Public opinion is a more effective weapon than fighting against the governments of small countries dependent on the goodwill of others. Urban guerrillas will get far more publicity than rural, because there are more newspapermen and cine cameras in towns. Some countries are more in the limelight than others. An unexploded hand grenade found in an Israeli backyard will be reported, major operations resulting in dozens killed in Burma, Thailand or the Philippines may not be reported. Hence the endeavor of urban terrorists to concentrate on eye-catching operations.

Propaganda is of particular importance in civil wars when the majority of the population, as is often the case, takes a neutral, passive attitude in the struggle between incumbents and insurgents. The apathy of the majority usually favors the guerrillas more than their enemies. No guerrilla movement has obtained its objectives solely through propaganda; equally none has succeeded by terrorism alone.

Terror is used as a deliberate strategy to demoralize the government by disrupting its control, to demonstrate one's own strength and to frighten collaborators. More Greeks were killed by EOKA than British soldiers, more Arabs than Jews in the Arab rebellion of 1936–1939, more Africans than white people by the Mau Mau. The terrorist element has been more pronounced in some guerrilla movements than in others; in "urban guerrillaism" it is the predominant mode of the armed struggle, in China and Cuba it was used more sparingly than in Vietnam, Algeria or in Greece. While few guerrilla movements have been opposed in principle to terror, some, for strategic reasons, have only seldom applied it because they thought it tactically ineffective or because they feared that it would antagonize large sections of the population. It is impossible to generalize about the efficacy of terror as a weapon; it has succeeded in some conditions and failed in others. It was used with considerable effect in Vietnam and Algeria; elsewhere, notably in Greece and in various Latin American countries, it had the opposite effect. Much depends on the selection of targets, how easy it is to intimidate political opponents, whether it is just a question of "liquidating" a few enemies, or whether the political power of the incumbents is widely diffused. Guerrilla war has been defined by insurgents and counterinsurgents alike as the struggle for the support of the majority of the people. No guerrilla movement can possibly survive and expand against an overwhelmingly hostile population. But in the light of historical experience the measure of active popular support required by a guerrilla movement need not be exaggerated.

10. The techniques and organizational forms of guerrilla warfare have varied enormously from country to country according to terrain, size and density of population, political constellation, etc. Thus, quite obviously, guerrilla units in small countries have normally been small whereas in big countries they have been large. In some countries guerrilla units gradually transformed themselves into regular army regiments and divisions (Greece) and yet failed, in others they won the war though they never outgrew the guerrilla stage (Cuba) or despite the fact that militarily they were beaten (Algeria). In some guerrilla movements the personality of the leader has been of decisive importance. One need recall only Shamyl and Abd el-Kader in the nineteenth century; the same goes for more recent guerrilla wars (Tito, Castro, Grivas). On other occa-

sions personalities have been of little consequence; the fact that the French captured some of the leaders of the Algerian rebellion did not decisively influence the subsequent course of the war. The leaders of the Vietnam Communists were expendable, Mao probably was not.

There are, by definition, no *Blitzkrieg* victories in guerrilla war, yet some campaigns succeeded within a relatively short period (two years) whereas others continued, on and off, for decades. Some involved a great deal of fighting, resulting in great losses, others were, on the whole, unbloody (Cuba, Africa). There has been a tendency to explain the defeats of guerrilla movements by referring to their strategic errors. Thus the Greek Communists have been blamed for their premature decision to adopt regular army tactics, and the Huks for not carrying the war to the cities. But this does not explain why other guerrillas succeeded, despite the fact that they made even graver mistakes. Success or failure of a guerrilla movement depends not only on its own courage, wisdom and determination but equally on objective conditions and, last but not least, on the tenacity and aptitude of the enemy. Castro won because he faced Batista and similarly the Algerians were dealing with the Fourth Republic, a regime in a state of advanced disintegration. The Greek partisans and the Huks, on the other hand, had the misfortune to encounter determined opponents in the persons of Papagos and Magsaysay. But beyond all these factors, subjective and objective, there is still the element of accident which cannot possibly be accounted for, which defies measurement and prediction. Objective conditions help or hamper guerrilla movements, they make success or failure more likely. Given a certain historical process such as decolonization, the victory of a guerrilla movement, however ineffectual, is almost a foregone conclusion. But decolonization has been concluded and the old rule no longer applies as the guerrillas confront native incumbents, nor is it true with regard to separatist movements. Guerrillas have succeeded even when "objective" conditions were adverse and they have been defeated even when everything pointed to their victory. The presence of a great leader is a historical accident: without Tito the Yugoslav partisans would probably not have taken to the mountains; but for Castro the invasion of Cuba would not have taken place. The same applies, of course, to the antiguerrilla camp. Under a more forceful, more farsighted and more gifted leader than Chiang Kai-chek, the KMT might have won the war; Mao was perfectly aware of this possibility. Other accidents can be decisive for the outcome of a guerrilla war, for instance the presence of a government spy high up in the guerrilla command. During the early period of insurgency the accidental death of a leader or his arrest could be a fatal setback. Thus, the Huks never recovered from the arrest in Manila, by accident, of most of the members of the Communist party leadership. On the other hand a small, isolated guerrilla movement may achieve a breakthrough early on in its struggle owing to sheer good fortune rather than superior strategy. On at least two occasions the fortunes of the Chinese Communists were decisively affected by sudden changes in the international political constellation. The political orientation of more than a few guerrilla movements has certainly been a matter of accident; it was not from historical necessity that the Ovambo (SWAPO) should turn to the Soviet Union, whereas the Herero and Mbanderu should study Chairman Mao's Little Red Book.

11. Urban terrorism in various forms has existed throughout history; during the past decade it has become more frequent than rural guerrilla warfare. Some modern guerrilla movements were predominantly city-based; for instance, the IRA, EOKA, IZL and the Stern Gang, others were part urban (Algeria). Neither the nineteenth-century anarchists nor the Russian pre-revolutionary terrorists regarded themselves as guerrillas; their assassinations were largely symbolic acts of "punishment" meted out to individual members of the forces of oppression — they were not usually part of an overall strategy. Whereas guerrilla operations are mainly directed against the armed forces of the enemy and the security services, as well as installations of strategic importance, modern urban terror is less discriminate in the choice of its targets. Operations such as bank robberies, hijackings, kidnappings, and, of course, assassinations are expected to create a general climate of insecurity. Such actions are always carried out by small groups of people; an urban guerrilla group cannot grow beyond a certain limit because the risk of detection increases with the growth in numbers. A successful urban guerrilla war is possible only if the strength of the establishment has deteriorated to the point where armed bands can move about in the city. Such a state of affairs has occurred only on very rare occasions and it has never lasted for any length of time, leading within a few days either to the victory of the insurgents or the incumbents. The normal use of "urban guerrilla" is a euphemism for urban terrorism which has a negative public relations image. Thus the Tupamaros always advised their members to dissociate themselves from "traditional terrorism" and only a few

fringe groups (Marighela, Baader-Meinhof) openly advocated terror. Urban terrorism can undermine a weak government, or even act as the catalyst of a general insurgency but it is not an instrument for the seizure of power. Urban terrorists cannot normally establish "liberated zones"; their operations may catch headlines but they cannot conduct mass propaganda nor build up a political organization. Despite the fact that modern society has become more vulnerable than in the past to attacks and disruptions of this kind, urban terrorism is politically ineffective, except when carried out in the framework of the overall strategy of a political movement, usually sectarian or separatist in character, with an already existing mass basis.

12. Guerrilla movements have frequently been beset by internal strife, within their own ranks or between rival groups. Internal dissension has been caused by quarrels about the strategy to be pursued (China, Greece) or by the conflicting ambitions of individual leaders (Frelimo, Columbia). The rivalry between the political and the military leadership, unless these were identical, has also been a frequent cause of friction. Contemporary Far Eastern and Southeast Asian Communist guerrilla movements have been relatively free of such internal struggle; elsewhere splits have been the rule rather than the exception. The Algerian rebels and the PAIGC succeeded in immobilizing their competitors early on in their struggle. In other countries as much effort has been devoted by the insurgents to fighting against their rivals as against the common enemy (IMRO, Yugoslavia, Albania, Greece, Angola). Sometimes a division of labor between rival organizations prevented open clashes while the struggle against the common enemy lasted. This was true for instance of Mexico, Mandatory Palestine, the Palestine Arab resistance and Ulster. But once the fight against the foreign enemy has been won the struggle for power frequently results in a free-for-all between rival guerrilla groups (Congo, Angola) or sets former comrades-in-arms against each other (the Irish Civil War).

THE FUTURE OF GUERRILLA WARFARE

The assessment of the future prospects of guerrilla warfare has to take historical experience as its starting point: in what conditions did such warfare occur, and why did it succeed or fail? The historical record shows, to repeat once again, that nineteenth-century guerrilla wars invariably failed to achieve their objectives except with the support of a regular army, domestic or foreign. During the Second World War guerrilla movements had limited successes against overextended enemy units; but they used the war to consolidate their power and in the political vacuum after the war they emerged as the chief contenders for power (China, Yugoslavia, Albania, Vietnam).

A powerful impetus was given to guerrilla war after 1945 with the disintegration of colonial empires. The colonial powers no longer had the will to fight, and even if guerrilla operations were militarily quite ineffective, to combat them became so costly that the imperial power eventually withdrew its forces.

Guerrilla war against domestic rulers has succeeded in the past — with one exception — only during a general war or immediately following it, with the collapse of central state power. Weakened as the South Vietnam regime was by Vietcong activities, the decisive assault was launched by a regular army. Separatist guerrilla movements have not so far scored decisive victories. Their future prospects will depend to a large extent on the amount of foreign aid they receive. If their political demands are limited in character (administrative-cultural autonomy) or if their secession would not decisively weaken the state they may succeed in certain cases. If on the other hand the loss would be unacceptable, their chances must be rated low, except at a time of general crisis such as war. The appeal of a separatist guerrilla movement is of necessity limited; its survival and success depends on the assumption that the authorities will not apply extreme measures ranging from resettlement on a massive scale to physical extermination.

The conditions conducive to the success of guerrillas have become much less promising with the virtual end of decolonization and the absence of general war. Could the Cuban example be emulated elsewhere? Could, in other words, a guerrilla movement succeed in peacetime in undermining an existing government to such an extent that its collapse became a distinct possibility? Certain developments favor insurrection: urban terrorism has become transnational, supported by foreign governments or by terrorist movements abroad. At the same time, the destructive power of the weapons used by terrorists has greatly increased. While the rifle, the machine gun and the hand grenade (or the bomb) were the classical weapons of the guerrilla during the last hundred years, the guerrilla of the future will have advanced weapons such as missiles at his disposal; he may be able to manufacture a crude atomic bomb or steal one.[17] But the political uses of nuclear blackmail by terrorist groups should not be exaggerated — it is not an instrument for the seizure of power.[18] In any case, the destructive power of the weapons in the hands of the state has grown even more and the outcome will depend in the last resort on the will and the ability of the government to apply this force. The military balance of power has shifted to the detriment of the guerrillas; they can seldom operate in the open country, and the scope of terrorist activities in urban centers is limited (the decline of hijacking).

It has been maintained that large-scale conventional wars have become so difficult and expensive that terrorists may be employed by foreign governments to engage in surrogate warfare and that terrorism may become the conventional war of the future. This seems unlikely for both military and political reasons. Recent technological advances such as precision-guided munitions provide more destructive energy in smaller packages than ever before and have revolutionized delivery accuracy. These new weapons however are effective above all against tanks and combat aircraft. But tanks and combat aircraft were never the guerrilla's worst enemies whereas in fighting in urban areas precision-guided munitions will be of strictly limited use. It is quite likely that in a future war massed forces will count for less and small forces with great firepower will be of considerable importance. There may well be a dispersal of forces, a return, on a higher level of technological development, to the partisan tactics of the eighteenth century with comparatively small, highly mobile units raiding the enemy's rear. But it is unlikely that guerrilla units operating in peacetime will derive much benefit from these innovations. They will not be able, as a rule, to retransfer their activities from the cities to the countryside, for means of detection in the open have greatly improved. If it is true that military power will become more diffuse, it is equally true that military power without a central command, close coordination, supply and logistics is ineffective.

In peace a determined army or police force will always be able to destroy the guerrillas and terrorists. The guerrillas have to rely on the government's inability to use the full power at its disposal, the constraints imposed by world opinion and public opinion at home. But this applies only to liberal-democratic regimes. Their number has been shrinking and guerrilla or terrorist activities could well hasten this process. What Regis Debray said about the Tupamaros applies *mutatis mutandis* to guerrillas and terrorists operating in democratic societies in general; that digging the grave of the "system" they dig their own grave, for the removal of democratic restraints spells the guerrilla's doom. Is it safe to expect that governments will be either so inefficient or so permissive as not to employ effective antiguerrilla or antiterrorist measures in an emergency? This is becoming less and less likely. The strategy of guerrilla war may be used between sovereign states with attacks launched from sanctuaries beyond state borders. But such war by proxy will usually be dangerous for it may lead to full-scale war; it is unlikely in time of peace that the Chinese will instigate guerrilla warfare in Siberia or vice versa. A guerrilla campaign may still be possible against a small country in certain circumstances if support by a major power discreetly (or not so discreetly) is provided to various separatist or opposition groups.

Democratic regimes always seem highly vulnerable to terrorist attack. The constitutional restraints in these regimes make it difficult to combat terrorism and such failure exposes democratic governments to ridicule and contempt. If, on the other hand, they adopt stringent measures they are charged with oppression, and the violation of basic human rights. If terrorists are put on trial they will try to disrupt the legal procedure and to make fair administration of justice impossible. Having been sentenced, terrorists and their sympathizers could then claim that they are victims of gross injustice. Up to this point, the media (always inclined to give wide publicity to acts of violence), are the terrorists' natural ally. But as terrorist operations become more frequent, as insecurity spreads and as wide sections of the population are adversely affected, there is a growing demand for tougher action by the government even if this should involve occasional (or systematic) infringements of human rights. The swing in popular opinion is reflected in the media focusing no longer on the courage and unselfishness of the terrorists but on the psychopathological sources of terrorism and the criminal element — sometimes marginal, at others quite prominent, but always present in "urban guerrilla" operations. Unless the moral fiber of the regime is in a state of advanced decay, and the political will paralyzed, the urban terrorists would fail to make headway beyond the stage of provocation, in which, according to plan, public opinion should have been won over to their cause, but is in fact antagonized.

Prospects for urban terror seem a little more promising in the Third World, but only in certain rare constellations, some of which have already been discussed. The security forces in these countries are less experienced and effective than in Communist regimes, but usually they will be capable of coping with challenges of this kind unless the rebels receive powerful support from abroad. Irrespective of how brutally a guerrilla movement is suppressed it will be next to impossible to mobilize foreign public opinion against an oil-producing country or one that has good relations with its neighbors and other Third World nations. World public opinion can be mobilized only against a relatively weak country that has powerful enemies among its neighbors, and few friends.

Even if the authority of the state is fatally undermined, even if a power vacuum exists, the prospects of guerrilla or terrorist victory have dimmed, for there is a stronger contender for power — the army. Military coups have become more and more frequent: they may in future become the normal form of political change in most parts of the world.

Latin American Communist leaders have noted that the revolutionary process largely depends on enlisting the "patriotic forces" among the military on the side of the Communists.[19] The same applies *mutatis mutandis* to the Middle East, Africa and parts of Asia. But such military coups can turn right as well as left. The slogans will be nationalist-populist in any case and the difference in policy between left- and right-wing military dictatorships may not always be visible to the naked eye. Those with a more pronounced left-wing bias will steal much of the guerrillas' thunder, those inclined more to the right will effectively suppress them. The army command seizes the key positions of the state apparatus and quite frequently establishes a state party. The help of civilian (or guerrilla) political activists may be accepted in this process but they are regarded at the same time as rivals and since the army officers have no desire to share power the civilians will be kept at a safe distance from the levers of power.

During the last fifteen years some hundred and twenty military coups have taken place whereas only five guerrilla movements have come to power; three of them as the result of the Portuguese military coup in 1974; Laos and Cambodia fell after the collapse of Vietnam. The military dictator may be overthrown but the challenge will again come from within the army. Not being over-extended and weakened by foreign wars, the army in Third World countries is in a strong position as a contender for domestic power.

The conditions that caused insurgencies have not disappeared — men and women are still exploited, oppressed, deprived of their rights and alienated. "Objective, revolutionary situations" still abound and will continue to exist. But the prospects for conducting successful guerrilla war in the postcolonial period have worsened, except, perhaps, to a limited extent in the secessionist-separatist context. Guerrilla war may not entirely disappear but, seen in historical perspective, it is on the decline, together with its traditional foes — colonialism on the one hand and liberal democracy on the other. Thus the function of guerrilla movements is reverting to what it originally was — that of paving the way for and supporting the regular army. In the past such assistance was military — today it is mainly political. It is holding the stirrup so that others may get into the saddle.

The transition from high to low tide in the fortunes of guerrilla war has been sudden. This is not to say that the conditions that once favored its rise may not recur — following a major war or a natural catastrophe or the weakening of the authority of the state for some other reasons. But at present the age of the guerrilla is drawing to a close. The retreat into urban terror, noisy but politically ineffective, is not a new departure but, on the contrary, the end of an era.

Notes

CHAPTER ONE: PARTISANS IN HISTORY

1. C. W. Abeli, *Savage Life in New Guinea* (London, 1901), 138; James Adair, *The History of the American Indians* (London, 1775), 382 *et seq*; H. H. Turney-High, *Primitive War, Its Practice and Concepts* (New York, 1971), 128.
2. Niese, ed., *Josephus Flavius* (Berlin, 1887), Book 2, chapter 19.
3. Dio Cassius, chapter 69, 12–13.
4. *Thucydides*, III, 94–98, quoted in F. E. Adcock, *The Greek and Macedonian Art of War* (Berkeley, 1957), 17.
5. The main sources for the Battle of Teutoburg Forest are Tacitus's *Annali*, Dio Cassius, and Suetonius.
6. Caesar, *De bello Gallico*, Book VII. See also Camille Jullian, *Vercingetorix* (Paris, 1921), and T. Rice Holmes, *Caesar's Conquest of Gaul* (London, 1899).
7. Antonio Garcia y Bellido, "Bandas y Guerrillas en las luchas con Roma," *Hispania*, V, No. 21, 548.
8. The main sources are Appian, Diodorus, Dio Cassius. The best short summary is Hans Gundel's "Viriathus" in Pauly and Wissowa, *Real Encyclopaedie d. klassischen Altertumswissenschaft*, 1893–.
9. Appian, *Roman History*, I, trans. Horace White (London, 1958), 258.
10. *Cambridge Ancient History*, VIII, 316.
11. "Su coincidencia con los practicados por las guerrillas en nuestra Guerra de la Independencia es absoluta." Bellido, "Bandas . . . ," 589.
12. The question whether Viriathus should be considered Portuguese or *Celtibero* remains in dispute.
13. Plutarch 12–13. See also the dissertation by Stahl, *De Bello Sertoriano* (Erlangen, 1907); Mommsen, *Römische Geschichte*, III, and Schulten, in Pauly and Wissowa, *Real Encyclopaedie*.
14. Mommsen, *Römische Geschichte*, III, 37.
15. See G. Köhler, *Die Entwicklung des Kriegswesens und der Kriegsführung in der Ritterzeit* (Breslau, 1886) II; Hans Delbrück, *Geschichte der Kriegskunst*, (Berlin, 1907) III; Emil Daniels, *Geschichte des Kriegswesen* (Leipzig, 1910).
16. *Oeuvres de Froissart*, VI, 32.
17. The main source for the Tuchins is M. Boudet, *La Jacquerie des Tuchins 1363–84* (Paris, 1895).
18. H. W. V. Temperley, *History of Serbia* (London, 1919), 125.

19. F. S. Stevenson, *History of Montenegro* (London, 1912), 123; G. Finlay, *History of the Greek Revolution* (London, 1861), I.
20. Lieutenant G. Arbuthnot, *Herzegovina or Omer Pasha and the Christian Rebels* (London, 1862), 152.
21. Ibid., 266.
22. *The Present State of the Ottoman Empire* (London, 1668).
23. G. Rosen, *Die Balkan Haiducken* (Leipzig, 1878), 26–27.
24. Finlay, *History* . . . , 32.
25. General Gordon, *History of the Greek Revolution*, I, 313.
26. J. B. Parsons, "Attitudes towards the late Ming Rebellions," in *Oriens Extremus*, VI (1959); Erich Hauer, "Li Tsu-cheng and Chang Hsien-chung," in *Asia Minor*, II (1925). For a short general description see Roland Mousnier, *Fureurs Paysannes* (Paris, 1967), 238–306.
27. Basing himself on Ewald (*Abhandlung*), Clausewitz refers to the many incidents in which American riflemen had abducted English generals (*Vorlesungen über den kleinen Krieg*, *Schriften*, I [Göttingen, 1966], 439). Frederick the Great, on the other hand, took a dim view of the importance of the lessons of the war in America. On one occasion he wrote: "The people who come back from America imagine they know all there is to know about war, and yet they have to start learning war all over again in Europe." (Quoted in Peter Paret, *York and the Era of Prussian Reform 1807–1815* [Princeton, 1966], 43).
28. Piers Mackesy, *The War for America* (London, 1964), 30, 366.
29. The most recent biography of Marion is Hugh F. Rankin, *The Swamp Fox* (New York, 1973). Of the earlier works K. T. Headley, *Washington and his Generals* (New York, 1847), II, and W. Gilmore Simms, *The Life of Francis Marion* (New York, 1846) should also be mentioned.
30. Rankin, op. cit., 173. See also Mark Boatner, ed., *Cassell's Biographical Dictionary of the American War of Independence, 1773–1783* (London, 1966).
31. A. K. Gregorie, *Thomas Sumter* (Columbia, 1931).
32. James Graham, *The Life of General Daniel Morgan* (New York, 1859).
33. Esmond Wright, *Washington and the American Revolution* (London, 1973), 135.
34. Jac Weller, "Irregular but Effective: Partisan Weapons Tactics in the American Revolution: Southern Theatre," *Military Affairs* (Fall 1967), 120.
35. W. D. James, *A Sketch of the Life of Brigadier General Francis Marion* (Charleston, 1824), 59.
36. Weller, loc. cit., 126.
37. Ibid., 119.
38. Ch. I. Chassin, *La Vendée Patriotique* (Paris, 1892), II, 293.
39. This account is based mainly on the works of Ch. I. Chassin, Emile Gabory, Joseph Clemenceau and Savary. Of the recent literature A. Montagnon, *Une guerre subversive* (Paris, 1959), Charles Tilly, *The Vendée* (London, 1964), and Peter Paret, *Internal War and Pacification: The Vendée 1789–1796* (Princeton, 1961) are the most important.
40. P. Paret, *Internal War* . . . , 34.
41. Quoted in Tilly, *The Vendée*, 333.
42. Chassin, *La pacification de l'ouest*, II.
43. Gabory relates the story of the peasant Guitton who killed twenty-seven soldiers after having found his wife and children dead after a punitive raid (*Napoleon et la Vendée* [Paris, 1914], 12).
44. Chassin, *La preparation*, III, 441.
45. Chassin, *La Vendée patriotique*, I, 439.
46. Tilly, *The Vendée*, 334.
47. Chassin, *La Pacification*, I, 187.
48. Paret, *Internal War* . . . , 33.
49. For a military analysis of the Vendean wars see A. Montagnon, *Une guerre subversive*.

50. Chassin, *La Pacification*, III, 219.
51. Some of their leaders, such as Bonchamp, had participated in the American War of Independence. But it would be wrong to attribute undue importance to this fact, just as it is no doubt accidental that some of the Vendeans (and their conquerors) subsequently saw service in Spain. It is possible to establish a genealogy of guerrilla warfare — from South Carolina to the Vendée, from there to Spain, from Spain to North Africa to the Fenians (John Devoy). But such exercises are of no great significance.
52. Raymond Carr, *Spain 1809–1939* (Oxford, 1966), *passim*.
53. Quoted in Geoffrey de Grandmaison, *L'Espagne et Napoleon* (Paris, 1931), IV, 219.
54. The main sources used in this account are Gomez de Arteche y Moro, *Guerra de la Independencia* (Madrid, 1868–1903), 14 vols.; A. Grasset, *La Guerre d'Espagne, 1807–1813* (Paris, Nancy, 1914); Toreno, *Histoire du Soulèvement de la Guerre et de la Révolution d'Espagne* (Paris, 1836–38), 5 vols.; Oman, *A History of the Peninsular War* (Oxford, 1902–22), 6 vols.; Geoffrey de Grandmaison, *Espagne et Napoleon* (Paris, 1931), 3 vols.; *Diccionario Bibliográfico de la Guerra de la Independencia Española* (Madrid, 1944–52), 3 vols. Most recently: Juan Priego López, *Guerra de la Independencia* (Madrid, 1973), 3 vols.
55. The main sources for Mina are *A Short Extract from the Life of General Mina published by himself* (London, 1825); *Memorias del General Don Francisco Espoz y Mina* (New Edition) (Madrid, 1962), I; I. M. Iribarren, *Espoz y Mina el guerrillero* (Madrid, 1965); Hermilio de Oloriz, *Navarra en la Guerra de la Independencia. Biografia del guerrillero Don Francisco Espoz y Mina* (Pamplona, 1955).
56. Oman, loc. cit., III, 489.
57. Mina, *Memorias*, 86–87.
58. *A Short Extract*, 31.
59. *Archives de la Guerre*, 20 April 1813. Quoted in Grandmaison, III, 246.
60. There is a great deal of literature, much of it apocryphal, on the Empecinado, some of it also in English, e.g., "Passages in the Career of the Empecinado" in *Peninsular Scenes and Sketches* by the author of "Student of Salamanca" [Frederick Hardman] (Edinburgh, 1846), 1–97. The following account is based mainly on French sources and on *The Military Exploits etc. etc. of Don Juan Marin Diaz the Empecinado who First Commenced and then Organized the System of Guerrilla Warfare in Spain* (London, 1823).
61. *The Military Exploits*, 153.
62. See documents quoted in Grandmaison, III.
63. *The Military Exploits*, 60.
64. *The Military Exploits*, 14.
65. C. F. Henningsen, *The Most Striking Events of a Twelve Months Campaign with Zumalacarregui* (London, 1836), I, 177.
66. On Somaten and Miqueletes, see Artèche, VII, 56; and Boucheman, "Aperçu sur l'organisation d'Armée espagnole et des corps de partisans de 1808–1814," *Le Spectateur Militaire*, XXII (1859).
67. About the subsequent fate of the guerrilla leaders see *Diccionario Bibliográfico;* E. Guillon, *Les Guerres d'Espagne sous Napoleon* (Paris, 1902), and Grandmaison.
68. Mina, *Memorias*, 112.
69. Toreno, III, 340.
70. Mina, *Memorias*, 169.
71. Jac Weller, "Wellington's Use of the Guerrillas," *Royal United Services Institute Magazine* (May 1963), 155.
72. Grandmaison, 219.
73. Toreno, III, 30.
74. *Aus dem Leben des Generals der Infanterie z.D. Dr. Heinrich von Brandt,*

(Berlin, 1870), 76, 212; *Soldats suisses au service étranger* (Geneva, 1909), II, 35; Artèche, VII, 64; Grandmaison, 246.

75. R. Wohlfeil, *Spanien und die deutsche Erhebung* (1965), 180 *et seq.*, 230 *et seq.*
76. The following account is based chiefly on the three main works on the Tyrolean rising: Josef Hirn, *Tirols Erhebung im Jahre 1809* (Innsbruck, 1909); Hans von Voltelini, *Forschungen und Beiträge zur Geschichte des Tiroler Aufstandes in Jahre 1809* (Gotha, 1909); Karl Paulin, *Das Leben des Andreas Hofer* (Innsbruck, 1959). A full bibliography is in Hans Hochenegg, *Beihefte zu Tiroler Heimat* (Innsbruck, 1960).
77. Hans Kramer, *Andreas Hofer* (Vienna, 1970), 40.
78. *Tirol und die Tiroler im Jahre 1809* (1810), 32.
79. F. Schulze, ed., *Die Franzosenzeit in deutschen Landen* (Leipzig, 1909), I, 244; *Helden der Ostmark* (Vienna, 1937), 135.
80. Kramer, 47.
81. The most important source is Denis Davydov, *Voennie Zapiski* (Moscow, 1940). There is much interesting material in *Russkaya Starina*, e.g., Löwenstern's *Zapiski*, serialized in 1900–1901. One of the earliest major analyses of the war of 1812–13 is Mikhailovski-Danilevski, *Opisanie otechestvennoi voini v 1812 godu* (St. Petersburg, 1839), 3 vols. and the same author's *Imperator Alexander I i evo spodvizhniki v 1812, 1813, 1814 i 1815 godakh* (St. Petersburg, 1849). Among Soviet accounts E. Tarle's *Napoleon's Invasion of Russia* (London, 1942) and V. A. Garin, *Izgnanie Napolona iz Moskvi* (Moscow, 1938) should be mentioned.
82. Davydov, op. cit., 209.
83. Ibid., 22.
84. Ibid., 158. See also *Mémoires du Général Löwenstern* (Paris, 1903), I, 296.
85. Mikhailovski-Danilevski, *Opisanie*, III, 132.
86. D. Cherviakov, "Partisanskie Otryadi v otechestvennoi voine 1812," *Voenno-Istoricheski Zhurnal*, 6–7 (1941), 54.
87. Davydov, 177.
88. *Sovremennik* 3, 1836. Pushkin wrote to him: "Your essay did not escape the red ink. Military censors wanted to show that they can read."
89. Davydov, 424.
90. Quoted in Tarle, 250.
91. Garin, 93.
92. Mikhailovski-Danilevski, *Opisanie*, III, 102.

CHAPTER TWO: SMALL WARS AND BIG ARMIES

1. On Tupac Amaru, L. E. Fish, *The Last Inca Revolt* (Norman, 1966); Daniel Valcaral, *La Rebelión de Tupac Amaru* (Mexico, 1947); idem, *Rebeliones indigenas* (Lima, 1946); German Arciniegas, *Los Comuneros* (Bogotá, 1959); Boleslao Lewin, *La Rebelión de Tupac Amaru* (Buenos Aires, 1963).
2. Fish, *Last Inca Revolt*, 214.
3. On Pumacahua, I. C. Bouroncle, *Pumacahua. La Revolución de Cuzco de 1814* (Cuzco, 1956); Juan José Vega, *La Emancipación frente el indio peruano* (Lima, 1958).
4. Pedro M. Arcaya, *Insurección de los negros de la serranía de Coro* (Caracas, 1949), 31–32.
5. C. L. R. James, *The Black Jacobins* (New York, 1963), 54, 116–117. For the military aspects of the campaign see A. Metra, *Histoire de l'expedition des Français à Saint Domingue* (Paris, 1825), and Lemmonier-Delafosse, *Seconde campagne de Saint Domingue* (Le Havre, 1846).
6. Oswaldo Diaz Diaz, *Los Almeydas* (Bogotá, 1962); Raul Rivera Serna, *Los Guerrilleros del centro en la emancipación peruana* (Lima, 1958).

7. Memoirs of General Miller, quoted in John Lynch, *The Spanish American Revolutions* (London, 1973), 181.
8. Robert L. Gilmore, *Caudillism and Militarism in Venezuela, 1810–1910* (Athens, [Ohio], 1964), 71; Paez, *Autobiografía* (New York, 1946), I, 7.
9. Gilmore, *Caudillism and Militarism*, 83.
10. Jasper Ridley, *Garibaldi* (London, 1974), 185.
11. The standard works in English are Hugh M. Hamill Jr., *The Hidalgo Revolt* (Gainesville, 1966), and Wilbert H. Timmons, *Morelos of Mexico* (El Paso, 1963). An important early work is F. Robinson's *Mexico and her Military Chieftains*, first published in 1847, reprinted in 1970.
12. J. A. Dabbs, *The French Army in Mexico* (The Hague, 1963), 70.
13. Julio C. Guerrero, *La Guerra de guerrillas* (La Paz, 1940), 89.
14. Hugh Thomas, *Cuba* (London, 1971), 254. See also Ramiro Guerra, *Guerra de los diez años* (Havana, 1960), and Antonio Pirala, *Anales de la guerra de Cuba* (Madrid, 1896).
15. The chief sources are Wyler's autobiographical account, *Mi mando en Cuba* (Madrid, 1910) 6 vols. See also Hugh Thomas, *Cuba*, and M. F. Almagro, *Historia política de la España contemporanea* (Madrid, 1959), II.
16. "A nossa Vendeia" is the title of two articles by Euclides da Cunha in *O estado de São Paulo*, March 17 and July 17, 1897, reprinted in *Canudos e ineditos* (São Paulo), 1967. Da Cunha's classic *Os sertões*, published in English under the title *Rebellion in the Backlands* (Chicago, 1944), is devoted to the campaigns against Canudos.
17. While da Cunha's epic presents a magnificent literary account, its historical accuracy has been disputed. The literature on Canudos is considerable; for a modern biography of Consilheiro see Abelardo Montenegro, *Antonio Conselheiro* (Fortaleza, 1954); for the general historical background, José Maria Bello, *A History of Modern Brazil 1889–1964* (Stanford, 1966); for a modern interpretation, Ralph della Cave, "Brazilian Messianism and National Institutions: A Reappraisal of Canudos and Joaseiro," *Hispanic American Historical Review* (August 1968); for the military aspects of the campaigns, U. Peregrino, *Os sertões como historia militar* (Rio de Janeiro, 1956).
18. Da Cunha, *Rebellion*, 149.
19. Ibid., 194.
20. Ibid., 475.
21. Edgar Holt, *The Carlist Wars in Spain* (London, 1967), is a recent historical study. Antonio Pirala, *Historia de la guerra civil* (Madrid, 1868), 6 vols., is the most detailed account. See also A. Risco, *Zumalacarreguy en campañà* (Madrid, 1935); T. Wisdom, *Estudio histórico militar de Zumalacarreguy y Cabrera* (Madrid, 1890); and most recently Roman Oyarzuni: *Vida de Ramon Cabrera* (Barcelona, 1961).
22. Holt, *Carlist Wars*, 117.
23. Franz von Erlach, *Die Freiheitskriege kleiner Völker gegen grosse Heere* (Bern, 1867), 323.
24. George Finlay, *History of the Greek Revolution* (London, 1861), I, 189. See also T. Gordon, *History of the Greek Revolution* (Edinburgh, 1832), 2 vols. C. W. Cranley, *The Question of Greek Independence* (Cambridge, 1930); C. M. Woodhouse, *The Philhellenes* (London, 1968), and his earlier *The Greek War of Independence* (London, 1952). For a modern Greek view of the war of independence, see G. K. Asporas, *Politika historia tes neoteras Hellados* (Athens, 1930).
25. Finlay, *Greek Revolution*, 194–195.
26. W. F. Reddaway *et. al.*, *Cambridge History of Poland* (London, 1941), II, 161. The standard (Polish) biography of Kosciusko is by T. Korzon (Cracow, 1906).
27. Friedrich von Smitt, *Geschichte des polnischen Aufstandes* (Berlin, 1839), II, 159.
28. William Ansell Day, *The Russian Government in Poland* (London, 1867), 131.

29. Smitt, *Polnischen Aufstandes*, 383.
30. Ludwik Mieroslawski, *Kritische Darstellung des Feldzuges vom Jahre 1831 und hieraus abgeleitete Regeln für National-Kriege* (Berlin, 1847), I, 302. See also the Mieroslawski biography (in Polish) by M. Zychowski (Warsaw, 1963).
31. Pisacane, quoted by J. Ridley, *Garibaldi*, 253.
32. G. M. Trevelyan, *Garibaldi's Defence of Rome* (London, 1933), 89. See also the *Autobiography of Giuseppi Garibaldi* (London, 1889), 3 vols.
33. Trevelyan, *Garibaldi and the Thousand* (London, 1933), 218.
34. Ridley, *Garibaldi*, 605.
35. H. d'Ideville, *Memoirs of Marshall Bugeaud* (London, 1884), I, 211. Of the many Bugeaud biographies, Lucas-Dubreton (1931), E. de Lamaze (1943), M. Andrieux (1951), and L. Morard (1947) should be mentioned.
36. Geo. Wingrove Cooke, *Conquest and Colonisation in North Africa* (London, 1860), 211–212. For a general account see also J. Pichon, *Abd el-Kader* (Paris, 1899), and A. Bellemore, *Abd el-Kader, sa vie politique et militaire* (Paris, 1863).
37. Count P. Castellan, *Military Life in Algeria* (London, 1853), I, 204.
38. d'Ideville, *Marshall Bugeaud*, I, 299.
39. Ibid., 252.
40. The literature on the Caucasian campaigns is immense. The most detailed Russian account is General Potto's *Kavkazkaya voina* (St. Petersburg, 1887–1897), 4 vols.; the standard English history is John F. Baddeley, *The Russian Conquest of the Caucasus* (London, 1908). Modern descriptions are Lesley Blanch, *The Sabres of Paradise* (London, 1960), and Paul Chavchavadze, *The Mountains of Allah* (London, 1953). An interesting account from a Turkish point of view is M. M. Zihni's *Seyh Samil* (Ankara, 1958).
41. Baddeley, *Conquest of the Caucasus*, 146.
42. Friedrich Wagner, *Schamil als Feldherr, Sultan und Prophet* (Leipzig, 1854), 95.
43. W. E. D. Allen and Paul Muratoff, *Caucasian Battlefields* (London, 1953), 51.
44. Compare N. I. Pokrovsky, "Miuridism u vlasti," in *Istorik-Marksist*, 2 (1934), with the debate in *Voprosy istorii*, II, 1947. The extensive literature on the subject is analyzed in Paul B. Henze, "The Shamil Problem" in W. Z. Laqueur, ed., *The Middle East in Transition* (London, 1958), 415–443. For a post-Stalinist appraisal of Shamil see N. A. Smirnov, *Miuridism na Kavkaze* (Moscow, 1963).
45. Baddeley, *Conquest of the Caucasus*, 393.
46. J. A. MacGahin, *Campaigning on the Oxus and the Fall of Khiva* (London, 1874), 378.
47. Byron Farwell, *Queen Victoria's Little Wars* (London, 1973), 169. Another recent account is Donald Featherstone, *Colonial Small Wars* (Newton Abbot, 1973).
48. A recent Indian analysis is Dharm Pal, *Tantia Topi* (New Delhi, 1957). The most detailed account is Kaye and Malleson, *History of the Indian Mutiny* (London, 1888–89), 6 vols.
49. Edgar Holt, *The Strangest War* (London, 1962), 150. The most detailed account is James Cowan, *The New Zealand Wars* (Wellington, 1922), 2 vols. For a recent account, Keith Sindar, *The Origins of the Maori Wars* (Wellington, 1957).
50. Frederick W. Turner III, *Geronimo: His Own Story* (London, 1974), introduction, 23; according to the same source, Ché Guevara found considerable inspiration in reading about Geronimo.
51. C. L. Alderman, *Osseola and the Seminole Wars* (New York, 1973), 72.
52. John B. Trussell, "Seminoles in the Everglades," *Army* (December 1961).
53. V. C. Jones, *Gray Ghosts and Rebel Raiders* (New York, 1956), introduction.
54. Carl W. Breiham, *Quantrill and his Civil War Guerrillas* (Denver, 1959), 42.
55. C. F. Holland, *Morgan and his Raiders* (New York, 1942); Jones, *Gray Ghosts;*

L. L. Butler, *John Morgan and his Men* (New York, 1960); James Williamson, *Mosby's Rangers* (New York, 1909); J. Scott, *Partisan Life with Col. J. S. Mosby* (New York, 1867); A. R. Johnson, *The Partisan Rangers* (New York, 1904).

56. Williamson, *Mosby's Rangers*, 23.
57. Stanley F. Horn, *The Army of Tennessee* (Indianapolis, 1941), 195.
58. Mark M. Boatner, *Cassel's Biographical Dictionary of the American Civil War* (London, 1973), 568.
59. Holland, *Morgan*, 170, 350.
60. Wyeth, *Forrest*, 635.
61. *The War of the Rebellion: A Compilation of the Official Records* (Washington, 1864–1927), series I, vol. 39, 121.
62. Quoted in Carl E. Grant, "Partisan Warfare, Model 1861–5," *Military Review*, (Nov. 1958), 45.
63. Lieber's comments were published under the title *Guerrilla Parties Considered with Reference to the Laws and Usages of War* (New York, 1862).
64. *La Guerre de 1870: la défense nationale en Provence; mesures générales d'organisation* (Paris, 1911), 553. See also Freycinet, *La Guerre en Provence* (Paris, 1871).
65. Ibid., 557.
66. Michael Howard, *The Franco-Prussian War* (London, 1961), 254. The most detailed histories of the war are Pierre Lehautcourt, *La Défense nationale* (Paris, 1893–1898), 8 vols.; and *Histoire de la guerre de 1870–71* (Paris, 1901–1908), 7 vols. The multivolume official German and French accounts provide comparatively little material about partisan warfare.
67. H. Genevois, *Les Coups de main pendant la guerre* (Paris, 1896), 111.
68. See for instance, Fritz Hönig, *Der Volkskrieg an der Loire* (Berlin, 1893–1897), 6 vols., *passim*, and A. Ehrhardt, *Kleinkrieg* (Potsdam, 1935), 49.
69. Georg Cardinal von Widdern, *Der Krieg an den rückwärtigen Verbindungen* (Berlin, 1893–1899), pt. II, 14.
70. Ibid., pt. III, 35.
71. Howard, *Franco-Prussian War*, 252–253.
72. Christian Rudolf de Wet, *Three Years War* (London, 1902), 78.
73. The main sources are the (semiofficial) history by General Frederic Maurice and Captain M. H. Grant in 4 vols. (London, 1906–1910), and the *Times History of the War in South Africa, 1899–1902* in six vols. (London, 1900–1909). In addition there are countless eyewitness reports, autobiographical and biographical accounts.
74. *History of the War in South Africa*, IV, 265.
75. Ibid.
76. Ibid., 397.
77. Sir Arthur Conan Doyle, *The Great Boer War* (London, 1903), 404.
78. The raid is described in the most vivid eyewitness account of the whole war in Denys Reitz's *Commando* (London, 1942), 199 *et seq.*
79. De Wet, *Three Years War*, 305.
80. Ibid., 321 *et seq.*
81. Reitz, *Commando*, 310.
82. J. C. Smuts, *Jan Christian Smuts* (London, 1952), 83.
83. De Wet, *Three Years War*, 279–282.
84. Ibid., 93.
85. Richard L. Maullin, *The Fall of Dumar Aljure, a Colombian Guerrilla and Bandit* (Santa Monica, 1968).
86. Constancio Bernaldo de Quiros, *El Bandolerismo en España y en Mexico* (Mexico, 1959), 296 *et seq.*; F. Lopez Leiva, *El Bandolerismo en Cuba* (Havana, 1930), 28 *et seq.* On the general phenomenon of banditry see E. Hobsbawm, *Bandits* (London, 1969), and with reference to Brazil, Maria Isaura Pereira de Queiroz, *Os congaceiros* (Paris, 1968).

87. Martin Luis Guzman, *Memoirs of Pancho Villa* (Austin, 1956), 7.
88. Jen Yu-wen, *The Taiping Revolutionary Movement* (New Haven, 1973), 67–68.
89. In 1951 this was watered down to "a very large proportion."
90. Quoted in Stuart R. Schram, *The Political Thought of Mao Tse-tung* (London, 1963), 176, 196.
91. Mark Seldon, "The Guerrilla Movement in North West China," *China Quarterly*, 28 (1965), 70.
92. C. MacFarlane, *Lives and Exploits of Bandits and Robbers* (London, 1837), 36–37.
93. Ibid., 49.
94. David Hilton, *Brigandage in South Italy* (London, 1864), I, 48.
95. About Gasparone (or Gasbaroni) who fought the Austrians and the local rulers in the 1820s see *Le Brigandage dans l'états ponteficaux. Memoirs de Gasbaroni: celèbre chef de bande de la province de Frosinone*, redigé par Pierre Masi (Paris, 1867).
96. Hilton, *Brigandage*, 50.
97. Ibid., 60.
98. Ibid., 67. This is based on Sacchinelli's biography of Cardinal Ruffo; he had been Ruffo's secretary.
99. Jen Yu-wen, *Taiping Revolutionary Movement*, 210.
100. See Bloch's lecture "The Transvaal War" in London, reported in the *Journal of the Royal Service Institution* (1901), 1341.
101. *Deutsche Revue* (July 1901).

CHAPTER THREE: THE ORIGINS OF GUERRILLA DOCTRINE

1. The only work which gives a short historical survey of partisan and guerrilla doctrine in the eighteenth and nineteenth century is Werner Hahlweg, *Guerilla. Krieg ohne Fronten* (Stuttgart, 1968). F. Tudman, *Rat protiv rat* (Zagreb, 1957) shows familiarity with the literature but does not discuss it in detail.
2. Jomini, *Précis de l'art de la guerre* (Paris, 1838), I, 72–75.
3. For instance the Marquis de Santa Cruz de Marzenado, *Réflexions militaires et politiques* (Paris, 1735), II, 233 *et seq.*, and before him the works of Bernardin de Mendoza and of Melzo.
4. An English translation by Robert Scott was published in London in 1816.
5. De la Croix's treatise was among the very first of its kind but even previously Folard's unpublished *De la guerre des partisans* and Basta's *Governo della cavalleria leggiera* had been written.
6. De la Croix, *Traité de la petite guerre* (Paris, 1752); see M. Jähns, *Geschichte der Kriegswissenschaften* (München, 1891), III, 2710.
7. Grandmaison, *La petite guerre* (Paris, 1756), 15.
8. The Count de Saxe, *Reveries or Memoirs upon the Art of War* (London, 1757), 173.
9. Grandmaison, op. cit., 149, 174.
10. De Jeney, *Le Partisan ou l'art de faire la petite guerre* (The Hague, 1749), 6.
11. The literature is discussed in Jähns, op. cit., 2712–2717.
12. Baron de Wüst, *L'Art militaire du partisan* (The Hague, 1768), 98–99.
13. The following quotations are from the German translation of Emmerich's book, *Der Partheygänger im Kriege* (Dresden, 1791).
14. Emmerich, op. cit., 5–6.
15. Ibid., 106–110.
16. Ibid., 55.
17. Ibid., 107.
18. Ibid., 111.

19. J. von Ewald, *Abhandlung von dem Dienst der leichten Truppen* (Schleswig, 1796), and see also his subsequent *Belehrungen* in three volumes.
20. Von Ewald, *Abhandlung*, op. cit., IX.
21. Ibid., 274–275.
22. Ibid., 280.
23. His study is quoted here on the basis of the 1829 edition which was incorporated into a larger book, *Die Lehre vom Kriege*.
24. Valentini, op. cit., 5–6.
25. Napoleon on the Cossacks, *Bulletin*, 21 December 1806; on the German Free Corps see letter to the Empress, 16 May 1813; on the Vendée see *Mémoires;* about Spain see *Bulletin*, 15 November 1808 and *Mémorial*.
26. Guibet, *Essai général de tactique*, first published in 1772, in *Oeuvres militaires du comte Guibet* (Paris, 1803), I, *passim*.
27. Bülow, *Neue Taktik der Neuern wie sie seyn sollte* (Leipzig, 1805), II, 24.
28. Bülow, *Militärische und vermischte Schriften* (Leipzig, 1853), 52 *et seq*.
29. Jomini, *Précis de l'art de la guerre* (Paris, 1838), I, 72.
30. J. B. Schels, *Leichte Truppen; kleiner Krieg* (Vienna, 1813–1814) and Felipe de San Juan, *Instrucción de guerrilla* (Santiago, 1823).
31. See for instance M. Tevis, *La Petite Guerre* (Paris, 1855).
32. Carl von Clausewitz, *Schriften — Aufsätze — Studien — Briefe*, ed. W. Hahlweg (Göttingen, 1966), 226–539.
33. Ibid., 240.
34. Ibid., 381.
35. Ibid., 309.
36. Ibid., 436.
37. Ibid., 413–414.
38. *Preussische Gesetzsammlung*, no. 184 (1813), 79.
39. Carl von Decker, *Der kleine Krieg im Geiste der neueren Kriegsführung* (1822) — I have quoted from the French edition *De la petite guerre . . .* (Paris, 1845) — and Karol Bogumil Stolzman, *Partyzantka czyli wojna dla ludow powstajacach najwlasciwcza* (Paris and Leipzig, 1844).
40. Le Mière de Corvey, *Des partisans et des corps irréguliers* (Paris, 1823).
41. Ibid., X.
42. Ibid., 105.
43. Ibid., 145.
44. Decker, op. cit., 318, 325.
45. Carl von Decker, *Algerien und die dortige Kriegsführung* (1844).
46. Ibid., II, 155.
47. Stolzman, op. cit., 192 *et seq*.
48. Ibid., 38 *et seq*.
49. General Chrzanowski, *Über den Parteigänger-Krieg* (Berlin, 1846), 4, 5, 19; the original Polish version of this book was not accessible.
50. J. M. Rudolph, *Über den Parteigängerkrieg* (Zürich, 1847), 12.
51. Gingens-La Sarraz, *Les Partisans et la défense de la Suisse* (Lausanne, 1861), 130.
52. Translated under the title *Total Resistance* (Boulder, Col., 1965).
53. Commandante J. I. Chacon, *Guerras irregulares* (Madrid, 1883), 2 vols., Matija Ban, *Pravila o cetnikoj vojni* (Belgrade, 1848).
54. W. Rüstow, *Die Lehre vom kleinen Krieg* (Zürich, 1864), 12–17.
55. Ibid., 320, 340.
56. The Brussels Declaration (1874) was a compromise between countries with great armies, such as Russia, urging that armed bands should not have the rights of belligerents and Spain, Belgium, Holland and Switzerland insisting that no limitation be placed on the right of inhabitants in occupied territory to defend their country. (L. Nurick and R. W. Barret, "Legality of Guerrilla Forces under the Laws of War," *American Journal of International Law* [1946], 565.)

57. Albrecht von Boguslawski, *Der kleine Krieg und seine Bedeutung für die Gegenwart* (Berlin, 1881), 23. Boguslawski was also the author of a history of the war in the Vendée.
58. Published in several installments in the *Journal des Sciences Militaires* (September–December 1880).
59. Devaureix, op. cit., 450–451. But on other occasions Napoleon said that Soult was one of his most gifted generals.
60. V. Charenton, *Les Corps francs dans la guerre moderne* (Paris, 1900), 172.
61. T. Miller Maguire, *Guerrilla or Partisan Warfare* (London, 1904). Maguire (1849–1920) was a barrister and a successful army "coach" lecturing and writing about strategy and great campaigns. Among his students were Allenby, Gough, Wilson and other military leaders of the First World War. He was elected Fellow of the Royal Historical Society in 1866. Maguire had previously published a series of articles in eleven installments in the *United Service Magazine* between May 1901 and March 1902. He had first written about the subject in 1896 but "was only turned into the utmost contempt by experts and politicians for my pains. General Lloyd did the same in Woolwich, only to be laughed at." (November 1901), 172.
62. Ibid.
63. R. F. Johnson, *Night Attacks* (London, 1886), and G. B. Malleson, *Ambushes and Surprises* (London, 1885).
64. Major L. J. Shadwell, *North West Frontier Warfare* (Calcutta, 1902), 2–5; Brigadier General C. C. Egerton, *Hill Warfare* (Allahabad, 1899), 16; see also Lieutenant Colonel A. R. Martin, *Mountain and Savage Warfare* (Allahabad, 1898), *passim*.
65. Callwell, *Small Wars* (London, 1899), 8, 104. Callwell was an artillery officer who had been seconded to serve in intelligence. He saw action in Afghanistan and South Africa and resigned from the army when passed over for promotion in 1909. He had published sketches from army life which had apparently offended his superior. He was recalled to duty in 1914, became chief of operations in the war office, was promoted to major-general and knighted. Callwell died in 1928.
66. Ibid., 108, and Maguire, op. cit., 106.
67. Maguire, op. cit., 59.
68. Ibid., 61.
69. The following quotations are from the first edition of Callwell's book, published by Her Majesty's Stationery Office, London in 1896, 107 *et seq*.
70. Ibid., 115.
71. Ibid., 108–109.
72. T. H. C. Frankland, "Notes on Guerilla [*sic*] Warfare," *The United Service Magazine*, New Series, vol. 33, 183.
73. A. T. Barteniev, *Voenno-Istoricheski Sbornik* (St. Petersburg, 1912), 137.
74. *Voennaia Entsiklopedia* (Petrograd, 1914), 303.
75. Colonel Vuich, *Malaia Voina* (St. Petersburg, 1850), viii, 239.
76. F. Gershelman, *Partisanskaia Voina* (St. Petersburg, 1885), 11.
77. Ibid., 17.
78. Ibid., 241.
79. "Ostpreussen und der Tartaren Ritt," *Allgemeine Militair Zeitung*, no. 92 (1883).
80. *Partisanskie Deistvia* (St. Petersburg, 1894).
81. C. Hron, *Der Parteigänger-Krieg* (Vienna, 1885).
82. Wlodimir Stanislaus Ritter von Wilczynski, *Theorie des grossen Krieges* (Vienna, 1869), 121. On occasion Wilczynski waxed lyrical about partisan warfare: "A war of this kind is the flower, it is the poetry of strategy . . . it is everything which fantasy can imagine and for this reason it is quite impossible to put down firm rules for it." (Page 4.)

83. Hron, op. cit., 6.
84. Ibid., 54.
85. Ibid.
86. A. Ehrhardt, *Kleinkrieg* (Potsdam, 1935), 73–75, and Major General Kerchnawe in *Militär-Wissenschaftliche Mitteilungen* (January–April 1929).
87. The French and Italian literature on Buonarroti is surveyed in Samuel Bernstein, *Buonarroti* (Paris, 1949), and Elizabeth Eisenstein, *The First Professional Revolutionist* (Cambridge, Mass., 1959).
88. *Della guerra nazionale d'insurrezione per bande applicata all'Italia. Trattato dedicato ai buoni Italiani da un amico del paese* (Italia, 1830), 2 vols. The book is now exceedingly rare; two copies have been located in Bologna and Milan. See Piero Pieri, "Carlo Bianco Conte di Saint Jorioz e il suo Trattato . . . ," in *Bolletino Storico Bibliografico Subalpino* (Torino, 1957–1958), 2 parts. On Carlo Bianco and Buonarroti see Alessandro Galante Garrone, *Filipo Buonarroti e i rivoluzionari dell'ottocento* (Einaudi, 1951), 333–342.
89. Bianco, op. cit., II, 14.
90. *Manuale pratico del Rivoluzionario Italiano desunto dal trattato sulle guerra d'insurrezione per bande* (Italia, 1833).
91. Anonimo, "Della guerra di parteggiani," *La Minerva napolitana* (February 1821); quoted in Egidio Liberti, ed., *Techniche della guerra partigiana nel Risorgimento* (Florence, 1972), 64–65.
92. "Ristrettissimi mezzi grandiosi risultamenti," E. Liberti, *Techniche* . . . , 166–168.
93. G. Pepe, *Memoria su i mezzi che menano all' Italiana indipendenza* (Paris, 1833), also *L'Italia Militare* (Paris, 1836). A summary is contained in E. Liberti, 171–181.
94. *Studii sulla guerra d'indipendenza scritti da un uffiziale italiano* (Torino, 1847); this is a shorter and modified version of the 1817 manuscript.
95. Enrico Gentilini, *Guerra degli stracorridori o guerra guerriata* (Capolago, 1848); reprinted in Liberti, 581 *et seq.* and (in part) in Gian Mario Bravo, *Les Socialistes avant Marx* (Paris, 1970), III; "Stracorridori" is an archaic military term relating to cavalry scouts. On Gentilini, Luigi Bulferetti, *Socialismo risorgimentale* (Turin, 1949), 176–194.
96. Guiseppe Budini, *Alcune idee sull'Italia* (London, 1843); E. Liberti, 181–188.
97. Michele N. Allemandi, "Del sistema militare svizzero applicabile al popolo italiano," *Italia del Popolo* (1850).
98. Giuseppe La Masa, *Della guerra insurrezionale in Italia* . . . (Turin, 1856).
99. Carlo Pisacane, *Saggi storici-politici-militari sull'Italia* (Milan, 1858–1860), IV, 143; among the writers of the 1860s G. B. Zafferoni, *L'Insurrezione armata* (Milan, 1868), should be singled out.
100. *Czy Polacy moga sie wybic na niepodleglosc?*
101. I. Bem, *O powstaniu narodowym w Polsce* (Paris, 1846–1848), republished in Warsaw in 1956; H. M. Kamienski, *Wojna Ludowa* (Paris, 1866), part III; see also his *O prawda zywotnich* . . . (Brussels, 1844), 283.
102. Bystrzonowski's book *Siec strategiczna* . . . appeared in French translation in Paris in 1842 *(Notice sur le réseau stratégique de la Pologne . . .).* Like Pisacane, Bystrzonowski had participated in the war against Abd el-Kader in Algeria and he refers to the experience gained there.
103. A. Jelowicki, *O powstaniu i wojnie partyzanskiej* (Paris, 1835).
104. W. Nieszokoc, *O systemie wojny partyzanskiej wzniesionym wsrod emigracji* (Paris, 1838), 3–30.
105. "Konspiratomania wloska, szczepiona na swawoli szlacheckiej polskiej," *Demokrata Polski* (February 8, 1845). Mieroslawski's views are discussed in detail in Marian Zychowski's massive biography, *Ludwick Mieroslawski 1814–1878* (Warsaw, 1963), 183 *et seq.* Further bibliographical details on the Polish partisan debate are found in L. Przemski, "Zagadnienie wojny party-

zanskiej w przededniu Wiosny Ludow," *Wiosna Ludow* (Warsaw, 1948), part I, 349–417.

106. *Guida pratica del perfetto partigiano.* The treatise was republished after the liberation of Rome in 1944: Liberti, 168–171.
107. Emilio Lussu, *Théorie de l'insurrection* (Paris, 1971), 26–27; "Instructions pour une prise des armes," *Militant Rouge* (December 1926–1928), and in *Archiv für die Geschichte der Arbeiterbewegung* (1930).
108. Samuel Bernstein, *Auguste Blanqui and the Art of Insurrection* (London, 1971), 198.
109. Maurice Dommanget, *Auguste Blanqui, Des origines à la Revolution de 1848* (Paris, 1967), 185 *et seq.*
110. Quoted from Auguste Blanqui, *Politische Texte* (Frankfurt, 1968), 157–163.
111. A vivid description of barricade fighting was given in Victor Hugo's *Les Misérables*, bearing out Blanqui's criticism.
112. Bakunin frequently referred to robbers as the most revolutionary element in society. See for instance his letter to Nechaev on 2 June 1870, published by M. Confino, *Cahiers du monde russe et soviétique* IV (1966), 652.
113. "Rules for the Conduct of Guerrilla Bands" in Mazzini, *Life and Writings* (London, 1864), I, 369.
114. Ibid., 372.
115. Ibid., 378. *Istruzione per le bande nazionali*, was published in Lausanne in 1853; but Mazzini's first writings on the subject date back to the early 1830s: "L'Istruzione generale per gli affratellati nella Giovine Italia" (1831); "Della guerra insurrezionale" published in 1832 in the fifth number of *Giovine Italia;* several introductions to new editions of his work first published in 1832; and lastly *Istruzione del condottiere delle bande nazionali* (1853).
116. Marx-Engels, *Werke* (Berlin [East], 1960), VIII, 95.
117. From the Abbé de Pradt's *Mémoires historiques sur la révolution d'Espagne* (1816).
118. First published in the *New York Tribune* (30 October 1845); quoted from Karl Marx and Frederick Engels, *Revolution in Spain* (New York, 1939), 55.
119. Engels, *Der bisherige Verlauf des Krieges gegen die Mauren* (January 1860), and Marx-Engels, *Werke*, XIII, 548 *et seq.*
120. Marx-Engels, *Werke*, XVII, 131, 169, 187.
121. Ibid., VI, 387.
122. Engels, "Die Aussichten des Krieges," *Pall Mall Gazette* (8 December 1870), and *Werke*, XVII, 197.
123. "Kriegsführung im Gebirge einst und jetzt," in Marx-Engels, *Werke*, XII, 115.
124. Engels, "Persien-China," first published in the *New York Daily Tribune* (5 June 1857) and Marx-Engels, *Werke*, XII, 214.
125. Marx-Engels, *Werke*, XVII, 177.
126. Engels, "Introduction" (1895) to Karl Marx, *Class Struggle in France* (New York, 1964), 21–25.
127. "Es lebe der Tyrannenmord," *Freiheit* (London, 19 March 1881). The newspaper later appeared in New York.
128. New York, 1884, p. 1. The German title of the opus was *Revolutionäre Kriegswissenschaft. Ein Handbuch zur Anleitung betreffend Gebrauch und Herstellung von Nitro-Glyzerin, Dynamit, Schiessbaumwolle, Knallquecksilber, Bomben, Brandsätzen, Giften usw. The Anarchist Cookbook* published in New York in 1971 was modeled after Most's *Revolutionary Warfare* and acknowledges its intellectual debt.
129. General de Brack, *Advanced Posts of Light Cavalry* (London, 1850).
130. Henry Lachouque, *Napoleon's Battles* (London, 1966), 454.
131. "Précis des guerres de Jules César," *Mémoires de Napoléon*, IV, 18.

CHAPTER FOUR: THE TWENTIETH CENTURY (I): BETWEEN TWO WORLD WARS

1. Gouverneur H. Schnee, *Deutsch-Ostafrika im Weltkriege* (Leipzig, 1919), 171.
2. General von Lettow-Vorbeck, *Heia Safari* (Leipzig, 1920), 88.
3. *Operations in Waziristan 1919/1920*, Catalogue C.W.4 (Calcutta, 1921), 139.
4. Charles Horden, *Military Operations. East Africa* (London, 1941), 514 (*History of the Great War based on official documents*, HMSO).
5. The first and most reliable was *Meine Erinnerungen aus Ostafrika* (Leipzig, 1920); the most successful written for the benefit of the young generation was *Heia Safari* (Leipzig, 1920) and many subsequent editions; the most recent was *Mein Leben* (Bibrach, 1957).
6. Horden, *Military Operations*; only the first volume leading up to September 1916 has appeared.
7. Brian Gardner, *German East* (London, 1963); Leonard Mosley, *Duel for Kilimanjaro* (London, 1963); J. H. Sibley, *Tanganyikan Guerrilla* (London, 1971).
8. T. E. Lawrence, *Revolt in the Desert* (London, 1927), 95.
9. G. Macmunn and C. Falls, *Military Operations. Egypt and Palestine* (London, 1928), I, 237–240.
10. Lawrence, *Revolt*, 44.
11. Ibid., 202, 264, 314.
12. Schnee, *Deutsch-Ostafrika*, 29.
13. Lettow-Vorbeck, *Mein Leben*, 85.
14. Lettow-Vorbeck, *Meine Erinnerungen*, 17.
15. R. Meinertzhagen, *Army Diary* (London, 1960), 96.
16. Ibid., 205.
17. Lettow-Vorbeck, *Meine Erinnerungen*, 17.
18. Lettow-Vorbeck, *Heia Safari*, 132.
19. *Partisanskoe dvizhenie v Zapadnoi Sibiri*, 22, quoted in A. M. Spirin, *Klassi i partii v grazhdanskoi voini v Rossii* (Moscow, 1968).
20. G. Stewart, *The White Armies of Russia* (New York, 1933), 141.
21. R. Luckett, *The White Generals* (London, 1971), 212.
22. On partisan activities in the Civil War, *Istoria grazhdanskoi voini* (Moscow, 1959), IV. In addition there are monographs on local guerrilla activities in Omsk (M. V. Naumov, 1960), Irkutsk (A. G. Solodyankin, 1960), and the Soviet Far East (S. S. Kaplin, 1960).
23. W. H. Chamberlin, *The Russian Revolution* (New York, 1965), II, 215–217.
24. M. Kubanin, *Makhnovshchina* (Leningrad, n.d.), *passim*; for an excellent appraisal of the Makhno movement in English see David Footman, *Civil War in Russia* (London, 1961), 245–305.
25. Luckett, *The White Generals*, 278.
26. Chamberlin, *The Russian Revolution*, 236.
27. Chamberlin, *The Russian Revolution*, 437–440. The story of the Antonov movement has been told in a novel by Nikolai Virta (*Odinochestvo*) and a collection of essays (*Antonovshchina*) by S. Evgenov and O. Litovski.
28. F. Novitski quoted in R. Pipes, *The Formation of the Soviet Union* (Cambridge, Mass., 1964), 179. See also J. Castagné, *Les Basmatchis* (Paris, 1925), and S. Ginsburg, "Basmachestvo v Fergane," *Novy Vostok* (1925), 10–11.
29. *Malaya Sovetskaya Entsiklopediya* (Moscow, 1958), I, 825.
30. "Borba s kontrrevoliutsionnim vosstaniam," *Voina i revoliutsiya*, 7–9 (1926).
31. Ibid., 9 (1926). For another interesting Soviet treatment of counterinsurgency see S. Dubrovski, "Grigorovshchina," *Voina i revoliutsiya*, 4 and 5 (1928).
32. Hagen Schulze, *Freikorps und Republik 1918–1920* (Boppard, 1969), 39. The main studies on the Freikorps are E. von Schmidt-Pauli, *Geschichte der Freikorps* (Stuttgart, 1936); F. W. von Oertzen, *Die deutschen Freikorps* (Munich, 1938); and R. G. L. Waite, *Vanguard of Nazism* (Cambridge, Mass., 1952).

33. Von Oertzen, *Die deutschen Freikorps*, 21; Schulze, *Freikorps und Republik*, 41. There is a very detailed description of military operations in *Darstellungen aus den Nachkriegskämpfen deutscher Truppen und Freikorps*, edited by the Forschungsanstalt für Kriegs- und Heeresgeschichte (Berlin, 1936–1940), 8 vols. It has however little to say about the spirit of the Freikorps.
34. Ernst Sontag, *Korfanty* (Kitzingen, 1954); S. Sopicki, *Wojciech Korfanty* (Katowice, 1935).
35. H. von Riekhoff, *German-Polish Relations 1918–33* (Baltimore, 1971), 47.
36. Schulze, *Freikorps und Republik*, 44.
37. F. Sieburg, *Es werde Deutschland* (Frankfurt, 1933), 20.
38. On the Freikorps spirit see Schulze, *Freikorps und Republik*, 54–66; E. von Salomon, *Die Geächteten* (Berlin, 1930), and his *Freikorpskämpfer* (Berlin, 1938); Arnolt Bronnen, *Rossbach* (Berlin, 1930). For the resentment against sections of the old conservative officers corps see Heimsoth, *Freikorps greift an* (Berlin, 1930), 80–81.
39. *Encyclopaedia Britannica* (1957), X, 950.
40. Lenin, *Selected Works* (New York, 1967), I, 581; the article first appeared in *Proletary* (29 August 1906).
41. Lenin, *Collected Works* (New York, 1962), XI, 213; the article was first published in *Proletary* (30 September 1906).
42. Ibid.
43. Ibid.
44. Lenin, *Werke*, XI, 159.
45. Lenin, *Collected Works*, XXII, 311.
46. See for instance *V. I. Lenin o voine, armii i voennoi nauke* (Moscow, 1965).
47. John Erickson, "Lenin as Civil War Leader," in *Lenin, the Man, the Theorist, the Leader*, L. Schapiro and P. Reddaway, eds. (London, 1967), 174.
48. Lenin, *Werke*, XXIX, 545. See also *Werke*, 247, 281, 514.
49. Lenin, *Ausgewählte Werke*, 2 vols. (Moscow, 1946–1947), II, 595.
50. Recent anthologies on Leninism and guerrilla warfare usually cover ground that has only tenuous guerrilla connection; one would look in vain in them for what Lenin really said and wrote about *partisanshchina*. See for instance W. J. Pomeroy, ed., *Guerrilla Warfare and Marxism* (New York, 1968).
51. *Military Writings by Leon Trotsky* (New York, 1971), 25, 54. Trotsky's denunciations of guerrillaism caused some headaches to his latter-day disciples, who argued that he was merely opposed to post-revolutionary guerrilla war, not to guerrilla war *per se*. Before 1917 he had been neither for nor against it; the question was then scarcely of consequence to him. Like Lenin, he had found nothing wrong with the Latvian insurrection of 1905, but he could not envisage guerrilla war playing an important role in revolutionary strategy in the industrially developed countries. He was not concerned with the rest of the world because he did not expect socialist revolutions in the colonies.
52. From a speech in April 1922, quoted in *Military Writings by Leon Trotsky*, 81.
53. Quoted in E. Wollenberg, *The Red Army* (London, 1938), 38.
54. Joseph Stalin, *Marxism and the National and Colonial Question* (London, n.d.), 154. Much importance is attached to this quotation by C. A. Dixon and O. Heilbrunn in *Communist Guerrilla Warfare* (London, 1954), 24.
55. Alexander Orlov, *Handbook of Intelligence and Guerrilla Warfare* (Ann Arbor, 1963), *passim*.
56. A. Neuberg, *Armed Insurrection* (London, 1970); the original German edition, *Der Bewaffnete Aufstand*, appeared in 1928. "A. Neuberg" was a collective pseudonym for O. Piatnitsky and other Soviet and foreign Communist leaders including Marshal Tukhachevski and Togliatti.
57. Neuberg, *Armed Insurrection*, 259.
58. For instance, A. Kolan, "Partisanskaya voina v okkupirovannikh rayonnakh Kitaya," *Kommunisticheskii Internatsional*, 6 (1940), 60 *et seq*.

59. James Connolly, "Street Fighting — Summary," published first in *Workers' Republic* (24 July 1915). Quoted from Connolly's *Selected Writings* (London, 1973), 230.
60. Major Henri le Carron, *Twenty-five years in the Secret Service* (London, 1892).
61. John Devoy, *Recollections of an Irish Rebel* (New York, 1929), 65.
62. Among the recent accounts of the Easter Rising are Desmond Ryan, *The Rising* (Dublin, 1957); James Gleeson, *Bloody Sunday* (London, 1962); Max Caulfield, *The Easter Rebellion* (London, 1964).
63. Tom Barry, *Guerrilla Days in Ireland* (Dublin, 1949), 7–11.
64. On Collins, see Piaras Beaslai, *Michael Collins and the Making of a New Ireland* (London, 1926), and Margery Forester, *Michael Collins. The Lost Leader* (London, 1971).
65. Barry, *Guerrilla Days*, 26.
66. T. P. Coogan, *The IRA* (London, 1970), 47.
67. Ibid., 274.
68. J. Swire, *Bulgarian Conspiracy* (London, 1939), 103. See also J. Perrigault, *Bandits d'Orient* (Paris, 1931); Stoyan Christowe, *Heroes and Assassins* (London, 1935); A. Doolard, *Quatre mois chez les comitadjis* (Paris, 1932).
69. F. Tudman, *Rat protiv rat* (Zagreb, 1957), 109.
70. L. Zarine quoted in Albert Londres, *Terror in the Balkans* (London, 1935), 171.
71. David S. Woolman, *Rebels in the Rif* (London, 1959), 80; Augusto Vivero, *El derrumbamiento* (Madrid, 1922), 161; and for a general account of the war, Carlos Hernandez de Huerrera and Tomas Garcia Figueras, *Acción de España en Marruecos* (Madrid, 1929), I.
72. Stanley G. Payne, *Politics and the Military in Modern Spain* (Stanford, 1967), 168.
73. Woolman, *Rebels in the Rif*, 155.
74. Major General C. W. Gwynn, *Imperial Policing* (London, 1936), 300.
75. There is no satisfactory detailed account of the rebellion of 1936–1939. John Marlowe, *Rebellion in Palestine* (London, 1946) is a brief and reliable survey; *Sefer Toldot Hahagana* (Jerusalem, 1964), II, pt. 2, is a survey with the emphasis on Jewish defense rather than Arab attack.
76. Marlowe, *Rebellion in Palestine*, 158.
77. *Sefer Toldot Hahagana*, 766.
78. Marlowe, *Rebellion in Palestine*, 190.
79. *Sefer Toldot Hahagana*, 765.
80. On the "Plan of Ayala," John Womack, *Zapata and the Mexican Revolution* (London, 1972), 175 *et seq*,; on Pancho Villa, Guzman, *Pancho Villa*, and Celia Herera, *Francisco Villa* (Mexico, 1964). For the general background, Robert E. Quirk, *The Mexican Revolution 1914/5* (Bloomington, 1960); Alfonso Taracena, *La tragedia Zapatista* (Mexico, 1931), and *Venustiano Carranza* (Mexico, 1963); F. Tannenbaum, *Peace by Revolution* (New York, 1933); Jose T. Melendez, ed., *Historia de la revolución Mexicana* (Mexico, 1936), 2 vols.
81. On the military aspects of the Zapatista operations see Juan Barragan Rodriguez, *Historia del ejército y de la revolución constitucionalista* (Mexico, 1946), and Jesus Silva Herzog, *Breve historia de la revolución Mexicana* (Mexico, 1960), 2 vols.
82. On Prestes's attitude to guerrillaísm see R. H. Chilcote, *The Brazilian Communist Party* (New York, 1974), 88–89.
83. For an excellent summary of the Prestes campaign see F. R. Allemann, *Macht und Ohnmacht der Guerilla* (Munich, 1974), 25–45; more detailed descriptions are Helio Silva, *1926: A grande marcha* (Rio de Janeiro, 1971); N. Werneck Sodré, *Historia militar do Brasil* (Rio de Janeiro, 1967); and L. M. Lima, *A Coluna Prestes: marchas e combates* (São Paulo, 1945).
84. The only detailed account is Neil Macauley, *The Sandino Affair* (Chicago, 1967); on his anti-Americanism, ibid., 207. For a military assessment R. W.

Peard, "The Tactics of Bush Warfare," *Infantry Journal* (September–October 1931).

85. E. Lister, "Lessons from the Spanish Guerrilla War 1939–51," *World Marxist Review* (February 1965). Guillen's criticism appears in his *Historia de la revolución española* (Buenos Aires, 1962).
86. Barton Whaley, *Guerrillas in the Spanish Civil War* (Detroit, 1969), 30. For the general background Hugh Thomas, *The Spanish Civil War* (London, 1961), and Pierre Broué and Émile Témime, *La révolution et la guerre d'Espagne* (Paris, 1961).
87. Whaley, *Guerrillas*, 67 *et seq*.
88. Payne, *Politics and the Military*, 391.
89. Ehrhardt, *Kleinkrieg*, 3rd ed. (Berlin, 1944), 111; the first edition appeared in 1935.
90. Ibid., 102.
91. Lieutenant Colonel C. E. Vickery, "Small Wars," *Army Quarterly* (July 1923), 307.
92. *Army Quarterly* (January 1927), 349. The same point was made at the time by an American officer: "We must never overlook the fact that behind and over us is that force known as Public Opinion in the United States. . . . In small wars we are at peace no matter how thickly the bullets are flying." Major H. H. Utley, "An Introduction to the Tactics and Technique of Small Wars," *Marine Corps Gazette* (May 1931), 51.
93. Ibid., 353.
94. K. F. Nowak, ed., *Die Aufzeichnungen des Generalmajors Max Hoffmann* (Berlin, 1928), II, 328, 373–377.

CHAPTER FIVE: THE TWENTIETH CENTURY (II): PARTISANS AGAINST HITLER

1. *Orlovskaya oblast v godi velikoi otechestvennoi voini* (Orel, 1960), quoted in John A. Armstrong, ed., *Soviet Partisans in World War II* (Madison, 1964), 37. A recent Soviet source gives the total number of partisans active in the Orel region as 16,300. V. N. Andrianov, *Voina v tylu vraga* (Moscow, 1974), I, 113.
2. *European Resistance Movements 1939–1945* (Oxford, 1960), 376.
3. The use of the term partisan was banned by Himmler in August 1942 "for psychological reasons."
4. Horst Rohde, *Das deutsche Wehrmachttransportwesen im zweiten Weltkrieg* (Stuttgart, 1971), 331.
5. J. Marianovitch and Pero Morache, *Nash oslobodilchki rat i narodna revolucia 1941–45* (Belgrade, 1958), 311; Walter Hubatsch, ed., *Kriegstagebuch des Oberkommandos der Wehrmacht* (Frankfurt, 1963), III, pt. 2, 1618.
6. Tito, *Vojna dela* (Belgrade, n.d.), I, 128.
7. According to V. Strugar the partisans diverted up to 55 enemy divisions (V. Strugar, *Der jugoslawische Volksbefreiungskrieg 1941 bis 1945* [East Berlin, n.d.], 300); according to German dates, the total strength of Axis forces dispatched in 1942 against Tito and Mikhailovich was one German division, nine battalions of the *Landwehr*, seven Italian divisions and some 44,000 Croat soldiers — altogether about 120,000 men (*Kriegstagebuch*, II, pt. 1, 138). This figure is identical with Tito's own estimate for 1943: 120,000 Axis soldiers faced 20,000 of his own men.
8. The high incidence of party officials in the partisan command is noted in the official Soviet history of the war, *Geschichte des grossen Vaterländischen Krieges der Sowjet Union* (East Berlin, 1963), II, 148; according to this source up to fifty percent of the members of the major partisan units were party members; according to other sources it was only fifteen to thirty percent. The higher figure probably refers to the early stage of the partisan movement, the lower to August 1943.

9. L. N. Bichkov, *Partisanskoe Dvizhenie* (Moscow, 1965), 419.
10. According to Soviet sources the number of Soviet partisans was 90,000 by the end of 1941; their number fell sharply with the beginning of winter, but rose again in April–May 1942. See Andrianov, *Voina v tylu vraga.*
11. Erich Hesse, *Der sowjetrussische Partisanenkrieg 1941–1944* (Göttingen, 1969), 134.
12. Ibid., 206. Hitler wrote in a circular letter in October 1942 that successes were achieved "only where the struggle against banditry has been carried out with utter disregard and brutality." Hubatsch, *Kriegstagebuch*, 237.
13. Gerald L. Weinberg in Armstrong's *Soviet Partisans*, 513; Hesse, *Partisanenkrieg*, 227.
14. Hesse, *Partisanenkrieg*, 229.
15. But even the largest Soviet partisan units were smaller than the guerrilla armies of the civil war. Mamontov's (counterrevolutionary) partisan army in the Altai had a strength of 30,000. There was, of course, not that much difference in equipment at the time between "regulars" and "partisans."
16. Andrianov, "Reidi partisan," in *Voenno — Istoricheski Zhurnal* 3 (1973).
17. Hesse, *Partisanenkrieg*, 248.
18. *Vsenarodnoe Partisanskie dvizhenie v Byelorussii* (Minsk, 1967), I, 146.
19. Among the best-known and most interesting accounts of partisan life are the following: A. Fyodorov, *Podpolnyi obkom deistvuyet* (Moscow, 1947); P. Ignatov, *Zapiski partisana* (Moscow, 1944); S. Kovpak, *Ot Putivlya do Karpat* (Moscow, 1945); G. Linkov, *Voina v tylu vraga* (Moscow, 1951); D. N. Medvedev, *Silnye dukhom* (Moscow, 1951); P. Vershigora, *Lyudi s chistoi sovestyu* (Moscow, 1951); and the collective volume *Sovetskye Partisany* (Moscow, 1960). Some of these books were translated into foreign languages; Medvedev's account was reissued in 1975 in an edition of 150,000.
20. Weinberg in Armstrong's *Soviet Partisans*, 361–388.
21. The Bryansk partisans had a tank battalion (Andrianov, *Voina v tylu vraga*, 110).
22. I. Vinogradov, *Doroga cherez front* (Leningrad, 1964), 7.
23. E. Klink, *Das Gesetz des Handelns: die Operation Zitadelle 1943* (Stuttgart, 1966), 130.
24. Bradley F. Smith and A. F. Peterson, *Heinrich Himmler: Geheimreden 1933 bis 1945* (Berlin, 1974), 163.
25. Hesse, *Partisanenkrieg*, 247.
26. Armstrong, quoted in *Soviet Partisans*, 38.
27. Ibid., 39. Kenneth Macksey has noted the inability of the partisans to hamper the Germans when they were winning. "Subtract what few partisans there were in operation before Stalingrad and little difference would have been made to the outcome. The Germans would have penetrated as fast and as far as they did regardless of the partisans." K. Macksey, *The Partisans of Europe in the Second World War* (New York, 1975), 255.
28. For a general survey of the state of Soviet partisan studies, Yu. P. Petrov, "Sostoianie i zadachi razrabotki istorii partisanskoi dvizheniya," *Voprosy Istorii* (1971), 5, 30 *et seq.*
29. F. W. Deakin, *The Embattled Mountain* (London, 1971), 100.
30. Smith and Peterson, *Himmler Geheimreden*, 242.
31. *Zbornik dokumentata i podataka o Narodnooslobiladskom ratu jugoslavenskih naroda* (Belgrade, 1950–1960), consists of 130 volumes. It is subdivided into several series; part 2 includes the documents of the general staff, the others contain the documents on a geographical basis, e.g., pt. I, Serbia, pt. 3, Montenegro, etc. There is a full bibliography, prepared by B. Dajovic and M. Radevic (Belgrade, 1969). See also Strugar, "Aperçu bibliographique," *Revue d'histoire de la deuxième guerre mondiale* (July 1972), 53–62.
32. *Zbornik*, section 6, III, 142; most of the "excesses" in Montenegro took place in July 1941.

33. Ph. Auty, *Tito* (London, 1970), 177; V. Dedijer, *Tito speaks* (London, 1953), is the semiofficial biography.
34. *Zbornik*, series 2, V, 187.
35. Tito, *Vojna Dela*, I, 129.
36. Hubatsch, *Kriegstagebuch*, pt. I, 139.
37. Ibid., III, part 2, 1253.
38. The Yugoslav Communists had themselves been in touch with the Germans. In March 1943 Velebit and Djilas, two of their highest-ranking commanders, traveled to Zagreb and, according to German documents discovered after the war, promised they would stop fighting the Germans if these would leave them alone in their bases in the Sanjak. "The partisan saw no reason for fighting our army — they added that they fought against German troops only in self-defence — but wished solely to fight the Chetniks." (Quoted in W. Roberts, *Tito, Mihailovic and the Allies* [Rutgers University Press, 1973], 108.) Kasche, the German minister in Zagreb, in his dispatches to Berlin advocated a German accommodation with Tito's partisans; militarily it would be useful if the partisans were given a free hand against the Chetniks. These negotiations were cut short by Hitler who said, "One does not negotiate with rebels, rebels must be shot."
39. Auty, *Tito*, 208.
40. There are no comprehensive statistics; seventy-five percent of the soldiers of one Slovene division were peasants; in other parts of Yugoslavia the peasants' share was perhaps even larger (K. Dincic, "La guerre de libération nationale en Yougoslavie," *Revue d'histoire de la deuxième guerre mondiale* [April 1960], 41).
41. W. Venohr, *Aufstand für die Tschechoslowakei* (Hamburg, 1969), 154.
42. V. Prevan, *Slovenské národné povstanie* (Bratislava, 1965). See also Gustav Husák, *Svedectvo o slovenskom národnom povstaní* (Prague, 1954).
43. T. Bor Komorowski, *The Secret Army* (London, 1951), and S. Korbonski, *Fighting Warsaw* (London, 1956), are the main accounts as seen from the Home Army. J. Kirchmayer, *Powstanie warszawskie* (Warsaw, 1959), gives the Communist version; H. v. Krannhals, *Der Warschauer Aufstand 1944* (Frankfurt, 1962), is the most detailed German monograph.
44. Krannhals, *Warschauer Aufstand*, 104.
45. The term "intellectuals" means no more in this context than the fact that they had received an education of sorts. This gave them a decisive advantage over their foes; Abas Kupi, for instance, was illiterate.
46. Julian Amery, *Sons of the Eagle* (London, 1948), 53. For an official history of the partisan movement see L. Kasneci, *Trempée dans le feu de la lutte* (Tirana, 1966). See also U.S. Army Intelligence Division: Resistance Factors and Special Forces Areas, Project No. A-229, *Albania* (Washington, 1957).
47. Hubatsch, *Kriegstagebuch*, III, pt. 2, 152.
48. A. Kedros, *La résistance grecque* (Paris, 1966), 237 *et seq.*; D. George Kousoulas, *Revolution and Defeat* (London, 1965), 160–169. See also Komninos Pyromaglou, *I ethniki antistasis* (Athens, 1975), and Heinz Richter, *Griechenland zwischen Revolution und Konterrevolution (1934–1946)* (Frankfurt, 1973).
49. Woodhouse, in *European Resistance Movements*, I, 382.
50. N. I. Klonis, *Guerrilla Warfare* (New York, 1972), 115.
51. Pyromaglou, in *European Resistance Movements* (Oxford, 1964), II, 317–318.
52. Zachariades, quoted in Kousoulas, *Revolution and Defeat*, 206.
53. Macksey, *The Partisans of Europe*, 191.
54. Henri Noguères, *Histoire de la résistance en France* (Paris, 1972), III, 162.
55. B. Ehrlich, *The French Resistance* (London, 1966), 165.
56. From the immense literature on the French resistance and the Maquis, the following special issues of the *Revue d'histoire de la deuxième guerre mondiale* should be singled out: 1, 30, 35, 47, 55, 61, 85 and 99.

57. The standard works on the Italian resistance after 1943 are those by Valiani (1947), Cadorna (1948), Salvadori (1955), Catalano (1956), Battaglia (1964), Bocca (1966), and the official Communist histories by Secchia-Frascati (1965) and Longo (1965).
58. G. A. Shepperd, *The Italian Campaign, 1943–45* (London, 1968), 302; A. Kesselring, *Soldat bis zum letzten Tage* (Bonn, 1953), 324, 330.
59. R. Battaglia, *Storia della resistenza Italiana* (Turin, 1964), 662.
60. On the major partisan republics see Hubertus Bergwitz, *Die Partisanen Republik Ossola* (Hanover, 1972); Anne Bravo, *La republica partigiana dell'alto Monferrato* (Torino, 1965), *passim;* G. Bocca, *Storia dell'Italia partigiana* (Bari, 1966), 458–503. For Longo's views on partisan tactics see his *Sulla via dell'insurrezione 1943–45* (Rome, 1954), 477–479.
61. Bocca, *Italia partigiana*, 569.
62. Guido Quazza, *La resistenza italiana* (Turin, 1966), 114; Bocca, *Italia partigiana*, 607.
63. Heinz Kuhnrich, *Der Partisanenkrieg in Europa 1939–1945* ([East] Berlin, 1968), 536–537.
64. Ibid.
65. The most authoritative Soviet work, *Istoria velikoi otechestvennoi voini*, III, 446, mentions a total of 120,000 Soviet partisans for 1943, which is lower than the German estimate. See Armstrong, III, 35–36.
66. Marcelle Adler-Bresse, "Témoignages allemandes sur la guerre des partisans," *Revue d'histoire de la deuxième guerre mondiale* (January 1964), 54.
67. F. O. Miksche, *Secret Forces* (London, 1950).
68. Lothar Rendulic, "Der Partisanenkrieg," *Bilanz des zweiten Weltkriegs* (Oldenburg, 1953), and Valdis Redelis, *Partisanenkrieg* (Heidelberg, 1958).
69. Henri Michel, *The Shadow War* (London, 1972), 290.
70. S. Hawes and R. White, *Resistance in Europe, 1939–1945* (London, 1975), 203.

CHAPTER SIX: THE TWENTIETH CENTURY (III): CHINA AND VIETNAM

1. Interview in *Communist International* (February 1938), 177.
2. When General Challe, one of the leaders of the right-wing conspiracy in Algiers, was put on trial he told the judges at great length about the wisdom of Mao. Peter Paret, *French Revolutionary Warfare from Indochina to Algeria* (New York, 1964), 112.
3. Mark Selden, *The Yenan Way in Revolutionary China* (Cambridge, Mass., 1971), 277.
4. A. M. Rumiantsev, *Istoki i evolutsia idei Mao Tse-tunga* (Moscow, 1972), 22–33.
5. Jean Chesneaux, *Peasant Revolts in China* (London, 1973), 78–81.
6. *Selected Works of Mao Tse-tung* (Peking, n.d.), I, 23.
7. Jerome Ch'en, *Mao and the Chinese Revolution* (London, 1965), 112; Maurice Meissner, *Li Ta-chao and the Origins of Chinese Marxism* (Cambridge, Mass., 1967), 81.
8. *Selected Military Writings of Mao Tse-tung* (Peking, 1968), 72, "A single spark can start a prairie fire."
9. S. B, Griffith, *The Chinese People's Liberation Army* (London, 1968), 29.
10. Robert Rothschild, *La chûte de Chiang Kai-chek* (Paris, 1972), 308.
11. Mao's report at the Sixth National Congress of the CCP, quoted in Brandt, Schwartz and Fairbank, *A Documentary History of Chinese Communism* (Cambridge, Mass., 1952), 162.
12. Ch'en, op. cit., 151; John E. Rue, *Mao Tse-tung in Opposition* (Stanford, 1966), 82 *et seq.*
13. J. Ch'en, "The Resolution of the Tsunyi Conference (January 1935)," *China Quarterly* (October 1969), 26.

14. Ch'en, *Mao,* loc. cit., 155–156.
15. The most detailed account of these events is in Dick Wilson, *The Long March* (London, 1971); see also *Veliki pokhod* (Moscow, 1959), and Anthony Garavente, "The Long March," *China Quarterly* (April 1965), 85 *et seq.*
16. John M. Nolan, "The Long March: Fact and Fancy," *Military Affairs* (Summer 1966), 81.
17. Detailed comprehensive studies of Communist guerrilla warfare during the Yenan period do not exist. The general literature is listed in the books by Ch'en, Griffith, Selden and Johnson, mentioned above. Of the eyewitness accounts Edgar Snow's is the most interesting inasmuch as the general background is concerned, whereas E. F. Carlson, *Twin Stars of China* (New York, 1940), is the most illuminating on military affairs. A great many theoretical analyses of Mao's strategy were published in later years. Among the more interesting are Katzenbach and Hanrahan, "The Revolutionary Strategy of Mao Tse-tung," *Political Science Quarterly* (September 1955); Chalmers A. Johnson, "Civilian Loyalties and Guerrilla Conflict," *World Politics* (1964), 287 *et seq*; Howard L. Boorman and Scott A. Boorman, "Chinese Communist Insurgent Warfare 1935–1949," *Political Science Quarterly* (June 1966).
18. Edgar Snow, *Red Star over China* (New York, 1961), 254.
19. Mark Selden, *The Yenan Way in Revolutionary China* (Cambridge, 1971), 66.
20. Lyman P. Van Slyke, *Enemies and Friends* (Stanford, 1967), 95.
21. Stuart Schram, ed., *Mao Tse-tung Basic Tactics* (London, 1967): this is a series of lectures which has not been included in Mao's *Selected Works*. It was virtually forgotten, to be rediscovered around 1970 and published — but only outside China.
22. "On Basic Lessons of Conventional War and the Conditions for Developing Guerrilla Warfare in Northern China," in Thomas W. Robinson, *A Politico-Military Biography of Lin Piao* (Santa Monica, 1971), part 1, 113 *et seq.*
23. *Selected Military Writings,* loc. cit., 138.
24. Ibid., 139.
25. Ibid., 141.
26. "Problems of Strategy in Guerrilla War against Japan" (May 1938), in *Selected Military Writings*, 157–165.
27. Ibid., 168.
28. Ibid., 181.
29. "On Protracted War," in *Selected Military Writings*, op. cit., *et seq.*
30. *Basic Tactics,* 55 *et seq.*
31. Robinson, loc. cit., 124.
32. Report given at the Seventh Congress of the Chinese Communist Party (April 1945) republished in *The Battle Front of the Liberated Areas* (Peking, 1962).
33. Chong-Sik Lee, *Counterinsurgency in Manchuria. The Japanese Experience 1931–1950* (Santa Monica, 1967), VII.
34. Griffith, op. cit., 74.
35. Jerome Ch'en, op. cit., 239–240.
36. Roy Hofheinz, Jr., in A. Doak Barnett, *Chinese Communist Politics in Action* (Seattle, 1967), 67.
37. Chalmers A. Johnson, *Peasant Nationalism and Communist Power. The Emergence of Revolutionary China 1937–1945* (Stanford, 1962), *passim.*
38. J. L. S. Girling, *People's War* (London, 1969), 79.
39. This passage from Mao's "On New Democracy" was deleted from subsequent editions. Stuart J. Schram, Introduction to *Basic Tactics,* op. cit., 29.
40. "On the New Stage," in Schram, *The Political Thought of Mao Tse-tung* (New York, 1963), 113–114.
41. Vo Nguyen Giap, *People's War, People's Army* (New York, 1962), 174.
42. Col. Roberts E. Biggs, "Red Parallel: The Tactics of Ho and Mao," *U.S. Combat Forces Journal* (January 1955).

43. Douglas Pike, *Vietcong* (Cambridge, 1966), 8.
44. Bernard B. Fall, *Viet-Nam Witness 1953–1966* (New York, 1968), 229; John T. McAlister, Jr., *The Beginnings of Revolution* (London, 1969), 206.
45. Joseph Buttinger, *Vietman: A Dragon Embattled* (New York, 1967), II, 760.
46. Jean Lacouture, *Ho Chi Minh* (New York, 1968), 14.
47. Bernard Fall, *Street without Joy* (Harrisburg, 1961), 24. For the general background of the history of the Communist and nationalist movement in Vietnam during this period, see Buttinger, op. cit.; Paul Mus, *Viet-Nam, Sociologie d'une guerre* (Paris, 1950); Philippe Devillers, *Histoire de Viet-Nam de 1940 à 1952* (Paris, 1952); and B. B. Fall, *Le Viet-Minh* (Paris, 1960).
48. Buttinger, II, 739; for firsthand accounts of Vietminh guerrilla warfare see the books by Bernard Fall and Wilfred G. Burchett.
49. Buttinger, II, 741.
50. Bernard B. Fall, *Truong Chih* (New York, 1963).
51. Ibid., 74.
52. Douglas Pike, *War, Peace and the Viet Cong* (Cambridge, Mass., 1969), 142 *et seq.*
53. Giap, *People's War, People's Army*, 48.
54. "The Big Victory, the Great Task," in Patrick J. McGarvey, *Visions of Victory. Selected Vietnamese Communist Writings 1964–68* (Stanford, 1969), 40.
55. Giap, op. cit., 108.
56. Ibid., 109.
57. McGarvey, op. cit., 41.
58. Robert J. O'Neill, *General Giap* (Sidney, 1969), 203.
59. McGarvey, op. cit., 15, 45.
60. George A. Carver, "The Faceless Viet Cong," *Foreign Affairs* (April, 1966), 360.
61. Carver, loc. cit.
62. J. J. Zasloff, *Origins of the Insurgency in South Vietnam, 1954–60: The Role of the Southern Vietminh Cadres,* RAND Memorandum RM-5163/2/ARPA (Santa Monica, 1967), 27.
63. Bernard B. Fall, *Last Reflections on a War* (New York, 1967), 219, 220.
64. Wesley R. Fishel, ed., *Anatomy of a Conflict* (Ithaca, 1968), 425.
65. For a study of Viet Cong political motivation see J. J. Zasloff's RAND Memo. RM 4703/2 2-ISA ARPA (August 1966).
66. John Gerassi, *Towards Revolution* (London, 1971), I, 107.
67. *Associated Press* (11 May 1975).
68. Douglas Pike, "How Strong is the NLF?" in Fishel, loc. cit., 412.
69. Duncanson, in Fishel, op. cit., 428.
70. Denis Warner, *The Last Confucian* (London, 1964), 32; Ton Tat Thien, "Vietnam, A Case of Social Alienation," *International Affairs* (July 1967).
71. Buttinger, op. cit., 984; Vietnam: W. Burchett, *Inside Story of the Guerrilla War* (New York, 1965), 84–89.
72. Pike, loc. cit., 418.
73. David Halberstam, *The Making of a Quagmire* (New York, 1965), 167.
74. Quoted in W. R. Fishel, 500–503.

CHAPTER SEVEN: NATIONAL LIBERATION AND REVOLUTIONARY WAR

1. Yehuda Bauer, *From Diplomacy to Resistance* (Philadelphia, 1970), 319.
2. There are many personal accounts of the anti-British struggle in Palestine between 1944 and 1948 but there is no comprehensive historical study. The paramilitary organizations involved have all published their official histories. *Sefer toldot ha-Hagana* (Tel Aviv, 1963), II, books 1 and 2; David Niv, *Ma'arakhot*

ha-Irgun ha-Zvai ha-Leumi (Tel Aviv, 1965–1973), III, IV; *Kovetz Lehi* (Tel Aviv, 1959); *Sefer ha-Palmach* (Tel Aviv, 1953). See also Natan Yalin Mor, *Lohame Herut Israel* (Tel Aviv, 1974).

3. M. Begin, *The Revolt* (London, n.d. [1951?]), 317.
4. Kousoulas, *Revolution and Defeat*, 236.
5. J. C. Murray, "The Anti-Bandit War" in Greene, *The Guerrilla*, 98.
6. Edgar O'Ballance, *The Greek Civil War 1944–1949* (London, 1966), 181. For a sympathetic though not uncritical account, Dominique Eudis, *The Kapetanios* (London, 1972).
7. The fullest account of the Indonesian struggle for independence is George McT. Kahin, *Nationalism and Revolution in Indonesia* (Ithaca, 1952). Among more recent works with a bearing on the period are J. M. van den Kroef, *The Communist Party of Indonesia* (Vancouver, 1965); B. Dahm, *Sukarno and the Struggle for Indonesian Independence* (Ithaca, 1969); idem, *History of Indonesia in the Twentieth Century* (London, 1971).
8. Arnold C. Brackman, *Indonesian Communism* (New York, 1963), 107.
9. J. H. Brimmell, *Communism in South East Asia* (London, 1959), 255–262; Ruth McVey, *The Calcutta Conference and the South East Asian Uprising* (Ithaca, 1958), *passim*.
10. Lucian W. Pye, *Guerrilla Communism in Malaya* (Princeton, 1956), was an early account of the fighting in Malaya. Others were Gene Hanrahan, *The Communist Struggle in Malaya* (New York, 1954), and V. Purcell, *Malaya Communist or Free?* (London, 1954). The fullest survey is Anthony Short, *The Communist Insurrection in Malaya 1948–1960* (London, 1975); the author was asked by the Malayan government to write the official history of the "emergency" and had access to almost all relevant sources. In the end the Malayan government refused however to give the book its blessing. Other important works are Robert Thompson, *Defeating Communist Insurgency* (London, 1966); Richard Clutterbuck, *The Long Long War* (London, 1967); idem, *Riot and Revolution in Singapore and Malaya* (London, 1973).
11. Short, *Insurrection in Malaya*, 39 *et seq*.
12. Ibid., 51.
13. Clutterbuck, *Riot and Revolution*, 271.
14. Sir Robert Thompson notes that the initial strength of the guerrillas in Malaya and Vietnam was about equal — 4,000 to 5,000 (*Defeating Communist Insurgency*, 47). But unlike the Vietcong, the guerrillas in Malaya had great difficulty in recruiting new cadres because they had no "popular," only a jungle base.
15. Short, *Insurrection in Malaya*, 319.
16. Clutterbuck, *Riot and Revolution*, 211.
17. Ibid., 272.
18. W. J. Pomeroy, ed., *Guerrilla Warfare and Marxism* (New York, 1968), 34–35.
19. Aguinaldo came from a well-to-do landowning family of mixed Chinese and Taganlog stock. He had been municipal captain of his home town and had a reputation as a proficient street fighter. On the struggle between the American army and Aguinaldo's forces, see J. A. Leroy, *The Americans in the Philippines* (New York, 1914), 2 vols.; T. M. Kalaw, *The Philippine Revolution* (Manila, 1925); W. Sexton, *Soldiers in the Sun* (Harrisburg, 1939); G. F. Zaide, *The Philippine Revolution* (Manila, 1954), and Leon Wolff, *Little Brown Brother* (London, 1961). The most recent study is Major Robert T. Yap-Diangco, *The Filipino Guerrilla Tradition* (Manila, 1971).
20. U. S. Baclagon, *The Huk Campaign in the Philippines* (Manila, 1960), 1.
21. Luis Taruc, *He Who Rides the Tiger* (London, 1967), 24.
22. B. T. Bashore in *Osanka*, 196.
23. B. Dasgupta, "Naxalite Armed Struggles and the Annihilation Campaign in Rural Areas," *Economic and Political Weekly*, nos. 4–6 (Bombay, 1973).
24. *Intercontinental Press* (2 June 1975), 741.

25. De Gaulle in conversation with Pierre Laffont, J. R. Tournoux, *La tragédie du Général* (Paris, 1967), 597.
26. The most detailed account of the Algerian war so far is Yves Courrière, *La guerre d'Algérie:* I, *Les fils de la Toussaint* (Paris, 1968); II, *Le temps des léopards* (Paris, 1969); III, *L'heure des colonels* (Paris, 1970).
27. William B. Quandt, *Revolution and Political Leadership. Algeria 1954–1968* (Cambridge, Mass., 1969), 91.
28. Edward Behr, *The Algerian Problem* (London, 1961), 60; Michael K. Clark, *Algeria in Turmoil* (New York, 1959), 58.
29. Quoted in C. and F. Jeanson, *L'Algérie hors la loi* (Paris, 1955), 298.
30. This applies, for instance, to G. Chaliand and A. Humbaraci; Quandt, *Revolution and Political Leadership*, 220.
31. Quoted in J. S. Ambler, *The French Army in Politics* (Columbus, 1966), 331.
32. Title of a book on Algeria by Mohammed Bessaoud (Paris, 1963).
33. *Case studies in Insurgency and Revolutionary warfare: Algeria 1954–62*, Special Operations Research Office (Washington, 1963), 19–29.
34. Dickey Chappelle, "How Castro won," reprinted in Osanka, 325.
35. Thomas, "The Origins of the Cuban Revolution," *The World Today* (October 1963), 490 *et seq.;* Theodore Draper, *Castroism, Theory and Practice* (New York, 1965), 103 *et seq.*
36. Malcolm Deas, "Guerrillas in Latin America: a Perspective," *The World Today* (February 1968), 74.
37. Thomas, *Cuba*, 791.
38. Ché Guevara, *Episodes of the Revolutionary War* (Havana, 1967), 13 *et seq.*
39. Draper, *Castroism*, 25.
40. For a discussion of the splits in the Palestinian resistance, the ideologies of the various groups and a bibliography see below, chapter 8.
41. Precise data have not been published, but occasional figures convey a glimpse of the magnitude of the sums involved. Thus according to a PLO spokesman, the arrears of the Arab states alone amounted to sixty million dollars in late March 1974. ("Voice of Palestine," Cairo Radio, 2 June 1974.)
42. For the general background of Kurdish-Arab relations see C. J. Edmond, *Turks and Arabs* (London, 1957); for descriptions of the Kurdish war, D. Adamson, *The Kurdish War* (London, 1964), and E. O'Ballance, *The Kurdish Revolt 1961–1970* (London, 1973); René Maunes, *Le Kurdistan ou la mort* (Paris, 1967); D. A. Schmidt, *Journey among Brave Men* (Boston, 1964).
43. J. Bowyer Bell, "Endemic Insurgency and International Order: The Eritrean Experience," *Orbis* (Summer 1974), 427–450. Originally there were both Muslims and Christians among the ELF cadres, but they split in 1971 along tribal and religious lines. Both the Saudis and Ghadafi resented the presence of Christians in the ELF but there seems to have been friction from the very beginning; it was difficult to get Christian and Muslim cadres even to eat together. Which did not stop the ELF's proclaiming itself a Marxist movement.
44. *Documents of the National Struggle in Oman and the Arabian Gulf* (London, 1974), 16; R. Fiennes, *Where Soldiers Fear to Tread* (London, 1975), is a firsthand account of the Oman war.
45. Ali Akbar Safayi Farahani, *What a Revolutionary Must Know* (London, 1973), 67. A former schoolteacher, Farahani fought with the Palestinians 1967–1969 and later participated in the Siahkal guerrilla movement in northern Iran. He was killed in 1970.
46. See *Organisations et combats du peuple de l'Iran* (n.p. [Paris?], n.d. [1974?]), *passim.*
47. P. Vielle and Abol Hassan Banisadr, *Pétrole et violence* (Paris, 1974), 107 *et seq.*
48. On *Dev Genc* and the various commando groups which evolved from it see J. M. Landau, *Radical Politics in Modern Turkey* (Leiden, 1974), 41–44.
49. Charles Foley, ed., *The Memoirs of General Grivas* (London, 1964), 135.

50. There is no detailed history of Grivas's campaign. The fullest account is still his own autobiography. See also Charles Foley, *Island in Revolt* (London, 1964); idem, *Legacy of Strife* (London, 1964); Robert Stephens, *Cyprus. A Place of Arms* (London, 1966).
51. For the background of the Mau Mau disorders see L. S. B. Leakey, *Defeating Mau Mau* (London, 1954), and F. D. Corfield, *Historical Survey of the Origins and Growth of Mau Mau* (London, 1960); C. G. Rosberg and J. Noltingham, *The Myth of Mau Mau* (London, 1966). For military aspects of the revolt, F. Kitson, *Gangs and Countergangs* (London, 1960), and Ian Henderson and Philip Goodhart, *The Hunt for Kimathi* (London, 1958). For a Mau Mau point of view, W. Itote, *Mau Mau General* (Nairobi, 1967); Donald L. Barnett and Karari Njama, *Mau Mau from Within* (London, 1966). See also *Life Histories from the Revolution, Mau Mau 1–3* (Richmond, Canada, 1974); these were written apparently by Barnett. (The problems with "autobiographies" of Asian or African guerrillas ghosted by Western well-wishers are manifold. To mention but one example: in the year 1953, *Born of the People*, an autobiography of Luis Taruc, was published in New York. When Taruc later surrendered and left the Communist party, he revealed that the book had been written "with the help of a friend" and edited by José Lava, general secretary of the Communist party; various chapters on theoretical subjects were inserted without his knowledge. Pomeroy, on the other hand, claimed that "this book was actually written by W. J. Pomeroy compiled from interviews with numerous Huk leaders.")
52. Guy Arnold, *Kenyatta and the Politics of Kenya* (London, 1974), 110.
53. D. L. Wheeler and R. Pélissier, *Angola* (London, 1971), 178–179.
54. Richard Gibson, *African Liberation Movements* (London, 1972), 281.
55. The main sources for the war in Guiné-Bissau are B. Davidson, *The Liberation of Guiné* (Harmondsworth, 1969), and Gerard Chaliand, *Lutte armée en Afrique* (Paris, 1967); Lars Rudebeck, *Guinea-Bissau: A Study of Political Mobilization* (New York, 1975), as well as Cabral's essays, *Unité et lutte* (Paris, 1975), all from a PAIGC point of view. The struggle in Angola and Mozambique is surveyed in R. H. Chilcote, *Portuguese Africa* (New York, 1967), and John A. Marcum, *The Angolan Revolution* (Cambridge, Mass., 1969). See also E. Mondlane, *The Struggle for Mozambique* (Harmondsworth, 1969).
56. Gibson, *African Liberation Movements*, 261.
57. K. W. Grundy, *Guerrilla Struggle in Africa* (New York, 1971), chapter 19.
58. "The campaign mounted by the enemy in claiming that the MPLA is a Communist organization can only be seen as propaganda intended to fool our people." A. Neto, *Messages to Companions in the Struggle* (Richmond, Canada, 1972), 27.
59. See chapter 8 below.
60. The best general account of the rise and fall of the Latin American guerrilla movements is Allemann, *Macht und Ohnmacht der Guerilla*. Also of interest are the earlier books by Richard Gott, *Rural Guerrilla in Latin America* (London, 1973), and Luis Mercier Vega, *Technique du contre état* (Paris, 1968). James Kohl and John Litt, *Urban Guerrilla Warfare in Latin America* (Cambridge, Mass., 1974), is a collection of texts with introductory comments. The most important work in Spanish is V. Bambirra, ed., *Dies años de insurrección en America Latina* (Santiago, 1971). For Venezuela see Luigi Valsalice, *Guerriglia e politica: L'esemplo de Venezuela 1962–1969* (Florence, 1973). The most up-to-date bibliography is *Bibliografia guerra revolucionaria y subversión en el continente* (Washington, 1973), published by the Library of the Inter-American Defense College.
61. Deas, "Guerrillas in Latin America," 74.
62. Anti–Mau Mau countergangs had first been used in Kenya and spread much confusion among the guerrillas. In Latin America right-wing terrorist groups emerged in many countries, frequently with the approval of the government or

the army. This applies to the Argentine National Orgainzation Movement (MANO), the Guatemalan NOA and MANO, the Brazilian *Escudrão da Morte*. The Spanish anti-Basque *Guerrilleros de Christo Rey* should also be mentioned in this context. For the urban guerrillas these were of course merely hired agents, just as in the eyes of the extreme right the Communists and Castroists were simply "bandits." Internal war is not the ideal period for detached political and social analysis.

63. For a comprehensive list, *Political Kidnappings 1968–1973*, Staff Study by the House of Representatives Committee on International Security (Washington, 1973). See also Brian M. Jenkins and Janera Johnson, *International Terrorism. A Chronology 1968–1974* (Santa Monica, 1975).
64. *Ulster*, by the *Sunday Times* Insight Team (London, 1972), 194 *et seq*. See also Martin Dillon and Denis Lehane, *Political Murder in Northern Ireland* (London, 1973).
65. See chapter 8 below.
66. D. V. Segré and J. H. Adler, "The Ecology of Terrorism," *Survival* (July–August 1973), 180.
67. B. M. Jenkins, *High Technology Terrorism and Surrogate War. The Impact of New Technology on Low-Level Violence* (Santa Monica, 1975), *passim*. For observations on the international character of terrorism J. Bowyer Bell, *Transnational Terrorism* (Washington, 1975).

CHAPTER EIGHT: GUERRILLA DOCTRINE TODAY

1. Boris Goldenberg, *Kommunismus in Lateinamerika* (Stuttgart, 1971), 361.
2. The basic texts of Latin American guerrilla writing are available in English, French and German. Among the more important general studies are the following: Vania Bambirra, ed., *Diez Años de Insurrección* (Santiago, 1971), 2 vols.; Richard Gott, *Guerrilla Movements in Latin America* (London, 1970); Luis Mercier Vega, *Guerrillas in Latin America* (London, 1969). Hugh Thomas, *Cuba or the Pursuit of Freedom* (London, 1971) and Theodore Draper, *Castroism, Theory and Practice* (London, 1965) are essential for the understanding of the Castro ideology. Some of the best studies on the subject are in German; this refers in particular to Boris Goldenberg, *Kommunismus in Lateinamerika* (Stuttgart, 1971) and Fritz René Allemann, *Macht und Ohnmacht der Guerilla* (München, 1974). The following are also of interest: Günter Maschke, *Kritik des Guerillero* (Frankfurt, 1973); Wolfgang Berner, *Der Evangelist des Castroismus–Guevarismus* (Köln, 1969); Richard E. Kiessler, *Guerilla und Revolution* (Bonn, 1975); Robert F. Lamberg, *Die castristische Guerilla in Lateinamerika* (Hanover, 1971). Of great help to students of the subject are the following bibliographies: Ronald H. Chilcote, *Revolution and Structural Change in Latin America: a Bibliography on Ideology, Development and the Radical Left (1930–1965)* (Stanford, 1970), 2 vols.; Anon., *Bibliografia: Guerra Revolucionaria y Subversión en el Continente* (Washington, 1973).
3. Interview with Andrew St. George, 4 February 1958; Ronald E. Bonachee and Nelson P. Valdes, *Revolutionary Struggle, 1947–1958;* Volume I of the *Selected Works of Fidel Castro* (Cambridge, Mass., 1972), 369.
4. The second Declaration of Havana, 4 February 1962 in M. Kenner and J. Petras, *Fidel Castro Speaks* (London, 1972), 164.
5. Ché Guevara, *Guerrilla Warfare* (London, 1969); the article was originally published in *Cuba Socialista* (September 1963), 1–17.
6. Guevara, *Guerrilla Warfare*, 14.
7. Debray, *Revolution in the Revolution* (New York, 1967), 104–106.
8. Guevara, *Guerrilla Warfare*, 19.
9. Regis Debray, *Strategy for Revolution* (London, 1973), 46–47.

10. Speech at the University of Havana, 13 March 1967, in Kenner and Petras, op. cit., 119.
11. *OLAS: Première Conférence de l'organisation latino-americaine de solidarité* (Paris, 1967), 72.
12. Debray, *Revolution*, 26.
13. Alberto Bayo, *150 Questions to a Guerrilla* (Boulder, 1963).
14. Guevara, *Guerrilla Warfare — a Method;* in Malin, op. cit., 276.
15. Guevara, *Guerrilla Warfare*, 121.
16. The romantic ("Byronic") inspiration of guerrilla leaders was noted first by Davydov and later by Maguire, the London lawyer who, around the turn of the century, was one of the first to present a systematic guerrilla doctrine.
17. "Frente a todos" Bohemia (8 January 1956), in Bonachee and Valdes, 1, 299.
18. T. Draper, *Castroism: Theory and Practice* (London, 1965).
19. Ibid., 55.
20. *Bohemia* (28 July 1957).
21. Liborio Justo, *Bolivia, la revolución derrotada* (Cochabamba, 1967), 261. Quoted in Goldenberg, op. cit.
22. Leo Huberman and Paul Sweezy, eds., *Regis Debray and the Latin American Revolution* (New York, 1969).
23. Douglas Bravo, "Cuba: Rectificación tactica o estrategia," French translation in *Temps Modernes* (July 1971).
24. Regis Debray, *Les Epreuves de Feu. La Critique des Armes* (Paris, 1974), II, 121–122.
25. Ibid., 123.
26. Interview with the Mexican newspaper *Sucesos*, in Vega, op. cit., 242–246.
27. Gott, op. cit., 262–265; Norman Gall, *Teodoro Petkoff: The crisis of the professional revolutionary*, part I, "Years of Insurrection." Field Staff Reports 1972, No. 1, 16 *et seq*; Robert J. Alexander, *The Communist Party of Venezuela* (Stanford, 1969), *passim*.
28. Hector Bejar, *Peru 1965: Notes on a Guerrilla Experience* (New York, 1970), 124.
29. Hugo Blanco, *El Camino de Nuestra Revolución* (Lima, 1964), *passim*; Robert J. Alexander, *Trotskyism in Latin America* (Stanford, 1973), 174–175.
30. For a representative selection of his writings see John Gerassi, ed., *Camilo Torres, Revolutionary Priest* (London, 1973). There is a recent biography: W. J. Broderick, *Camilo Torres* (New York, 1975).
31. Anon., *La Guerrilla por dentro* (Bogotá, 1971), *passim*; the author of this book was the former guerrilla leader Jaime Arenas. Conrad Dentrez, *Les mouvements révolutionnaires en Amerique Latine* (Brussels, 1972); Allemann, loc. cit., 272–274.
32. G. Lora, *Neubewertung der Guerilla* (Berlin, 1973), 142.
33. Alexander Craig, "Urban Guerrilla in Latin America," *Survey* (Summer 1971), 124.
34. The writings of Carlos Marighela have been widely translated; the books and articles of Abraham Guillen are not readily available even in Spanish. A comprehensive bibliography has been supplied by Russell, Miller and Hildner; "The Urban Guerrilla in Latin America," *Latin American Research Review* (Spring 1974). Among the few general studies on the subject, the following ought to be mentioned: Robert Moss, *Urban Guerrillas* (London, 1972); James Kohl and John Litt, *Urban Guerrilla Warfare in Latin America* (Cambridge, Mass., 1974). Ernesto Mayans, ed., *Tupamaros: antologia documental* (Cuernavaca, 1971) is an excellent collection of the main documents on the urban guerrilla in Uruguay.
35. A. Guillen, *Estrategia de la guerrilla urbana* (Montevideo, 1966), 63; quoted in Donald C. Hodges, ed., *Philosophy of the Urban Guerrilla* (New York, 1973), 236.

36. A. Guillen, *El pueblo en armas: estrategia revolucionaria* (unpublished, 1972), quoted in Hodges, 257–258.
37. Hodges, loc. cit., 241.
38. Hodges, loc. cit., 263–277.
39. Carlos Marighela, *For the Liberation of Brazil* (London, 1971), 178–182.
40. Ibid., 47.
41. R. Moss, op. cit., 195.
42. Joao Quartin, *Dictatorship and Armed Struggle in Brazil* (New York, 1971), 194–195.
43. *Minimanual*, in Marighela, op. cit., 81.
44. More books and articles have been written about the Tupamaros than about any other Latin American guerrilla movement. The most important are, in addition to Mayans's collection of documents mentioned above: A. Mercader and Jorge de Vega, *Tupamaros: estrategia y acción* (Montevideo, 1969); Alain Labrousse, *The Tupamaros* (London, 1973); Maria Esther Gilio, *The Tupamaros* (London, 1972).
45. Originally published in the Chilean journal *Punto Final* and frequently reprinted. Quoted here from Kohl and Litt, op. cit., 227–236.
46. Interview with "Urbano," Kohl and Litt, op. cit., 268.
47. Debray, *Les Epreuves de Feu*, loc. cit., 277. This is a variation on one of Debray's favorite theses, first pronounced in the 1960s about the revolutionarization by the revolutionaries of the counterrevolution.
48. *Freedom Struggle*, "By the Provisional IRA" (n.p., 1973), 11.
49. Patxi Isaba, *Euzkadi Socialiste* (Paris, 1971), 98. See also Ortzi, *Historia de Euskadi* (Paris, 1975).
50. Pierre Vallières, *Nègres blancs d'Amerique* (Montreal, 1969).
51. A. Schubert, ed., "Das Konzept Stadtguerrilla," *Stadtguerrilla* (Berlin, 1971), 111. See also *Holger, der Kampf geht weiter, Dokumente und Beitraege zum Konzept Stadtguerrilla* (Gaiganz, 1975).
52. Kollektiv RAF, *Uber den bewaffneten Kampf in Westeuropa* (Berlin, 1971), 47.
53. Schubert, op. cit., 137.
54. H. J. Müller-Borchert, *Guerilla im Industriestaat* (Hamburg, 1973), 108.
55. *Erklärungen von Horst Mahler* (Rote Hilfe, Berlin, 1974), 8.
56. Jerry Rubin, *Do it. Scenario of the Revolution* (New York, 1970), 125.
57. "Communiqué No. 1" in Harold Jacobs, ed., *Weatherman* (Berkeley, 1970), 125.
58. *Scanlans* (January, 1971), 15; *The Berkeley Barb* (15 February 1974).
59. "Communiqué No. 4" in Jacobs, op. cit., 518.
60. *Scanlans*, loc. cit., 14.
61. Philip S. Foner, ed., *The Black Panther Speaks* (New York, 1970), 107, 122.
62. Eldridge Cleaver, *On the Ideology of the Black Panther Party* (n.p., n.d.), 11.
63. *Break de Chains* (New York, 1973), 14.
64. *Break de Chains*, op. cit., 11–12.
65. Frantz Fanon, *Pour la révolution Africaine* (Paris, 1969), 186.
66. F. Fanon, *The Wretched of the Earth* (London, 1967), 64.
67. Irene L. Gendzier, *Frantz Fanon* (New York, 1973), 203.
68. Fanon, *The Wretched of the Earth*, 109.
69. Jack Woddis, *New Theories of Revolution* (London, 1972), 174.
70. Nguyen Nghe, "Fanon et les problèmes de l'indépendence," *La Pensée* (February 1963).
71. Ronald H. Chilcote, "The Political Thought of Amilcar Cabral," *Journal of Modern African Studies*, 3 (1968), 386.
72. Amilcar Cabral, *Revolution in Guinea* (London, 1969), 51.
73. Speech in Havana, January 1966, reprinted in *L'Arme de la Théorie*, I (Paris, 1975) and in *Portuguese Colonies: Victory or Death* (Havana, 1971), 133.
74. Gerard Chaliand, *Armed Struggle in Africa* (New York, 1969), 114.

75. A. Cabral, *Unité et lutte*, II, *La pratique révolutionnaire* (Paris, 1975), 195 *et seq.*
76. Havana speech, loc. cit.: see also Amilcar Cabral, *Die Revolution der Verdammten* (Berlin, 1974), 88.
77. Amar Ouzegane, *Le meilleur combat* (Paris, 1962), 300.
78. *El Moudjahid* (15 November 1957), quoted in André Mandouze, ed., *La révolution algérienne par les textes* (Paris, 1961), 132.
79. This refers, for instance, to Mustafa Talas, *Harb al isabat* (Damascus, n.d.), which was published in several editions. Talas later became the chief of staff of the Syrian army. His book was dedicated to Guevara.
80. Y. Harkabi, *Fedayeen Action and Arab Strategy* (London, 1968), 14.
81. Harkabi, op. cit., 18.
82. Hisham Sharabi, *Palestine Guerrillas* (Washington, 1970), 32.
83. *Min muntalaqat al amal al fidai* (Amman, 1967), 67.
84. Walid Kazziha, *Revolutionary Transformation in the Arab World* (London, 1975), 54.
85. The basic ideological texts of the various groups are readily available in many editions; they were systematically reproduced in the *Journal of Palestine Studies*, Beirut. A convenient collection is Bichara et Naim Khader, ed., *Textes de la révolution palestinienne* (Paris, 1975). Among the more important descriptive accounts are Gerard Challiant, *La résistance palestinienne* (Paris, 1970); John K. Cooley, *Green March, Black September* (London, 1973); Ehud Yaari, *Strike Terror* (New York, 1970); Edgar O'Ballance, *Arab Guerrilla Power (1967–1972)* (London, 1973). The central theoretical issues are discussed in books by Naji Alush, Elias Murgus and Anis Qasim (in Arabic) and Y. Harkabi (in Hebrew).
86. Rolf Tophoven, *Fedayin, Guerrilla ohne Grenzen* (Bonn, 1973), 109. Fatah alone received 80–85 million in 1973; the Libyan government gave 30 million.
87. Peter Paret and John W. Shy, "Guerrilla Warfare and U.S. Military Policy," in T. N. Greene, *The Guerrilla and How to Fight Him* (New York, 1962), 37.
88. Kwame Nkrumah, *Handbook of Revolutionary Warfare* (London, 1968).
89. Abdul Haris Nasution, *Fundamentals of Guerrilla Warfare* (London, 1965), 55, 73.
90. Ibid., 17.
91. General Grivas, *Guerrilla Warfare and Eoka's Struggle* (London, 1964), 73.
92. Ibid., 74.
93. *World Marxist Review* (May 1964).
94. William J. Pomeroy, op. cit., 34.
95. Lin Piao, "Long live the Victory of People's War," *People's Daily* (August 1966); *Peking Review* (3 September 1966).
96. Peter van Ness, *Revolution and Chinese Foreign Policy* (Berkeley, 1971), 7.
97. *The Polemic on the General Line of the International Communist Movement* (Peking, 1965), 15.
98. Roger Trinquier, *Modern Warfare* (London, 1964), 6.
99. Otto Heilbrunn, *Partisan Warfare* (London, 1962), 40.
100. Kenneth W. Grundy, *Guerrilla Struggle in Africa* (New York, 1971), 42.
101. R. Trinquier, *La guerre moderne* (Paris, 1959); G. Bonnet, *Les guerres insurrectionelles et révolutionnaires* (Paris, 1958), as well as the books and articles by Chassin, Souyris, Nemo, Lacheroy, Rocquigny, Hogard and others.
102. Peter Paret, *French Revolutionary Warfare from Indochina to Algeria* (New York, 1964), 7.
103. Captain Souyris in *Revue militaire d'information* (October, 1958), 38.
104. For a discussion of the revolutionary war doctrine, J. S. Ambler, *The French Army in Politics* (Columbus, 1966), 308–336.
105. Ximenes (pseud.) in *Revue militaire d'information* (August–September 1958), 27–40.

106. E. Behr, *The Algerian Problem* (London, 1961), 140; Ambler, op. cit., 324.
107. David S. Sullivan and Martin J. Sattler, eds., *Revolutionary War and Western Response* (New York, 1971), 7 *et seq*.
108. E. L. Katzenbach, Jr, "Time, Space and Will: The Politico-Military View of Mao Tse-tung," in T. N. Greene, ed., loc. cit., 19; Virgil Ney, "Guerrilla Warfare and Modern Strategy," in F. M. Osanka, ed., *Modern Guerrilla Warfare* (Glencoe, 1962), 38.
109. G. Lichtheim, *Imperialism* (New York, 1971), 164.
110. Dept. of the Army. Operations against Irregular Forces, *Field Manual 31–15* (1961), 5.
111. Julian Paget, *Counter-Insurgency Campaigning* (London, 1967), 23.
112. John J. McCuen, *The Art of Counter Revolutionary War* (London, 1966), *passim.*
113. Frank Kitson, *Low Intensity Operations* (London, 1971), 32.
114. J. Baechler, in Sullivan and Sattler, eds, op. cit., 79.
115. Robert Thompson, *Revolutionary War in World Strategy, 1945–1969* (London, 1970), 11.
116. J. L. S. Girling, *People's War* (London, 1969).
117. David Galula, *Counterinsurgency Practice* (New York, 1964), *passim.*
118. Eqbal Ahmad, loc. cit., 4.
119. Ibid., 15.
120. Richard Clutterbuck, *Protest and the Urban Guerrilla* (London, 1973), 13 *et seq.*

CHAPTER NINE: A SUMMING UP

1. For instance Otto Heilbrunn, *Partisan Warfare* (London, 1962), 40, and many other authors.
2. E. J. Hobsbawm, *Revoltionaries* (London, 1973), 165.
3. G. Fairbarn, *Revolutionary Guerrilla Warfare* (London, 1974), 16.
4. M. Elliot-Bateman rightly notes that "people's war" is not a new form of war but that it was forgotten or repressed. On the other hand it is far-fetched, to put it mildly, to consider Lawrence's exploits in Arabia a case of "people's war" as he does. See "The Form of People's War," *Army Quarterly* (April 1970), 38.
5. Juana Azurduy de Padilla, one of the chief guerrilla leaders in the Andes, was made *Teniente Coronel* in 1816 (Joaquin Gautier's biography [La Paz, 1973], 199.) In the La Plata wars of the early nineteenth century the guerrilla portiòn of the campaign (*guerra de recursos*) was almost entirely entrusted to the women of Paraguay by Brigadier General Eliza Lynch and Lieutenant-Colonel Margaret Ferreira. See also Julio Diaz Arguedas, *Guerrilleros y Heroinas de la Independencia* (La Paz, 1974), 13–15.
6. The term "internal war" in fact antedates "guerrilla"; it was used in the eighteenth century, but it appears here in its specific modern sense. "People's war," needless to say, is not a new expression either but, following Mao, it has acquired a different specific meaning. The use of "partisan" as a military term can be traced back in English to the early eighteenth century, and in French and Italian to the late sixteenth.
7. The following is a fairly representative but by no means exhaustive sample for the discussion of guerrilla theories: Henry Bienen, *Violence and Social Change* (Chicago, 1968), 40–65; Lucien Pye and others in Harry Eckstein, ed., *Internal War* (New York, 1964); Harry Eckstein, "On the Etiology of Internal Wars," *History and Theory* 2 (1965), 133–163; Chalmers Johnson, "Civilian Loyalties and Guerrilla Conflicts," *World Politics* (July 1963); Samuel P. Huntington, "Guerrilla Warfare in Theory and Practice" in Osanka, op. cit; J. K. Zawodny,

Russell Rhyne, Klaus Knorr and other contributors to the special issue of *The Annals* (May 1962); Franklin A. Lindsay, "Unconventional Warfare," *American Scholar* (Summer 1962); D. F. Robinson, "Irregular Warfare," *Army Quarterly and Defence Journal* (July 1974); B. Singh and Ko Wang Mei, *Theory and Practice of Modern Guerrilla Warfare* (New York, 1971); M. Elliot-Bateman, "The Form of People's War," *Army Quarterly* (April 1970); P. Kecskemeti, *Insurgency as a Strategic Problem* (Santa Monica, 1967); Charles Wolf, Jr., *Insurgency and Counterinsurgency: New Myths and Old Realities* (Santa Monica, 1965); Eqbal Ahmad, E. R. Wolf and M. Gelden in N. Miller and R. Aya, eds., *National Liberation, Revolution in the Third World* (New York, 1971); Roger Darling, "Analyzing Insurgency," *Military Review* (February 1974); Nathan Leites and Charles Wolf, *Rebellion and Authority —An Analytical Essay on Insurgent Conflicts* (Chicago, 1970).

8. Kecskemeti, op, cit., 15.
9. Pye in Eckstein, ed., op. cit., 162.
10. Eckstein in *History and Theory*, op. cit., 153.
11. *Revue Militaire Génerale* (January 1957).
12. L. Oppenheim, *International Law* (London, 1940), Section 254.
13. Jürg H. Schmid. *Die völkerrechtliche Stellung der Partisanen im Kriege* (Zürich, 1956); Carl Schmitt, *Theorie des Partisanen* (Berlin, 1963); Alfred Bopp, *Moderner Krieg und Kriegsgefangenenrecht* (Würzburg, 1970); Charles Zorgbibe, *La Guerre Civile* (Paris, 1975); J. Siotis, *Le droit de la guerre et les conflits armés d'un caractère non international* (Geneva, 1958); M. Venthey, *La guérrilla: le problème du traitement des prisonniers* in *Annales d'Etudes Internationales* (Geneva, 1972); F. Kalshoven, "The Position of Guerrilla Fighters under the Law of War," *International Society for Military Law* (Leyden, 1969); J. R. Rosenau, ed., *International Aspects of Civil Strife* (Princeton, 1964); R. Pinto, "Les règles du droit international concernant la guerre civile," *Revue des Cours de l'Academie de Droit International* (1965), vol. 114.
14. Bienen, loc. cit., 105.
15. Darling, op. cit.
16. Huntington in Osanka, op. cit., XVI.
17. B. M. Jenkins, *High Technology Terrorism and Surrogate War: The Impact of Surrogate War on Low-Level Violence* (Santa Monica, 1975).
18. This is, in any case, part of a wider problem, that of individuals blackmailing society. A terrorist "movement" will not be needed to engage in nuclear extortion; a small group of madmen or criminals, or perhaps a single individual will be equally effective, perhaps even more so, because the smaller the group the more difficult to identify and combat it.
19. Luis Padilla in *World Marxist Review* (April 1975) and T. Timofeev in *Kommunist* (April 1975).

Chronology of Major Guerrilla Wars

(Including general wars in which guerrilla operations played a significant role)

United States War of Independence	1775–1783
Vendée	1792–1796
South Italy (Ruffo)	1799
Spain	1809–1812
Tyrol	1809
Russia	1812
Latin America (Wars of Independence)	ca. 1810–1821
Greece	1821–1832
Carlist Wars	1834–1839
Poland	1831, 1863
North Africa (Abd el-Kader)	1830–1847
Caucasus (Shamyl)	1834–1859
Italy	1848–1849 etc.
Colonial Wars Such As:	
Anglo-Burmese Wars	1824–1826, 1852, 1885
Sikh Wars	1845–1849
Maori Wars	1845–1870
Kaffir War	1851–1852
Ashanti Wars	1863–1874
Bhutan	1865
Zulu War	1879
Sudan	1883–1885
Tonkin Uprisings	1883–1895
Madagascar	1884, 1895
Waziristan	1919–1923
American Indian Wars	1850–1890
Russia in Central Asia	1837–1884
Taiping Rebellion	1851–1865
United States War of Secession	1861–1865
Mexico	1862–1867
Cuba	1868–1878, 1895–1898
Franco-Prussian War	1870–1871

Bosnia	1878–1879
Macedonia (IMRO)	1893–1934
Boer War	1899–1902
Philippines	1899–1902
Mexico	1911–1919
East Africa (Lettow-Vorbeck)	1914–1918
Ireland	1916, 1919–1922
Arabia (Lawrence)	1916–1917
Russian Civil War	1918–1921
Soviet Union (Basmatchis)	1919–1930
North Africa (Abd el-Krim)	1921–1927
Brazil (Prestes)	1926–1927
Nicaragua (Sondino)	1927–1933
China	1927–1945
Palestine	1936–1939
Second World War:	
Soviet Union	1941–1944
Yugoslavia	1941–1945
Albania	1941–1944
Poland	1944
Slovakia	1944
Greece	1942–1944
France	1941–1944
Italy	1943–1944
Palestine	1944–1948
Iraq (Kurds)	1945–1975
Philippines (Huk)	1946–1956, 1972–
Vietnam	1946–1952
Burma	1947–
Greek Civil War	1947–1949
Indonesia	1947–1949
Malaya	1948–1956
Mau Mau (Kenya Emergency)	1952–1956
Algeria	1954–1960
Sudan	1955–1972
Cyprus	1955–1959
Cuba	1957–1959
Eritrea	1958–
Vietnam, Laos, Cambodia	1959–1975
Angola	1961–1975
Mozambique	1961–1975
Venezuela	1962–1965
Peru	1962–1965
Guinea-Bissau	1963–1974
Rhodesia	1963–
Guatemala	1964–1967
Colombia	1964–
Chad	1965–
Israel	1965–
Oman	1965–
Thailand	1965
Brazil	1968–1970
Uruguay	1968–1972
India (Naxalites)	1969–1972
Ulster	1969–
Argentina	1972–

Abbreviations

(The figures indicate approximate date of foundation or activity.)

AK	*Armia Kraiowa* (Home Army)	Poland	World War II
AL	*Armia Ludowa* (People's Army)	Poland	World War II
ALN	*Ação Libertadora National*	Brazil	1968
ALN	*Armée de la Libération Nationale*	Algeria	1954–1960
CNR	*Conseil National de la Resistance*	France	World War II
EDES	*Ethnikos Dimokratikos Ellinikos Syndemos* (National Republican Greek League)	Greece	World War II
ELAS	*Ethnikos Ellinikos Laikos Apelephterikos Stratos* (National Popular Liberation Army)	Greece	World War II
ELF	Eritrean Liberation Front	Ethiopia	1958–
ELN	*Ejército de Liberación Nacional*	Bolivia	1967
ELN	*Ejército de Liberación Nacional*	Columbia	1965
ELN	*Ejército de Liberación Nacional*	Peru	early 1960s
EOKA	*Ethniki Organosis Kypriakou Agoniston* (National Organization of Cypriot Fighters)	Cyprus	1954–
EOKA II		Cyprus	1972–1974
ERP	*Ejército Revolucionario del Pueblo*	Argentina	1970
ERP	*Ejército Revolucionario del Pueblo*	San Salvador	
ETA	*Euzkadi ta Askatasuna*	Spain-Basque	1966
FALN	*Fuerzas Armadas de Liberación Nacional*	Venezuela	1962

FAR	*Fuerzas Armadas Revolucionarias*	Argentina	1973
FAR	*Fuerzas Armadas Rebeldes*	Guatemala	1962
FARC	*Fuerzas Armadas Revolucionarias de Columbia*	Columbia	1966
FATAH	Palestinian Arabs	(Exile)	
FFI	*Forces Francaises de l'Interieur*	France	1942–1945
FLCS	Front for the Liberation of Coastal Somalia	Somalia	1974
FLQ	*Front de Libération de Quebec*	Canada	early 1970s
FRELIMO	*Frente de Libertação de Moçambique*	Mozambique	1962
FROLINAT	*Front de Libération National du Tchad*	Chad	1966
FROLIZI	Front for the Liberation of Zimbabwe	Rhodesia	1971
FSLN	*Frente Sandinista de Liberacion National*	Nicaragua	*ca.* 1960
FTP	*Franctireurs Partisans*	France	World War II
GRAE/FLNA	*Governo Revolucionario de Angola no Exilio/Frente Nacional de Libertação de Angola*	Angola	late 1950s
HRB	*Hrvatsko Revolucionarno Bratsvo*	Croatia (exile)	1960s
IRA–Officials	Irish Revolutionary Army	Ulster	
IRA–Provisionals	Irish Revolutionary Army	Ulster	
IZL	*Irgun Zvai Leumi* (National Military Organization)	Palestine	1940–1948
LEHI	*Lohame Herut Israel* (Fighters for the Freedom of Israel)	Palestine	1940–1948
MIR	*Movimiento de Izquierda Revolucionaria*	Bolivia	1974
MIR	*Movimiento de Izquierda Revolucionaria*	Chile	1965
MIR	*Movimiento de Izquierda Revolucionaria*	Peru	early 1960s
MIR-13	*Movimiento Revolucionario de Noviembre 13*	Guatemala	1963
MLN	*Movimiento de Liberación Nacional* (Tupamaros)	Uruguay	1963–1972
MNLA	Malayan National Liberation Army	Malaya	World War II and after
Monteneros		Argentina	1968
MPLA	*Movimento Popular para a Libertação de Angola*	Angola	1960
NPA	New People's Army	Philippines	1968–1969
PAIGC	*Partido Africano de Independencia da Guinea 'Portuguesa' e das Ilhas de Cabo Verde*		1956

PDFLP	Popular Democratic Front for the Liberation of Palestine	(Exile)	1968
PFLP	Popular Front for the Liberation of Palestine	(Exile)	1968
PFLP General Command	Popular Front for the Liberation of Palestine (General Command)	(Exile)	1968
PFLO	Popular Front for the Liberation of Oman (formerly PFLOAG)	Oman	1963
PLO	Palestine Liberation Organization	(Exile)	1965
RAF	*Rote Armee Fraktion* (Baader-Meinhof)	Germany	1970
SIAKHAL		Iran	1970
SWANU	South West African National Union		1960
SWAPU	South West African People's Organization		1960
TPLA	Turkish People's Liberation Army	Turkey	1970
UDA	Ulster Defence Association	Ulster	
UNITA	*União Nacional para a Independencia Total de Angola*	Angola	1966
UVF	Ulster Volunteer Force	Ulster	
ZANU	Zimbabwe African National Union	Rhodesia	1963
ZAPU	Zimbabwe African People's Union	Rhodesia	1961–1962

Bibliography

BIBLIOGRAPHICAL NOTE

There is no comprehensive bibliography on guerrilla literature. The wars in the Vendée and Andreas Hofer's insurrection in Tyrol have been studied in great detail; whereas the literature on the Spanish guerrillas, on the Russian partisans (1812), and Cardinal Ruffo's expedition in 1799 is relatively sparse.

Of the early theoreticians of partisan warfare Grandmaison, de Jeney, Emmerich, Ewald and von Valentini whose works were published between 1750 and 1800 are the most important; with the exception of von Valentini they are largely anecdotal in character. Denis Davydov's account of the Russian partisans is the most vivid.

The books of Decker (1821) and Le Mière de Corvey (1823) encompass the experience gained in the Napoleonic wars; far more systematic than their predecessors they were copied or paraphrased in countless subsequent works.

Clausewitz and Jomini refer to people's wars and the former taught a course on the technique of "small war." His notes on the subject, published in 1966, have not yet appeared in English.

The writings of the Polish and Italian authors on the technique of military insurrection are of paramount importance because of their preoccupation with the politics and the strategy of wars of national liberation. Some of them thought that partisan units would be transformed sooner or later into a regular army, others were in favor of "pure guerrillaism." E. Liberti, *Techniche della guerra partigiana nel Risorgimento* (Florence, 1972) is an excellent, very detailed survey of the Italian nineteenth-century literature; his volume also reproduces Gentilini's *Stracorridori* (1848) and Carlo Bianco's *Manuale Pratico* (1833) though not his more important *Trattatto* (1830). There is no such introduction into the equally interesting Polish literature of the period. Marian Zychowski's biography of Mieroslawski (in Polish; Warsaw, 1963) contains probably the best bibliographical guide available. Bem's and Stolzman's works were republished in Warsaw after 1945, most other works of this period have become exceedingly rare. Mazzini's and Pisacane's writings are readily available today, but not those of other Italian writers of the 1840s and 1850s such as Cesare Balbo, Guglielmo Pepe, La Masa *et al.*

The German and Austrian nineteenth-century literature (Schels, Boguslawski, Rüstow, Hron) deals mainly with partisan operations in the enemy's rear in collaboration with the regular army; the same refers to Russian authors such as Gershelman and Klembowski. There are very many British and French accounts of colonial campaigns in the nineteenth century but only very few studies of a general character. The most important are those by Callwell (1899) and Maguire (1904). Devaureix (1881) and Charenton (1900) on the other hand deal only with the European experience. T. E. Lawrence and Lettow-Vorbeck described their activities during the First World War in considerable detail. The various Latin American guerrilla wars during the nineteenth century and early twentieth have been covered in absorbing detail but there were few attempts to generalize on the basis of this rich experience.

Socialist authors with a very few exceptions were not interested in partisan warfare. Marx and Engels referred to it on rare occasions, Lenin did so even more infrequently. Blanqui's *Instructions* were published only after the First World War. Johannes Most provided the first do-it-yourself manual for terrorists but even he was not interested in the wider aspects of urban terror. Thus it was left to Mao to rediscover the theory and practice of guerrilla warfare, and in his wake, to Giap, Guevara, Debray and the others. Recent, i.e., post–World War II, literature is readily available; the theoretical works are by necessity didactic and repetitive, the personal accounts are sometimes fascinating. The only two works dealing with guerrilla war in historical perspective are Professor Werner Hahlweg's *Guerrilla* (1968), a short survey with an excellent bibliography, and F. Tudman's much more voluminous *Rat protiv rat* (War against War) (in Serbo-Croat, Zagreb, 1957), heavily preoccupied with theBalkan roots and manifestations of guerrilla war.

There have been several popular histories of guerrilla warfare from Percy Cross Standing's *Guerrilla Leaders of the World* (London, 1912) to Robert B. Asprey's *War in the Shadows* (New York, 1975). The anthologies such as those by Osanka (1962) and Pomeroy (1968) cover almost exclusively contemporary guerrilla warfare.

The following selective list includes works of historical or topical importance. Books on specific guerrilla wars, terrorism and counterinsurgency have, as a rule, not been included. Readers may find the references in the footnotes of the present volume of some help in their search for further literature.

K. Adaridi. *Freischaren und Freikorps. Auf Grund von Kriegserfahrungen.* Berlin, 1925.

F. R. Allemann. *Macht und Ohnmacht der Guerilla,* Munich, 1974.

T. Argiolas. *La guerriglia. Storia e dottrina.* Florence, 1967.

J. A. Armstrong, K. de Witt, eds. *Soviet Partisans in World War II.* Madison, 1964.

General de Brack. *Advanced Posts of Light Cavalry.* London, 1850.

V. Bambirra, ed. *Diez Años de Insurrección.* 2 vols. Santiago, 1971.

A. Bayo. 150 *Questions to a Guerrilla.* Boulder, 1963.

H. Bejar. *Peru 1965, Notes on a Guerrilla Experience.* New York, 1970.

J. Bowyer Bell, *Myth of the Guerrilla.* New York, 1971.

Carlo Bianco di St. Jorioz. *Della guerra nazionale d'insurrezione per bande applicata all'Italia.* Italy, 1830.

H. Blanco. *El camino de nuestra revolución.* Lima, 1964.

A. von Boguslawski. *Der kleine Krieg und seine Bedeutung für die Gegenwart.* Berlin, 1881.

G. Bonnet. *Les guerres insurrectionelles et révolutionnaires de l'antiquité à nos jours.* Paris, 1958.

H. von Brandt. *Der kleine Krieg in seinen verschiedenen Beziehungen.* Berlin, 1850.

G. Budini. *Alcune idee sull'Italia.* London, 1843.

D. von Bülow. *Militärische und vermischte Schriften.* Leipzig, 1853.

C. E. Callwell. *Small Wars, Their Principles and Practice.* London, 1899.

A. Cabral. *Revolution in Guinea.* London, 1969.

———. *Unité et Lutte,* 2 vols. Paris, 1975.

G. Cardinal von Widdern. *Der kleine Krieg und der Etappendienst. Aus dem deutsch-französischen Krieg 1870–71.* Leipzig, 1892–1897.

F. Castro. *Selected Works: Volume I, Revolutionary Struggle.* Cambridge, Mass., 1971.

L. M. Chassin. *La conquête de la Chine par Mao Tse-tung (1945–1949)*. Paris, 1952.
Chizzolini. *Della guerra nazionale*. Milan, 1863.
General W. Chrzanowski. *O wojnie partyzanckiej*. Paris, 1835.
C. von Clausewitz. *Schriften — Aufsätze — Studien — Briefe*, ed. W. Hahlweg. vol. I. Munich, 1966.
J. Connolly. *Revolutionary Warfare*. Dublin, 1968.
De la Croix. *Traité de la petite guerre*. Paris, 1752.
H. von Dach. *Der totale Widerstand*. Bern, 1966.
D. Davydov. *Voennie zapiski*. Moscow, 1940.
R. Debray. *Revolution in the Revolution*. London, 1968.
———. *Strategy for Revolution*. London, 1973.
———. *La critique des armes*. 2 vols. Paris, 1974.
C. von Decker. *Der kleine Krieg im Geiste der neueren Kriegsführung*. Berlin, 1821.
G. Desroziers. *Combats et partisans*. Paris, 1883.
A. Devaureix. *De la guerre de partisans*. Paris, 1881.
H. Eckstein, ed. *Internal War*. New York, 1964.
A. Ehrhardt. *Kleinkrieg*. Potsdam, 1935.
A. Emmerich. *The Partisan in War*. London, 1789.
———. *Der Parteigänger im Kriege oder der Nutzen eines Corps leichter Truppen für eine Armee*. Dresden, 1791.
F. Engels. *Ausgewählte militärische Schriften*. 2 vols. Berlin, 1958, 1964.
J. von Ewald. *Abhandlungen über den kleinen Krieg*. Kassel, 1785.
G. Fairbarn. *Revolutionary Guerrilla Warfare*. London, 1974.
F. Fanon. *Les damnés de la terre*. Paris, 1961.
D. Galula. *Counterinsurgency, Warfare, Theory and Practice*. London, 1964.
Vo Nguyen Giap. *People's War, People' Army*. Hanoi, 1962.
———. *The Military Art of People's War*. New York, 1970.
E. Gentilini. *Guerra degli stracorridori a guerra guerriata*. Capolago, 1848.
F. Gershelman. *Partisanskaia Voina*. St. Petersburg, 1885.
A. Gingins-La Sarraz. *Les partisans et la défence de la Suisse*. Lausanne, 1861.
B. Goldenberg. *Kommunismus in Latein Amerika*. Stuttgart, 1970.
R. Gott. *Guerrilla Movements in Latin America*. London, 1970.
Grandmaison. *De la petite guerre ou traité du service des troupes legères en campaigne*. Paris, 1756.
T. N. Greene, ed. *The Guerrilla and How to Fight Him. Marine Corps Gazette*, 1962.
G. Grivas. *General Grivas on Guerrilla Warfare*. New York, 1965.
C. Grosse. *Kurzgefasste Geschichte der Parteigängerkriege in Spanien 1833–1836*. Leipzig, 1837.
E. (Ché) Guevara. *Guerrilla Warfare*. London, 1969.
W. Hahlweg. *Krieg ohne Fronten*. Stuttgart, 1968.
B. Liddell Hart. *T. E. Lawrence in Arabia and After*. London, 1934.
C. Helmuth. *Der kleine Krieg*. Magdeburg, 1855.
D. C. Hodges, ed. A. *Guillen: Philosophy of the Urban Guerrilla*. New York, 1973.
Ho Chi-Minh. *Selected Works: Volumes I and II*. Hanoi, 1961.
K. Hron. *Der Parteigänger-Krieg*. Vienna, 1885.
M. Jähns. *Geschichte der Kriegswissenschaften*. Munich, 1891.
A. Jelowicki. *O powstaniu i wojnie partyzanskiej*. Paris, 1835.
M. de Jeney. *Le partisan, ou l'art de faire la petite guerre avec succès, selon le génie de nos jours*. The Hague, 1759.
Chalmers Johnson. *Autospy on People's War*. Berkeley, 1973.
R. F. Johnson. *Night Attacks*. London, 1886.
H. de Jomini. *Précis de l'art de la guerre*. Paris, 1838.
H. M. Kamienski. *Wojna Ludowa*. Paris, 1866.
R. E. Kiessler. *Guerilla und Revolution*. Bonn, 1975.
V. N. Klembovski. *Partisanskie Deistviia*. St. Petersburg, 1894.

P. Klent. *Partizanska Taktika*. Belgrade, 1965.
J. Kohl, J. Litt. *Urban Guerrilla Warfare in Latin America*. Cambridge, Mass., 1974.
Kollektiv RAF. *Über den bewaffneten Kampf in Westeuropa*. Berlin, 1971.
H. Kühnrich. *Der Partisanenkrieg in Europa 1939–1945*. Berlin, 1965.
J. Kunisch. *Der Klein Krieg*. Frankfurt, 1973.
T. E. Lawrence. *Revolt in the Desert*. London, 1927.
———. *The Seven Pillars of Wisdon*. London, 1935.
E. Liberti, ed. *Techniche della guerra partigiana nel Risorgimento*. Florence, 1972.
Lin Piao. *Long Live the Victory of People's War*. Peking, 1965.
E. Lussu. *Teoria dell' insurrezione*. Milan, 1969.
G. B. Malleson. *Ambushes and Surprises*. London, 1885.
Mao Tse-tung. *On Guerrilla Warfare*. London, 1961.
———. *Selected Works: Volume I*. Peking, 1965.
———. *Basic Tactics*. London, 1967.
C. Marighela. *For the Liberation of Brazil*. London, 1971.
A. R. Martin. *Mountain and Savage Warfare*. Allahabad, 1898.
Marx-Engels. *Werke*. [East] Berlin, 1960.
G. la Masa. *Della Guerra insurrezionale in Itala*. Turin, 1856.
G. Maschke. *Kritik des Guerilleros*. Frankfurt, 1973.
E. Mayans, ed. *Tupamaros: antologia documental*. Cuernavaca, 1971.
J. J. McCuen. *The Art of Counter-Revolutionary War*. London, 1966.
A. Mercader, J. de Vega. *Tupamaros: estrategia y acción*. Montevideo, 1969.
Le Mière de Corvey. *Des partisans et des corps irréguliers*. Paris, 1823.
L. v. Mieroslawski. *Kritische Darstellungen des Feldzuges vom Jahre 1831 und hieraus abgeleitete Regeln für Nationalkriege*. 2 vols. Berlin, 1847.
F. O. Miksche. *Secret Forces: The Technique of Underground Movements*. London, n.d.
T. Miller Maguire. *Guerrilla or Partisan Warfare*. London, 1904.
R. Moss. *Urban Guerrillas*. London, 1972.
J. Most. *Revolutionäre Kriegswissenschaft*. New York, 1884.
H. J. Müller Borchert. *Guerrilla im Industriestaat*. Hamburg, 1973.
A. H. Nasution. *Fundamentals of Guerrilla Warfare*. London, 1963.
A. Neuberg. *Armed Insurrection*. London, 1970.
W. Nieszokoc. *O Systemie wojny partyzanskiej wzniesionym wsrod emigracji*. Paris, 1835.
A. Orlov. *Handbook of Intelligence and Guerrilla Warfare*. Ann Arbor, 1963.
F. M. Osanka, ed. *Modern Guerrilla Warfare*. Glencoe, 1962.
J. Paget. *Counter-Insurgency Campaigning*. London, 1967.
P. Paret. "French Revolutionary Warfare from Indochina to Algeria." *Princeton Studies in World Politics*, No. 6, 1964.
P. Paret, J. W. Shy. "Guerrillas in the 1960s." *Princeton Studies in World Politics*, No. 1, 1962.
G. Pepe. *Memoria su i mezzi che menano all Italiana indipendenza*. Paris, 1833.
G. Pisacane. *Saggi storici-politici-militari sull' Italia*. New ed., Milan, 1957.
W. J. Pomeroy, ed. *Guerrilla Warfare and Marxism*. New York, 1968.
Ray de Saint Genies. *L'officier partisan*. 6 vols. Paris, 1769.
De la Roche-Aymon. *Essay sur la petite guerre*. Paris, 1770.
F. W. Rüstow. *Die Lehre vom Kleinen Krieg*. Zürich, 1864.
Felipe de San Juan. *Instrucción de Guerrilla*. La Paz, 1846.
J. B. Schels. *Der Kleine Krieg*. Vienna, 1848.
———. *Leichte Truppen, kleiner Krieg*. 2 vols. Vienna, 1813–1814.
C. Schmitt. *Theorie des Partisanen*. Berlin, 1963.
K. B. Stolzman. *Partyzanka czyli wojna dla ludow powstajacych najwlasciwsza*. Paris, 1844.
R. Taber. *The War of the Flea*. New York, 1965.
M. Talas. *Harb al isabat*. Damascus, 1966.

E. V. Tarle, ed. *Partisanskaia Voina*. Moscow, 1943.
L. Taruc. *Born of the People*. New York, 1953.
M. Tevis. *La petite guerre*. Paris, 1855.
C. W. Thayer. *Guerrillas*. London, 1964.
R. Thompson. *Defeating Communist Insurgency*. London, 1966.
J. Broz Tito. *Selected Military Works*. Belgrade, 1966.
R. Trinquier. *La guerre moderne*. Paris, 1961.
F. Tudman. *Rat Protiv Rat*. Zagreb, 1957.
T. Mitev Urkovachev. *Partisanskata Voina*. Sofia, 1966.
Freiherr W. von Valentini. *Abhandlung über den kleinen Krieg*. Berlin, 1799.
Ritter de Ville. *Von Parteyen*. Breslau, 1755.
Colonel Vuich. *Malaia Vonia*. St. Petersburg, 1850.
W. St. Ritter von Wilczynski. *Theorie des grossen Krieges mit Hilfe des kleinen oder Partisanen-Krieges bei theilweiser Verwendung der Landwehr*. Vienna, 1869.
M. le Baron de Wüst. *L'art militaire du partisan*. The Hague, 1768.
G. S. Zafferoni. *L'insurrezione armata*. Milan, 1868.

Index